FAIRNESS 2012: FAIR NExt Generation of ScientistS 2012

Journal of Physics: Conference Series Volume 426

Hersonissos, Greece
3-8 September 2012

ISBN: 978-1-62748-303-2
ISSN: 1742-6588

Printed from e-media with permission by:

Curran Associates, Inc.
57 Morehouse Lane
Red Hook, NY 12571

CURRAN ASSOCIATES INC. proceedings.com

Some format issues inherent in the e-media version may also appear in this print version.

Printed by Curran Associates, Inc. (2013)

For permission requests, please contact the Institute of Physics
at the address below.

Institute of Physics
Dirac House, Temple Back
Bristol BS1 6BE UK

Phone: 44 1 17 929 7481
Fax: 44 1 17 920 0979

techtracking@iop.org

Additional copies of this publication are available from:

Curran Associates, Inc.
57 Morehouse Lane
Red Hook, NY 12571 USA
Phone: 845-758-0400
Fax: 845-758-2634
Email: curran@proceedings.com
Web: www.proceedings.com

Papers

PREFACE

FAIRNESS 2012 was the first in a series of workshops designed to bring together excellent international young scientists with research interests focused on physics at FAIR (Facility for Antiproton and Ion Research) and was held on 3–8 September 2012 in Hersonissos, Greece. The workshop covered a wide range of topics, both theoretical developments and current experimental status, that concentrated around the four scientific pillars of FAIR. FAIR is a new accelerator complex with brand new experimental facilities, that is currently being built next to the existing Helmholtzzentrum for Schwerionenforschung close to Darmstadt, Germany.

The spirit of the conference was to bring together young scientists, e.g. advanced PhD students and postdocs and young researchers without permament position to present their work and to foster active informal discussions and the build-up of networks. Every participant at the meeting, with the exception of the organizers, gave an oral presentation and all sessions were followed by an hour long discussion period. During the talks questions were collected anonymously in a circulating box to stimulate these discussions. Since the physics program of FAIR is very broad, this was reflected in the wide range of topics covered at the conference:

- Physics of hot and dense nuclear matter, QCD phase transitions and critical point
- Nuclear structure, astrophysics and reactions
- Hadron Spectroscopy, Hadrons in matter and Hypernuclei
- Special emphasis is put on the experiments CBM, HADES, PANDA, NuSTAR, as well as NICA and the RHIC low beam energy scan
- New developments in atomic and plasma physics

In each of these different areas one invited speaker was selected to give a longer introductory presentation. The write-ups of the talks presented at FAIRNESS 2012 are the content of this issue of *Journal of Physics: Conference Series* and have been refereed according to the IOP standard for peer review. This issue constitutes therefore a collection of the forefront of research that is dedicated to the physics of FAIR.

February 2013,

Organizers of FAIRNESS 2012: Almudena Arcones, Marcus Bleicher, Miriam Fritsch, Tetyana Galatyuk, Diana Nicmorus, Hannah Petersen, Claudia Ratti and Laura Tolos

Support for holding the conference was provided by:

HIC for FAIR
Helmholtz International Center

EMMI

FAIR

FIAS Frankfurt Institute for Advanced Studies

Peer review statement

Abstract Metrics

All papers published in this volume of *Journal of Physics: Conference Series* have been peer reviewed through processes administered by the proceedings Editors. Reviews were conducted by expert referees to the professional and scientific standards expected of a proceedings journal published by IOP Publishing.

Transport coefficients from field theory and FAIR physics

Daniel Fernández-Fraile

Institut für Theoretische Physik, Johann Wolfgang Goethe-Universität, Max-von-Laue-Str. 1, 60438 Frankfurt am Main, Germany

E-mail: danfer@th.physik.uni-frankfurt.de

Abstract. I review recent progress in the understanding of hydrodynamics and transport coefficients in field theory. This subject plays an important role in the future heavy-ion program at the FAIR facility in Darmstadt. One of the main goals of the experiment is obtaining a better understanding of the QCD phase diagram, in particular to determine whether there is a critical point at intermediate baryon densities and temperatures. Transport coefficients are essential inputs to describe the space-time evolution of the plasma created in heavy-ion experiments. Moreover, their characteristic behavior near the critical point may also help to obtain experimental proof of its existence.

1. Hydrodynamics

Hydrodynamics is an effective quantum[1] field theory which describes *any* physical system at long-enough distances and times. The degrees of freedom are (long-living) massless modes corresponding to fluctuations from the equilibrium values of hydrodynamical variables (e.g., charge density, momentum density, ...) [1].

Let us consider a microscopic quantum field theory in which the only conserved currents are the energy-momentum tensor $T_{\mu\nu}$ and J_μ (a current associated to some internal continuous symmetry).[2] Hydrodynamic equations are the conservation laws $\partial_\mu\langle T^{\mu\nu}\rangle = 0$ and $\partial_\mu\langle J^\mu\rangle = 0$, supplemented with information about dissipative processes which take place in the fluid due to the interaction between the microscopic particles. These dissipative processes enter into the hydrodynamic equations as a gradient expansion of the hydrodynamic variables around their equilibrium values. Since hydrodynamics applies at long-enough distances and times, variations of the hydrodynamic quantities happen at much larger distances than the mean-free path for the collision between particles.

In order to simplify the hydrodynamical equations, instead of using $\langle T_{\mu\nu}\rangle$ and $\langle J_\mu\rangle$ as the hydrodynamic variables, it is convenient to express them in terms of the local temperature $T(x)$, the local chemical potential $\mu(x)$, and the local fluid velocity $\boldsymbol{u}(x)$. This can be done

[1] The temperature plays the role of $\hbar$, and the (UV) cutoff of the theory is played by the mean free path of the particles $\lambda_{\rm mfp}$. In the language of effective field theories, the perturbative expansion is performed in powers of $k\lambda_{\rm mfp}$, where k represents the momentum or frequency scale of the process under consideration.

[2] In a more general situation, hydrodynamic modes will also be phases of symmetry-breaking condensates (near second-order phase transitions, not only the phase but also the magnitude of the condensate) and Abelian gauge fields of unbroken $U(1)$ symmetries [2].

systematically by expanding $\langle T_{\mu\nu}\rangle$ and $\langle J_\mu\rangle$ in gradients of these new hydrodynamic variables. To linear order in gradients, one obtains in the Landau frame the *constitutive relations* [1]:[3]

$$\langle T^{\mu\nu}\rangle = \epsilon u^\mu u^\nu + p\Delta^{\mu\nu} - \eta\Delta^{\mu\alpha}\Delta^{\nu\beta}\left(\partial_\alpha u_\beta + \partial_\beta u_\alpha - \frac{2}{d}\eta_{\alpha\beta}\partial_\mu u^\mu\right) - \zeta\Delta^{\mu\nu}\partial_\lambda u^\lambda + \mathcal{O}(\partial^2)\ , \quad (1)$$

$$\langle J^\mu\rangle = nu^\mu - \sigma T\Delta^{\mu\nu}\partial_\nu(\mu/T) + \mathcal{O}(\partial^2)\ , \quad (2)$$

where the metric is $\eta_{\mu\nu} = \mathrm{diag}(-1,1,\ldots,1)$ and $\Delta^{\mu\nu} \equiv \eta^{\mu\nu} + u^\mu u^\nu$ is a projector operator. ϵ, p and n are the energy-density, pressure, and charge density respectively, and they are related by the equation of state in equilibrium. It can be shown by arguments of non-negative divergence of the entropy current (or, alternatively, using symmetry properties of Green functions involving $T_{\mu\nu}$ and J_μ), that the *transport coefficients* obey $\eta \geq 0$ (*shear viscosity*), $\zeta \geq 0$ (*bulk viscosity*), and $\sigma \geq 0$ (*charge conductivity*). Since charge diffusion is related to heat dissipation, instead of using the charge conductivity, sometimes it is used either the *diffusion constant* D or the *heat conductivity* κ. These three quantities are related by the Einstein-like relation[4]

$$\sigma = \frac{\partial n}{\partial \mu} D = \frac{T^2 n^2}{(\epsilon+p)^2}\kappa\ , \quad (3)$$

where μ is the chemical potential.

But this is not the full story regarding linear hydrodynamics; recently there has been an intense activity, motivated by the AdS/CFT duality [3], in exploring the influence on hydrodynamics from effects in the microscopic quantum field theory such as anomalies [4], inclusion of external fields [5], or parity-violation [6]. Moreover, if the expansion parameter $k\lambda_{\rm mfp}$ is not small, one in principle has to consider higher orders in the derivative expansion as well (see for instance [7]). Second-order hydrodynamics was also introduced originally to cure the problems of non-causal propagation,[5] instabilities,[6] and the lack of a well-posed initial value problem for rotating fluids in the first-order approximation. Nevertheless, second-order hydrodynamics breaks down at the "classical" level because of thermal fluctuations [1]. This is due to the effective field theory nature of hydrodynamics; thermal fluctuations are analogous to loops in Feynman diagrams of field theory. Another recent advance in the formulation of hydrodynamics has been the finding that hydrodynamical equations (not limited only to the linear order) for certain classes of strongly-interacting conformal field theories in d dimensions can be mapped (i.e., they are dual) to the long-wave limit of Einstein's equations in $d+1$ dimensions. This has been called the *fluid-gravity correspondence* [8], and it is currently also attracting much attention.

Although the perfect fluid approximation (no dissipation) can be a good approximation in many physical situations, quantum mechanics actually prevents (by the Uncertainty Principle) from the existence of perfect fluids in Nature [9,10,11,12]. In particular, viscous effects are crucial for understanding the physics of the heavy-ion programs currently running at RHIC and LHC, and for the future program at FAIR. In the next section I will review some of the recent studies concerning the linear transport coefficients of QCD.

[3] The word 'frame' in this context is used in a different sense from Lorenz transformations. It means a particular choice within the arbitrariness that exist for defining the quantities $T(x)$, $\mu(x)$, and $\boldsymbol{u}(x)$ in a non-equilibrium situation. With a different choice of frame, the arrangement of the different transport coefficients in the constitutive equations could change. For instance, the bulk viscosity could appear contributing to the charge current. Nevertheless, the value of the transport coefficients is frame-independent.

[4] Sometimes the heat conductivity is defined with an extra power of T multiplying σ. With our definition, both κ and σ have the same dimensions.

[5] Due to the non-hyperbolic nature of the linear hydrodynamic equations, high-k modes would propagate with a super-luminar velocity, which represents a problem for numerical simulations.

[6] The equilibrium state around which the gradient expansion is performed would be unstable against small perturbations.

Journal of Physics: Conference Series **426** (2013) 012001 doi:10.1088/1742-6596/426/1/012001

2. Transport coefficients of QCD: kinetic theory vs field theory

As we have seen in the previous section, transport coefficients are, together with the equation of state, the essential inputs to describe the space-time evolution of the fluid by means of the hydrodynamic equations. Their value has to be determined using the underlying theory, in our case QCD. However, the calculation of physical observables in QCD at intermediate energies (which is the situation in FAIR) is a challenging task because of the lack of a small parameter in the theory to perform perturbation theory with. Traditionally, transport coefficients have been calculated using kinetic theory (KT) (e.g., for an analysis in high-temperature QCD see [13]). This is a reasonable approach if the system is dilute enough, which in the case of QCD only happens at very high and very low energies. At intermediate energies (strong couplings), KT is no-longer applicable and one should rely on field-theory methods. In field theory, transport coefficients are calculated as the limit when the external momentum and frequency go to zero (in this order) of spectral functions corresponding to correlators of conserved currents.[7] For instance, in the case of the shear viscosity [14]:

$$\eta = \frac{1}{20} \lim_{\omega \to 0^+} \lim_{|\boldsymbol{k}| \to 0^+} \frac{\partial \rho_\eta(\omega, |\boldsymbol{k}|)}{\partial \omega} \ , \quad \rho_\eta(\omega, |\boldsymbol{k}|) = \int \mathrm{d}^4 x \ \mathrm{e}^{-\mathrm{i}\boldsymbol{k}\cdot\boldsymbol{x} + \mathrm{i}\omega t} \langle [\pi_{ij}(x), \pi^{ij}(0)] \rangle \ , \tag{4}$$

where $\pi_{ij} \equiv T_{ij} - \frac{1}{3}\delta_{ij} T^k_k$ (the indices correspond to spatial coordinates). To my knowledge, so far the only accurate approaches within field theory to calculate transport coefficients at strong-coupling are the large-N expansion (for vector groups) (see for instance [15]) and the AdS/CFT correspondence [16]. None of these two methods deal with QCD, but with field-theory models which resemble QCD is some aspects, so one expects at least to obtain some insight about the qualitative behavior of transport coefficients in QCD from them. The approach based on the AdS/CFT correspondence is specially promising because there is currently a very intense activity to extend it to tackle a larger class of theories, and eventually QCD. Another (and also very promising) strategy consists in performing a numerical calculation on the lattice. However, the statistical errors are still large even at zero chemical potential,[8] and calculating at finite densities is even more difficult because of the sign problem [19].

Besides the methods described above, which apply rigorously in the strong-coupling regime, some KT- or perturbative-based approaches try to extend their domain of applicability by including additional degrees of freedom, collision processes, or resummations of higher-order diagrams. For instance, in the quark-gluon plasma (QGP) phase, higher-order contributions to some transport coefficients have been calculated [20,21,22], which allow a better extrapolation down to temperatures near Λ_{QCD}. On the other hand, in the hadronic-phase one in principle expects that the pionic component dominates the dynamics at very low temperatures.[9] Since the interaction between hadrons weakens at low energies (in the massive case, cross sections

[7] It is important to mention that even the calculation of transport coefficients in the regime of temperatures where perturbation theory is applicable is highly non-trivial due to the presence of "pinching" singularities. In order to obtain the leading order in terms of the coupling constant it is necessary to carry out a resummation of an infinite set of diagrams. It turns out that, after performing the resummation, the result can be identified with the result one would obtain from an effective KT [14]. Therefore, when perturbation theory is applicable, it is more convenient to use from the beginning KT instead of field-theory.

[8] Transport coefficients are obtained from correlation functions in real time, whereas on the lattice these functions are calculated in Euclidean time. Both quantities are related through a convolution [17] and, in order to obtain the real-time correlator at low frequencies from a finite set of values of the Euclidean correlator, one has to make further assumptions (use an ansatz) for the shape of the correlation function in real time and then find the parameters of the ansatz from the best fit to the imaginary-time correlator. This is the so-called Maximum Entropy Method (see for instance [18] and references therein).

[9] The case of the bulk viscosity of a multi-component gas deserves further study. Because of the exponential increase of ζ at low temperatures for a one-component fluid (controlled by the mass of the field [14]), it is not obvious whether states heavier than the pion might dominate the bulk viscosity at low temperatures.

tend to a constant value), KT is applicable in that case [23].[10] A way of extending these low-temperature calculations consists in including additional hadronic states [24,25,26]. However, near the cross-over transition, the number of relevan hadronic degrees of freedom becomes large and it is still unclear how to describe all their interactions. Performing resummations in Chiral Perturbation Theory is another approach to try to obtain insight on the qualitative behavior of transport coefficients near the transition temperature [27]. Using this technique it was shown that unitarity makes the pion gas to fullfil the KSS bound (read below) [28].

There are still no accurate results available for the linear transport coefficients (shear and bulk viscosities) of QCD near the cross-over temperature at zero baryon chemical potential. The shear viscosity has been the most studied one. Most of the analyses point to a value of the quotient η/s (which can be considered as a measure of how "perfect" a fluid is) very close to the proposed KSS lower bound $1/4\pi$ obtained using the AdS/CFT conjecture [10]. It is by now acknowledged that this does not constitute a universal lower bound, but a bound which is obeyed by a large class of gauge theories with gravity duals (for a result violating the bound see for instance [29]). The fact that QCD seems to have a value of η/s very close to the KSS bound is a remarkable result, which may also indicate the existence of a gravity dual formulation of the theory at least in the limit of large number of colors, something that string theorists are trying to find with great dedication.

The bulk viscosity has been much less studied than the shear viscosity. This is probably, I believe, because in the high temperature regime it is found to be negligible with respect to the shear viscosity [30] (the bulk viscosity vanishes in the conformal limit). Moreover, initial analyses of this transport coefficient in the hadronic phase also obtained a very small ζ [23]. However, more recent studies have shown that it could be non-negligible near the transition temperature [31,32,33,34,35]. It is not a general rule that the bulk viscosity has to be smaller than the shear viscosity, in fact, some systems present a bulk viscosity which is several orders of magnitude larger [36]. It has been also recently conjectured a lower bound for the bulk viscosity [37], but there are already examples where this bound is violated [38].

At zero baryon chemical potential, another linear transport coefficient which has been recently studied is the electrical conductivity [18,39,40]. This corresponds to dissipation due to the introduction of an external electric field coupled to quarks (and at low temperatures its effect can be implemented through its coupling to pions). The frequency-dependent value of the electrical conductivity is directly related to the photon spectrum emitted from the QCD plasma, and the results of the calculations seem to be compatible (within the same order of magnitude) with experimental measurements of the spectrum at low energies.

2.1. Behavior of transport coefficients near phase transitions

For many systems which undergo a phase transition (and it is also the case of QCD with non-zero quark masses and zero baryon chemical potential), the quotient η/s reaches its minimum value at or close to the transition temperature [41,42]. The v_2 spectrum measured in heavy-ion collisions (measure of elliptic flow) indeed seems to indicate a small value of the shear viscosity for intermediate temperatures [43]. On the other hand, a rapid increase of the bulk viscosity near the transition temperature due to non-perturbative effects[11] could also produce observable consequences in the spectrum of produced particles due to clusterization of the plasma [44].

[10] At intermediate energies, non-perturbative effects such as resonances can appear. KT is still applicable at those energies provided some conditions are fulfilled [14]: (i) we include the non-perturbative effects and other thermal corrections into the scattering amplitudes and dispersion relations of the asymptotic hadrons, (ii) the resulting mean-free path of the asymptotic states remains much larger than their thermal de Broglie wave length.

[11] In [31], it was proposed an increase due to a direct correlation between the bulk viscosity and the trace anomaly, which lattice calculations have shown to have a peak near the transition temperature. However, analyses in other field theories indicate that this correlation may not be direct [38].

For the heavy-ion program at FAIR, it is of special importance the physics of the QCD critical point [45]. It belongs to the dynamic universality class of the Model H [2] (the same as that of the liquid-gas phase transition, according to the classification of [46]). For this universality class, the linear transport coefficients present the (divergent) critical behavior near T_c [46,47]: $\eta \sim t^{-0.05}$, $\zeta \sim t^{-2.8}$, and $\kappa \sim t^{-0.95}$ (thermal conductivity), with $t \equiv |T - T_c|/T_c$. Therefore, the bulk viscosity would probably cause a stronger experimental effect [48,49,50], whereas the shear viscosity is essentially insensitive to the critical point. For a recent analysis of the critical behavior of the thermal conductivity and its phenomenological consequences see [51].

Acknowledgments
I want to thank the organizers of FAIRNESS 2012 for the kind invitation to participate in the workshop. My work is supported by an Alexander von Humboldt fellowship. I also thank partial financial support from the Spanish Research projects FIS2008-01323 and FPA2011-27853-C02-02.

References

[1] Kovtun P 2012 *J. Phys.* A **45** 473001
[2] Son D T and Stephanov M A 2004 *Phys. Rev.* **D70** 056001
[3] Aharony O, Gubser S S, M Maldacena J, Ooguri H and Oz Y 2000 *Phys. Rept.* **323** 183
[4] Son D T and Surowka, P 2009 *Phys.Rev.Lett.* **103** 191601
[5] Jensen K, Kaminski M, Kovtun P, Meyer R, Ritz A and Amos Y 2012 *Phys. Rev. Lett.* **109** 101601
[6] Jensen K, Kaminski M, Kovtun P, Meyer R, Ritz A and Amos Y 2012 *JHEP* **1205** 102
[7] Baier R, Romatschke P, Son D T, Starinets, A O and Stephanov M A 2008 *JHEP* **04** 100
[8] Hubeny V E, Minwalla S and Rangamani M 2011 The fluid/gravity correspondence *Preprint* arXiv:1107.5780
[9] Danielewicz P and Gyulassy M 1985 *Phys. Rev.* **D31** 53
[10] Kovtun P, Son D T and Starinets A O 2005 *Phys. Rev. Lett.* **94**
[11] Kovtun P, Moore G D and Romatschke P 2011 *Phys. Rev.* **D84** 025006
[12] Torrieri G 2012 *Phys. Rev.* **D85** 065006
[13] Arnold P, Moore G D and Yaffe, L G 2000 *JHEP* **11** 001
[14] Jeon S 1995 *Phys. Rev.* **D52** 3591
[15] Aarts G and Martinez Resco J M 2004 *JHEP* **02** 061
[16] Son D T and Starinets A O 2007 *Ann. Rev. Nucl. Part. Sci.* **57** 9
[17] Aarts G and Martinez Resco J M 2002 *JHEP* **04** 053
[18] Ding H-T, Francis A, Kaczmarek O, Karsch F, Laermann E and Soeldner W 2011 *Phys. Rev* **D83** 034504
[19] de Forcrand P 2009 *PoS* **LAT2009** 010
[20] Arnold P, Moore G D and Yaffe L G 2003 *JHEP* **05** 051
[21] Xu Z and Greiner C 2008 *Phys. Rev. Lett.* **100** 172301
[22] Chen J-W, Dong H, Ohnishi K and Wang Q 2010 *Phys. Lett.* **B685** 277
[23] Prakash M, Prakash M, Venugopalan R and Welke G 1993 *Phys. Rept.* **227** 321
[24] Muronga A 2004 *Phys. Rev.* **C69** 044901
[25] Demir N and Bass S A 2009 *Phys. Rev. Lett.* **102** 172302
[26] Noronha-Hostler J, Noronha J and Greiner C 2009 *Phys. Rev. Lett.* **103** 172302
[27] Fernandez-Fraile D and Gomez Nicola A 2009 *Eur. Phys. J.* **C62** 37
[28] Fernandez-Fraile D and Gomez Nicola A 2007 *Eur. Phys. J.* **A31** 848
[29] Buchel A, Myers R C and Sinha A 2009 *JHEP* **0903** 084
[30] Arnold P, Dogan C and Moore G D 2006 *Phys. Rev.* **D74** 085021
[31] Kharzeev D and Tuchin K 2008 *JHEP* **09** 093
[32] Meyer H B 2008 *Phys. Rev. Lett.* **100** 162001
[33] Fernandez-Fraile D and Gomez Nicola A 2009 *Phys. Rev. Lett.* **102** 121601
[34] Chen J-W and Wang J 2009 *Phys. Rev.* **C79** 044913
[35] Lu E and Moore G D 2011 *Phys. Rev.* **C83** 044901
[36] Kogan A B and Meyer H 1998 *J. Low Temp. Phys.* **110** 899
[37] Buchel A 2012 *Phys. Rev.* **D85** 066004
[38] Fernandez-Fraile D 2011 *Phys. Rev.* **D83** 065001
[39] Gupta S 2004 *Phys. Lett.* **B597** 57
[40] Fernandez-Fraile D and Gomez Nicola A *Phys. Rev.* **D73** 045025

[41] Csernai L P, Kapusta J I and McLerran L D 2006 *Phys. Rev. Lett.* **97** 152303
[42] Dobado A, Llanes-Estrada F J and Torres-Rincon J M 2009 *Phys. Rev.* **D80** 114015
[43] Romatschke P and Romatschke U 2007 *Phys. Rev. Lett.* **99** 172301
[44] Torrieri G, Tomasik B and Mishustin I 2008 *Phys. Rev.* **C77** 034903
[45] Friman B, Höhne C, Knoll J, Leupold S, Randrup J, Rapp R and Senger P 2011 *The CBM physics book* (Berlin: Springer)
[46] Hohenberg P C and Halperin B I 1977 *Rev. Mod. Phys.* **49** 435
[47] Onuki A 2004 *Phase transition dynamics* (Cambridge: Cambridge University Press)
[48] Karsch F, Kharzeev D and Tuchin K 2008 *Phys. Lett.* **B663** 217
[49] Kapusta J I 2008 Viscous properties of strongly interacting matter at high temperature *Preprint* arXiv:0809.3746
[50] Moore G D and Saremi O 2008 *JHEP* **9** 015
[51] Kapusta J I and Torres-Rincon J M 2012 Thermal conductivity and chiral critical point in heavy ion collisions *Preprint* arXiv:1209.0675

Journal of Physics: Conference Series **426** (2013) 012002 doi:10.1088/1742-6596/426/1/012002

Azimuthal correlations and collective effects in a heavy-ion collisions at the LHC energies

Ilya Selyuzhenkov

Research Division and ExtreMe Matter Institute EMMI,
GSI Helmholtzzentrum für Schwerionenforschung, Darmstadt, Germany

E-mail: ilya.selyuzhenkov@gmail.com

Abstract. Recent highlights from the anisotropic flow and the azimuthal correlation measurements in a heavy-ion collisions at the LHC are presented. Various flow harmonics measured for the charged and identified particles versus transverse momentum, pseudo-rapidity, and the collision centrality are reported. New experimental probes of the local parity violation at the LHC energies using the charge dependent azimuthal correlations are also discussed.

1. Introduction

An azimuthal anisotropic flow which describes collectivity among particles produced in heavy-ion collision is recognized as a key observable used to infer information about the early time evolution of the nuclei interaction. This FAIRNESS 2012 Conference proceedings highlight recent results by the LHC experiments from the anisotropic flow measurements in relativistic heavy-ion collisions at the TeV energy scale. Experimental findings from the search for effects of the local parity violation in strong interaction using the charge dependent azimuthal correlations with respect to the reaction plane at the LHC energies are also discussed.

2. Anisotropic flow fluctuations

It is commonly understood that the even-by-event fluctuations in the initial energy density of a heavy-ion collision plays an important role in the development of the azimuthal asymmetries in the momentum distribution of the produced particles. Since fluctuations are mapped by the interaction among constituents of the expanding medium into the final momentum space asymmetry, the measurements of the anisotropic flow provide a unique experimental information about the properties of the created medium, its evolution, and profile of the initial conditions in a heavy-ion collision. Recently a number of new high statistics experimental results from LHC which help to understand details of flow fluctuations become available. Figure 1(a) shows the results of the multi-particle correlation analysis of the directed v_1, elliptic v_2, and triangular v_3, flow measured for Pb-Pb collisions at $\sqrt{s_{\rm NN}} = 2.76$ GeV by the ALICE Collaboration [1]. Observed non-zero odd flow harmonics represent a significant effect of the initial energy fluctuations especially for the most central collisions where all harmonics become comparable to each other. Agreement between the anisotropic flow estimates with 4, 6, and 8-particle correlations put stringent constraints on the shape of the event-by-even flow distribution.

Differential measurements of the anisotropic flow versus pseudorapidity η and transverse momentum $p_{\rm T}$ provide additional information on the extend of flow fluctuations to forward

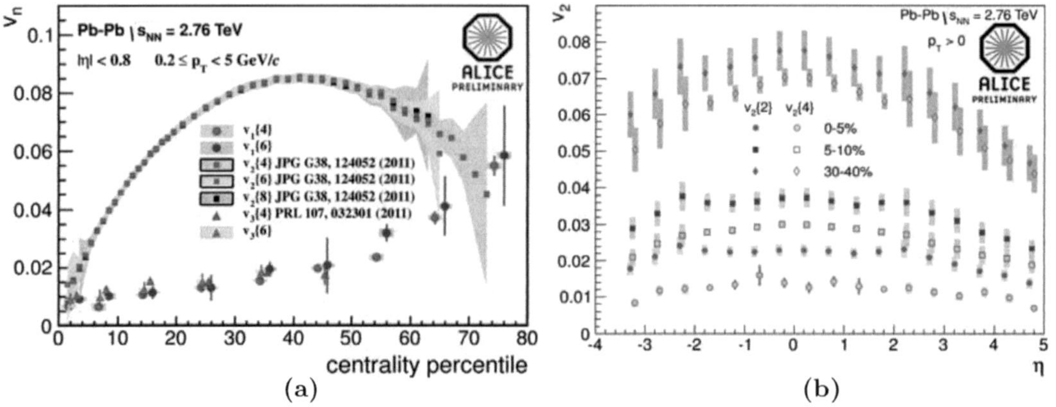

Figure 1. Charged particle v_1, v_2, and v_3 measured by ALICE [1, 5] in Pb-Pb collisions at $\sqrt{s_{\mathrm{NN}}} = 2.76$ TeV. (a) Multi-particle estimates of v_1, v_2, and v_3 at midrapidity vs. centrality. (b) Two- and four-particle estimates of v_2 vs. pseudorapidity.

rapidity and higher transverse momenta. Such measurements performed by the ATLAS [2], CMS [3], and ALICE [4] Collaborations are shown in Fig. 1(b) and Fig. 2. The flow estimates

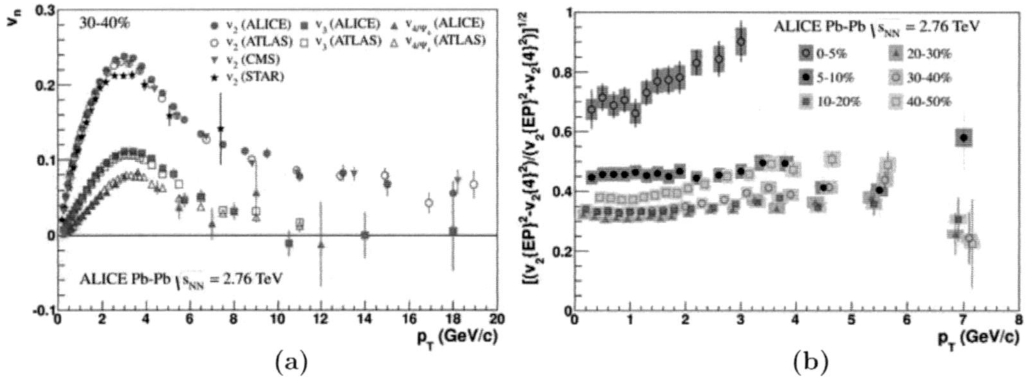

Figure 2. (a) Charged particle v_2, v_3, and v_4 measured for Pb-Pb collisions at $\sqrt{s_{\mathrm{NN}}} = 2.76$ TeV by the ATLAS [2], CMS [3], and ALICE [4] Collaborations. (b) Estimate of the relative v_2 flow fluctuations as a function of the transverse momentum for different collision centrality classes.

obtained with the event plane, two and four-particle measurement techniques [3, 4, 5] suggest that relative flow fluctuations weakly depend on p_{T} and η. Figure 2(b) shows that fluctuations extend to higher p_{T} up to about 8 GeV/c, while Fig. 1(b) indicates a similar amount of relative fluctuations up to $|\eta| \sim 5$.

Further constraints on the effects of the initial energy fluctuations is provided by the measurements of the multi-particle mixed harmonic correlations and the correlation between the different order symmetry planes. Figure 3 shows corresponding results by the ATLAS [6] and ALICE [1] Collaborations in comparison with the expectations from the event-by-event hydrodynamic model calculations using two different profiles of the initial condition [7]. Observed correlations between the two and three planes of symmetry and their qualitative agreement with the hydrodynamic model calculations points on their strong sensitivity to the

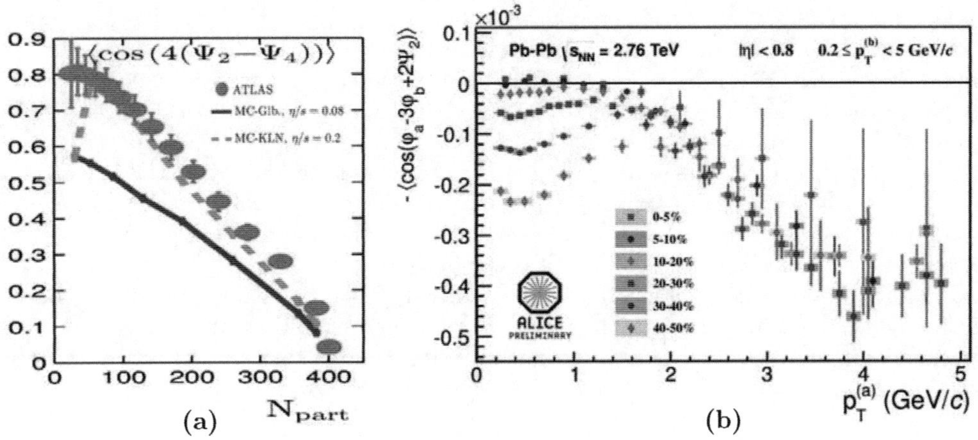

Figure 3. (a) 2nd and 4th order event plane correlation vs. the number of participants (centrality) measured by ATLAS [6] for Pb-Pb collisions at $\sqrt{s_{\rm NN}}$ = 2.76 TeV. Results are compared with the event-by-event hydrodynamic model calculations. (b) Three-particle azimuthal correlation proposed in [8] measured by ALICE [1] for the same collision system.

details of the fluctuating initial energy profile. Among other promising developments in the fluctuations studies are the shape measurements of the event-by-event flow distributions (see e.g. [9]), and the measurements of physics observables for event classes selected based on their azimuthal shapes. This new type of study suggested in [10] was successfully applied in the real data analysis by the ALICE Collaboration [11, 12].

3. Elliptic flow of identified particles

Viscous hydrodynamics is considered to be a relevant model to describe a thermalized phase in the time evolution of the system created in a heavy-ion collision. Applicability of the

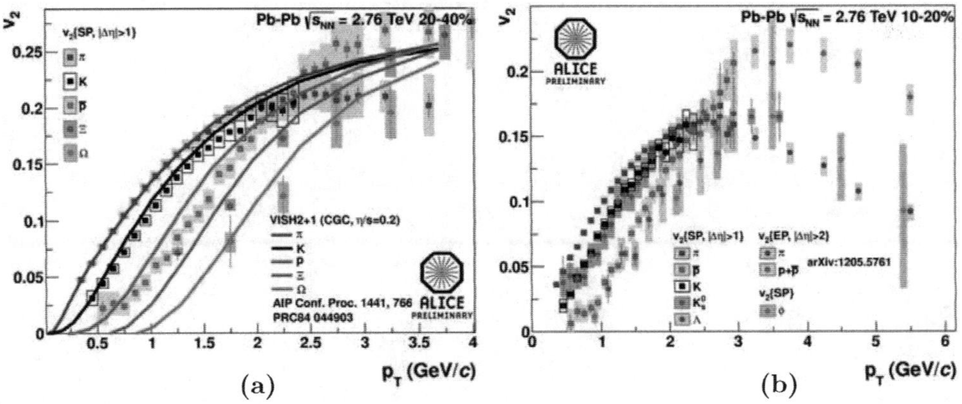

Figure 4. Identified particle $v_2(p_{\rm T})$ measured by ALICE [13] for Pb-Pb collisions at $\sqrt{s_{\rm NN}}$ = 2.76 TeV. Charged pion, kaon, and proton v_2 compared to that of (a) Ξ and Ω and the hydrodynamic model calculations [14], and (b) with v_2 of ϕ-meson, K_s^0 and $\Lambda/\bar{\Lambda}$ particles.

hydrodynamic description at the LHC energies can be tested by measuring the particle mass dependence of the elliptic and triangular flow at small transverse momenta, $p_T < 2-3$ GeV/c. Figure 4 shows $v_2(p_T)$ of pions and anti-protons in comparison to that of strange (K^0_s, $\Lambda/\bar{\Lambda}$) and multi-strange (ϕ, Ξ, and Ω) particles. The main trends of the observed mass splitting of v_2 at low transverse momenta is reproduced by viscous hydrodynamic model calculations [14] with a color glass condensate initial condition (see Fig. 4(a)). The flow of heavier particles (proton, ϕ-meson, Ξ and Ω) is more sensitive to the hadronic rescattering phase and the agreement with data improves when adding this additional phase into the model calculations (see Fig. 2 in [15]). v_2 of ϕ-meson which is shown in Fig. 4(b) follows the mass splitting expected from the hydrodynamics at lower transverse momenta, while its magnitude is close to v_2 of other mesons (pions) in the intermediate region of $p_T \sim 3-6$ GeV/c. Such behavior suggest the constituent quark number (n_q) scaling of v_2 which is expected for mesons and baryons produced via quark coalescence in a phase of the deconfined quarks and gluons. This scaling may indeed holds at the LHC within 20% at the $p_T/n_q \sim 1.2$ GeV/c (see Figs. (4) and (5) in [13]).

4. Probes of local parity violation in strong interaction

The parity symmetry violation by the strong interactions remains one of the open fundamental questions about the quantum chromodynamics. It is argued [16] that the parity symmetry can be locally violated in a heavy-ion collisions what will result in the experimentally observable separation of charges along the extreme magnetic field generated by the moving ions, the so called chiral magnetic effect (CME). As an experimental probe of the CME it was proposed [17] to use the azimuthal correlations with respect to the collision reaction plane which is perpendicular to the magnetic field generated in the collision. The STAR [18] at RHIC and the ALICE [19] at the LHC observed a clear charge dependence of the two-particle correlation with respect to the reaction plane. The measured observable is parity even and thus is sensitive to effect unrelated to the parity symmetry violation. Recently the ALICE Collaboration extended [20] the

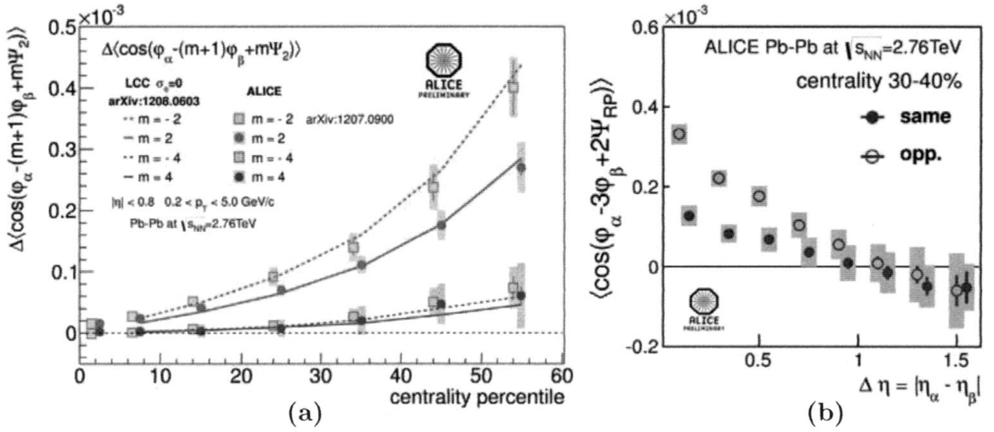

Figure 5. Charge dependent correlations measured by ALICE [20] for Pb-Pb collisions at $\sqrt{s_{NN}} = 2.76$ TeV. (a) The same minus the opposite charge correlations compared to the local charge conservation model [25]. (b) Correlation $\langle\cos(\phi_\alpha - 3\phi_\beta + 2\Psi_2)\rangle$ vs. rapidity separation [20].

measurements with the new set of the mixed harmonic correlators $\langle\cos(\phi_\alpha - (m+1)\phi_\beta + m\Psi_m)\rangle$ (for $m = \pm2, \pm4$) and the double harmonic correlator $\langle\cos(2\phi_\alpha + 2\phi_\beta - 4\Psi_4)\rangle$, where $\phi_{\alpha,\beta}$ is the azimuthal angle, α, β is the charge of the particle, and Ψ_m is the m-th order collision symmetry plane angle. Figure 5(a) shows the correlation difference $\Delta\langle\cos(\phi_\alpha - (m+1)\phi_\beta + m\Psi_m)\rangle$ between the same and the opposite charge combinations. The expected CME contribution to the correlator $\langle\cos(\phi_\alpha + \phi_\beta - 2\Psi_2)\rangle$ ($m = -2$ in Fig. 5) at the LHC energies is based on the extrapolation of the RHIC results. Depends on the model assumptions, it varies from about 20% [21] up to the full strength [16, 22, 23] of the correlations observed at the LHC if one assumes that magnetic field flux is independent of the collision energy. It is noted [24] that effects of local charge conservation modulated by the anisotropic flow can be responsible for a significant part of the measured correlations. The statement is also supported by model calculations [25] shown in Fig. 5(a). The charge independent part of the correlations can be dominated by effects of flow fluctuations [8], although the differential study of the charge dependent correlation shown in Fig. 5(b) reveals strong dependence on the pair separation in pseudorapidity which is not typical for flow fluctuations. At the moment none of the models is able to reproduce simultaneously the charge dependence and the charge insensitive baseline of the measured correlations.

5. Summary and outlook

The anisotropic flow and the multi-particle azimuthal correlation measurements at the LHC help to clarify the role of the initial energy fluctuations in the multiple particle production in a heavy-ion collision. The particle type and the mass dependence of the anisotropic flow suggest that the system produced in a heavy-ion collision is strongly coupled and evolved via the deconfined phase of quarks and gluons. The new measurements of the mixed harmonic charge dependent azimuthal correlations helps to constrain possible background sources in the search for effects of the parity symmetry violation in the QCD.

Acknowledgements

Supported by the Helmholtz Alliance Program of the Helmholtz Association, contract HA216/EMMI "Extremes of Density and Temperature: Cosmic Matter in the Laboratory".

References

[1] A. Bilandzic [ALICE Collaboration], arXiv:1210.6222 [nucl-ex].
[2] G. Aad *et al.* [ATLAS Collaboration], Phys. Rev. C **86**, 014907 (2012) [arXiv:1203.3087 [hep-ex]].
[3] S. Chatrchyan *et al.* [CMS Collaboration], arXiv:1204.1409 [nucl-ex].
[4] B. Abelev *et al.* [ALICE Collaboration], arXiv:1205.5761 [nucl-ex].
[5] A. Hansen [ALICE Collaboration], arXiv:1210.7095 [nucl-ex].
[6] J. Jia [ATLAS Collaboration], arXiv:1208.1427 [nucl-ex].
[7] Z. Qiu and U. Heinz, Phys. Lett. B **717**, 261 (2012) [arXiv:1208.1200 [nucl-th]].
[8] D. Teaney, L. Yan, Phys. Rev. **C83**, 064904 (2011).
[9] J. Jia [ATLAS Collaboration], arXiv:1209.4232 [nucl-ex].
[10] J. Schukraft, A. Timmins and S. A. Voloshin, arXiv:1208.4563 [nucl-ex].
[11] A. Dobrin [ALICE Collaboration], arXiv:1211.5348 [nucl-ex].
[12] L. Milano [ALICE Collaboration], Proceedings of the "Quark Matter 2012" conference.
[13] F. Noferini [ALICE Collaboration], arXiv:1212.1292 [nucl-ex].
[14] C. Shen, U. Heinz, P. Huovinen, H. Song, Phys. Rev. **C84**, 044903 (2011).
[15] F. Noferini [ALICE Collaboration], Acta Phys. Polon. Supp. **5**, 329 (2012).
[16] D. E. Kharzeev, L. D. McLerran and H. J. Warringa, Nucl. Phys. A **803**, 227 (2008).
[17] S. A. Voloshin, Phys. Rev. **C70**, 057901 (2004).
[18] B. I. Abelev *et al.* [STAR Collaboration], Phys. Rev. Lett. **103**, 251601 (2009) [arXiv:0909.1739 [nucl-ex]].
[19] B. Abelev *et al.* [ALICE Collaboration], arXiv:1207.0900 [nucl-ex].
[20] Y. Hori [ALICE Collaboration], arXiv:1211.0890 [nucl-ex].
[21] V. D. Toneev and V. Voronyuk, Phys. Atom. Nucl. **75**, 607 (2012) [arXiv:1012.1508 [nucl-th]].
[22] A. R. Zhitnitsky, Nucl. Phys. A **853**, 135 (2011) [arXiv:1008.3598 [nucl-th]].
[23] A. R. Zhitnitsky, Nucl. Phys. A **886**, 17 (2012) [arXiv:1201.2665 [hep-ph]].
[24] S. Schlichting, S. Pratt, Phys. Rev. **C83**, 014913 (2011).
[25] Y. Hori, T. Gunji, H. Hamagaki and S. Schlichting, arXiv:1208.0603 [nucl-th].

Journal of Physics: Conference Series **426** (2013) 012003 doi:10.1088/1742-6596/426/1/012003

Phenomenology of quarkyonic percolation at FAIR

Giorgio Torrieri, Stefano Lottini

FIAS, J. W. Goethe Universität, Frankfurt Am Main, Germany.

Abstract. We will give an introduction to the concept of quarkyonic matter, presenting an overview of what is meant by this term in the literature. We will then argue that the quarkyonic phase, as defined in the original paper, is a percolation-type phase transition whose phase transition line is strongly curved in $\rho_B - N_c$ space, where N_c is the number of colors and ρ_B the baryon density. With a toy model estimate, we show that it might be possible to obtain a percolating but confined phase at $N_c = 3, N_f = 2$ at densities larger than one baryon per one baryon size. We conclude by discussing how this phase can be observed at FAIR.

A concept recently receiving quite a lot of theoretical attention is that of "quarkyonic matter". Introduced in [1], it can be best defined as matter where temperature is below deconfinement, $T < T_c$, but quark chemical potential is of the order of the confinement scale (inverse of the baryon size), $\mu_Q \geq \Lambda_{\mathrm{QCD}}$, resulting in a density of a baryon per baryonic volume or higher. This regime is not easily accessible through controlled theoretical expansions: Effective hadronic models, since they expand in powers of Q/Λ_{QCD}, where Q is the momentum exchange, can be reasonably expected to diverge. Lattice calculations are affected by the sign problem, which generally can only be controlled at $\mu_Q \ll T$.

Hence, expectations for matter in this regime are very model-specific. Popular ideas included a confined yet chirally restored phase [2, 3], chirally inhomogeneus phases [4, 5, 6], a deconfined but chiral-broken plasma of constituent quarks [7] and a large-N_c generalization of nuclear matter [8, 9, 10]. Ref. [1] has generically argued that in this regime confinement coexists

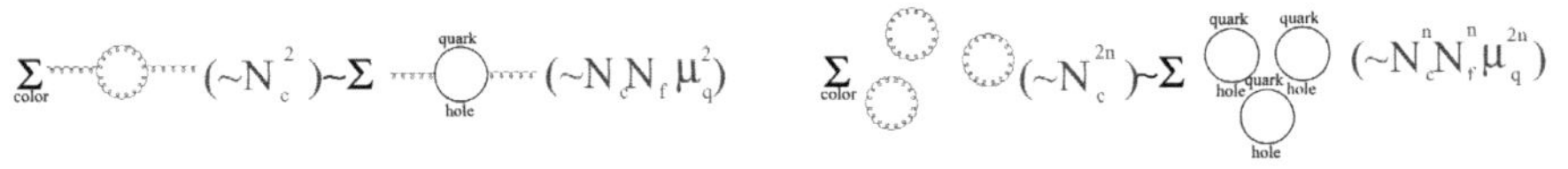

Figure 1. Diagrams at arbitrary N_c responsible for gluon antiscreening and quark-hole screening.

with pQCD and asymptotic freedom. The essence of this argument is rather simple and qualitative: Assuming that at finite chemical potential deconfinement appears when the quark-hole screening overpowers the gluon-gluon antiscreening, we can obtain how the chemical potential required for deconfinement varies with N_c. It turns out that at leading order (Fig. 1), $\mu_Q^{deconfinement} \simeq \sqrt{N_c/N_f}\Lambda_{\mathrm{QCD}}$. At higher order ($n$ loops) the N_c dependence is parametrically unchanged. Thus, looking at momentum space, we expect $\mu_Q \simeq \Lambda_{\mathrm{QCD}}$ matter to be firmly confined.

In configuration space, on the other hand, we have a baryon with N_c quarks per baryonic size. Hence, quarks of neighboring baryons are distant by $N_c^{-1/3} \to 0$. Thus, at least at large N_c, asymptotic freedom should somehow coexist with confinement. A natural picture of this, found in [1], is an asymptotically free gas of quarks moving in a mean field of baryons. Alternatively, the bulk of the Fermi surface is asymptotically free but its excitations are confined.

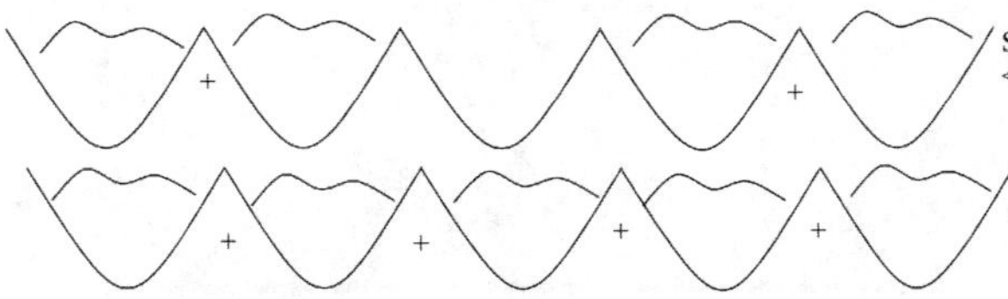

Figure 2. The influence of percolation of quark wavefunctions inside a dense distribution of hadrons (shown here as potential wells).

How relevant is this reasoning to our $N_c = 3, N_f = 2$ world? As we have already known for a few decades, the $N_c \to \infty$ expansion is qualitatively remarkably good for a lot of QCD phenomenology: That mesons are weakly-interacting quasi-particles, baryons look like semi-classical "skyrmion" states of mesonic fields, and planar diagrams dominate are all features of the large N_c expansion. On the other hand, some features of QCD thermodynamics are not even qualitatively well reproduced: At large N_c, deconfinement is a first order phase transition rather than a cross-over. At large chemical potential, binding energy of nuclear matter is *of the order of the baryon mass* $\sim N_c\Lambda_{\text{QCD}}$ (rather than $\sim 10^{-3}N_c\Lambda_{\text{QCD}} \ll m_\pi$, as in our world).

Thus, the existence of phase transitions in N_c, as well as T and μ_Q, is virtually assured. Such transitions can not obviously be studied through experimental data, but nevertheless remain pretty interesting theoretically: The large-N_c technique is one that has been extensively used for phenomenology as well as theory. The identification of the large-N_c limit with a classical dual gauge theory makes this topic more interesting still, since the phase transition in N_c is a quantum-gravitational one in its dual description. Unfortunately, as confinement shows, determining that a phase transition exists from a fundamental theory can be a very subtle issue.

In the rest of this work, we shall use percolation [11, 12] as a "toy model" to investigate a transition between normal nuclear matter and the quarkyonic phase which bears a lot of similarity with that described in [1]. Percolation is a simple yet surprisingly deep and universal paradigm: It requires an infinite network of "links", each with a probability, invariant in space, to be "on" or "off". The crucial question is the typical length of a connected cluster of "on" links as a function of the individual link probability.[1] Somewhat subtly, this link's dependence on probability is not monotonic: At the critical "percolation" probability (typically not very high, $\mathcal{O}(1)/N_{neighbours}$), this length diverges. Considering this as a correlation length, it is clear percolation defines a second order phase transition, an insight which is confirmed by the appearance of critical behavior at the percolation threshold. The abstractness and generality of percolation makes it very appealing as a toy model, since *any* local propagator in an extended medium should give rise to percolation. Specifically, by analogy with the well-known conductor-insulator transition in QED, percolation signals the regime where one can not consider quark wavefunctions to be effectively localized to a baryon since quantum tunneling will have the potential to make quarks propagate through arbitrary distances at equilibrium (Fig. 2).

To investigate the curvature in ρ_B-N_c space, we posutulate a generic interaction propagator, respecting what we know about Yang-Mills theory: Due to confinement, it has to fall off

[1] Excluding the infinite component, if any.

Journal of Physics: Conference Series **426** (2013) 012003 doi:10.1088/1742-6596/426/1/012003

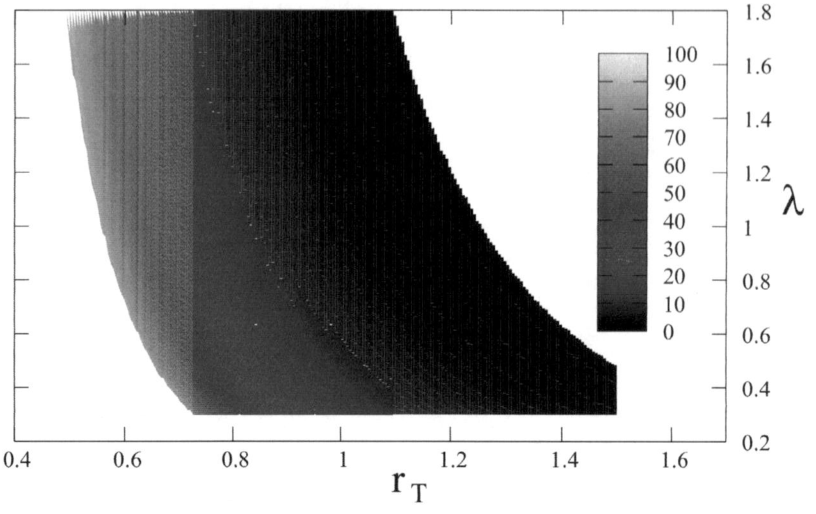

Figure 3. The dependence of the percolation critical number of colors N_c on λ, r_T at a density of one baryon per baryon size and hexagonal packing [11].

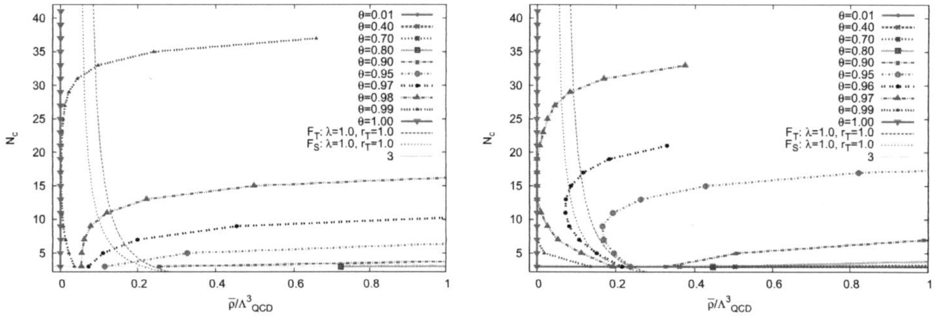

Figure 4. Two-dimensional plots in the (baryon density,N_c) plane of the percolation curves (lines without data points) vs. the deconfinement curves at various temperatures ($\theta = T/T_c$), for $N_f = 1, 3$ (resp. left and right plots).

at distances higher than $\Lambda_{\rm QCD}^{-1}$ in configuration space, and the interaction probability has to $\sim \lambda/N_c$:

$$f(x) = \frac{\lambda}{N_c} f\left(\frac{x}{r_T}\right) \quad , \quad f\left(\frac{x}{r_T} \geq 1\right) \to 0 \quad ; \tag{1}$$

in units of $\Lambda_{\rm QCD}$, it is natural that $r_T \sim \mathcal{O}(1)$. Since at that scale λ becomes strong, it is natural that it is $\lambda \sim \mathcal{O}(1)$ too. As long as these requirements are satisfied, the shape of $f(x)$ (Θ-function in position space, Gaussian, etc) is irrelevant. Fig. 3 [11] shows the dependence of the critical number of colors N_c for a regular hexagonal packing on λ, r_T at fixed baryon density. It is clear that at the chemical potential of $\mu_Q \simeq \Lambda_{\rm QCD}$, indicated in [1], $N_c = 3$ is not enough to trigger a percolation transition, which requires $N_c \simeq \mathcal{O}(10)$. However, the possibility of quarkyonic matter at $1 \leq \mu_Q/\Lambda_{\rm QCD} \leq \sqrt{N_c/N_f}$ remains open if the effective number of flavors $N_f < 3$. Note that, since the mass of the strange quark $\simeq \Lambda_{\rm QCD}$, it is far from clear whether the

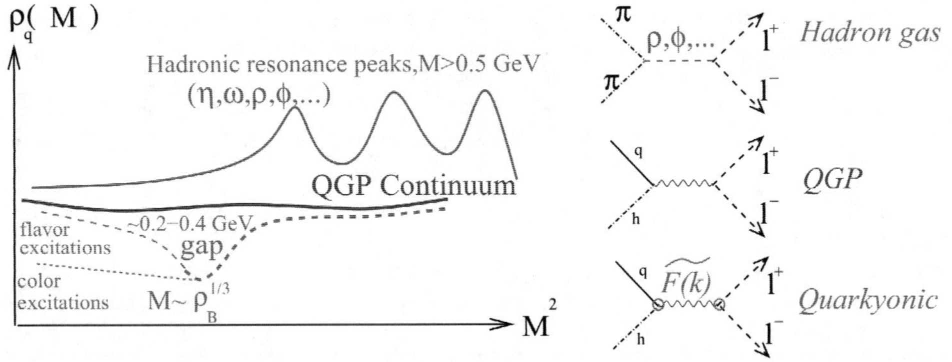

Figure 5. Spectral function of dileptons in a hadron gas, in a pQCD, and in quarkyonic matter.

above hyerarchy can be maintained, since it is uncertain whether the strange flavor is "light" for our purposes.

To further investigate whether the quarkyonic phase can appear, we varied density as well as N_c [12]. Note that, while this is an obvious extension, technically it is somewhat difficult because the definition of "nearest neighbor" is somewhat sloppy with varying density (the nearest neighbor at one density can be farther than the next-to-nearest neighbor at the other). To obtain the critical percolation probability, we integrated out physical baryons into a square lattice, and performed a renormalization-group analysis. The details of this procedure are available in [12].

Unlike percolation, baryonic density necessary for deconfinement goes up with N_c because of the $\mu_Q \sim \sqrt{N_c/N_f}\Lambda_{\mathrm{QCD}}$ requirement. Hence, the "true high-N_c limit" is one where deconfinement happens at higher μ_Q than percolation. In this limit, there will be both a confined percolating phase and a deconfined phase. At the low N_c limit, percolation would require a higher μ_Q than deconfinement. Since percolation requires baryonic wavefunctions to be present, this means percolation *never* occurs. Where is our world in this hyerarchy? In Fig. 4 [12] we plotted, as a function of N_c, μ_Q, T, the density of an ideal baryon gas at deconfinement. As can be seen, a sliver of percolating but confined phase diagram remains. While the "ideal" approximation for baryon gas density is rough in this limit, this calculation shows percolation is *possible* in our world. Hence, an experimental search for it might be necessary.

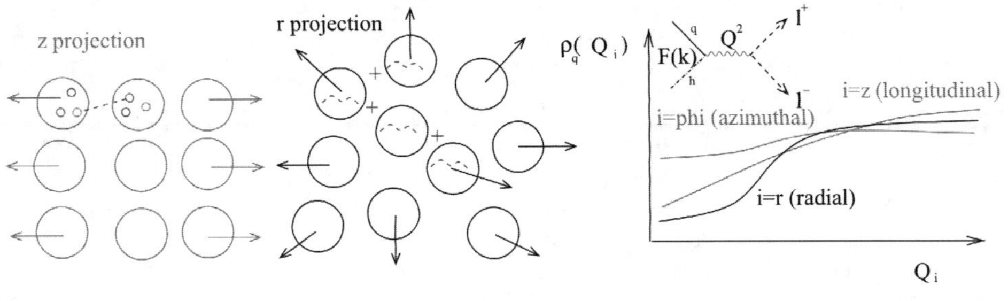

Figure 6. Influence of the baryonic dynamics on the decomposed quark-hole spectral function in quarkyonic matter.

What would be an experimental signature for the production of "quarkyonic matter"? This question was not really answered by [1]; however, the coexistence, in quarkyonic matter, of pQCD degrees of freedom with baryonic mean fields makes exploration of the spectral function by dilepton detection a promising avenue to investigate. Consider a reaction such as $q + \text{hole} \to \gamma^* \to l^+ l^-$ whose rate Γ will be given by

$$\Gamma(M) = \int \mathcal{M}\left((p_1 - p_2)^2\right) f(p_1) f(p_2) d^3p_1 d^3p_2 \delta\left((p_1^\mu - p_2^\mu)^2 - M^2\right) \quad ; \tag{2}$$

the assumption that quark and hole degrees of freedom be quark-like means $\mathcal{M}(...)$ should not be too different from the quark-quark scattering matrix elements calculated within pQCD.

The distribution functions $f(p_{1,2})$ will not however be those of a free gas but will contain a mean field reflecting the distribution of baryons in the system. Equivalently, the Eigenstates of the unperturbed quarkyonic matter will not be free particles (i.e., the spectral function will not be flat in M), but rather Eigenstates of the Hamiltonian

$$\hat{H} = \hat{k}^2 + m_q^2 + V_{\text{mf}}(\vec{x}) \quad , \quad V_{\text{mf}}(\vec{x}, \vec{y}) = \sum_i^{(\text{baryons})} \Phi_{\text{mf}}(x - x_i) \tag{3}$$

where the mean field baryonic potential Φ_{mf} will contain dips centered around the position of the baryons in configuration space (see Fig. 2). If baryons were immobile and at regular intervals, as we assumed in Fig. 3 and 4, we would recover Bloch's theorem for QED conductors: the spectral function of dileptons would experience a dip at $M \simeq \rho_B^{1/3}$, as shown in Fig. 5, due to the disappearance of quark wavefunctions having a momentum $\simeq \rho_B^{1/3}$. Such dips would be distinctive for quarkyonic matter, since a hadron gas would give *peaks* for the production of resonances and a perturbative QGP would give a relatively flat spectral function (Fig. 5).

Density of inhomogeneities and the movement of baryons are likely to at least somewhat spoil this picture. However, the dynamics of baryons is calculable by semiclassical dynamical models such as uRQMD [13], and shows regularities such as transverse and longitudinal expansion. Also, experimentally we can access not just the invariant mass of the spectral function but its full vector decomposition $M_{y,r,\phi}$ in rapidity-transverse-azimuthal space. Since in quarkyonic matter quarks are not point particles but non-trivial Eigenfunctions of the distribution of baryons, the shape of the $M_{y,r,\phi}$ distribution is likely going to be non-trivial, and reflecting macroscopic dynamics. To calculate it, one would need to solve the Dirac equation for dips centered around a classical distribution of baryons (whose positions are given by RQMD), and then fold the solutions with pQCD interaction matrices. This work is currently in progress, and we expect to get quantitative predictions for dilepton distributions from quarkyonic matter.

In conclusion,we have used a percolation-based model to argue for the existence of a dense "quarkyonic" phase in which quark wavefunctions can percolate to arbitrary distances, yet confinement persists and baryons are well-defined objects. We have shown that this phase is almost certain to exist in the limit $N_c \gg N_f$, and is not excluded in our $N_c = 3, N_f = 2$ world. We have proposed dilepton spectral functions as a possible experimental verification of this phase, and await developments at low-to-intermediate energy heavy ion collisions, such as the FAIR program. G. T. acknowledges the financial support received from the Helmholtz International Center for FAIR within the framework of the LOEWE program (Landesoffensive zur Entwicklung Wissenschaftlich-Ökonomischer Exzellenz) launched by the State of Hesse.

[1] L. McLerran and R. D. Pisarski, Nucl. Phys. A **796**, 83 (2007) [arXiv:0706.2191 [hep-ph]].
[2] L. McLerran, K. Redlich and C. Sasaki, Nucl. Phys. A **824**, 86 (2009) [arXiv:0812.3585 [hep-ph]].
[3] L. Y. Glozman, V. K. Sazonov and R. F. Wagenbrunn, arXiv:1111.0949 [hep-ph].
[4] S. Carignano, D. Nickel and M. Buballa, Phys. Rev. D **82**, 054009 (2010) [arXiv:1007.1397 [hep-ph]].
[5] T. Kojo, Y. Hidaka, L. McLerran and R. D. Pisarski, Nucl. Phys. A **843**, 37 (2010) [arXiv:0912.3800 [hep-ph]].
[6] K. Miura, T. Z. Nakano and A. Ohnishi, Prog. Theor. Phys. **122**, 1045 (2009) [arXiv:0806.3357 [nucl-th]].
[7] P. Castorina, R. V. Gavai and H. Satz, arXiv:1003.6078 [hep-ph].
[8] O. Bergman, G. Lifschytz and M. Lippert, JHEP **0711**, 056 (2007) [arXiv:0708.0326 [hep-th]].

[9] J. de Boer, B. D. Chowdhury, M. P. Heller and J. Jankowski, arXiv:1209.5915 [hep-th].
[10] G. Torrieri and I. Mishustin, Phys. Rev. C **82**, 055202 (2010) [arXiv:1006.2471 [nucl-th]].
[11] S. Lottini and G. Torrieri, Phys. Rev. Lett. **107**, 152301 (2011) [arXiv:1103.4824 [nucl-th]].
[12] S. Lottini and G. Torrieri, arXiv:1204.3272 [nucl-th].
[13] H. Petersen, M. Bleicher, S. A. Bass and H. Stocker, arXiv:0805.0567 [hep-ph].

Journal of Physics: Conference Series **426** (2013) 012004 doi:10.1088/1742-6596/426/1/012004

Role of hyperons in black hole formation

Sarmistha Banik[(a,b)]

[(a)]Saha Institute of Nuclear Physics, APC Division, Kolkata, INDIA
[(b)]BITS Pilani, Hyderabad Campus, Sameerpet Mondal, Hyderabad, INDIA

E-mail: sarmistha.banik@saha.ac.in

Abstract. A phase transition from hadronic to exotic phases might occur in the early post-bounce phase of a core collapse supernova. We investigate the role of strange hyperons in the dynamical collapse of a non-rotating massive star to a black hole using 1D General relativistic simulation *GR1D*. We follow the dynamical formation and collapse of a protoneutron star (PNS) from the gravitational collapse of a $40M_{solar}$ progenitor of Wooseley, adopting Shen hyperonic EoS. We also study the neutrino signals that may be used as a probe to core collapse supernova. We compare the results with those of Shen nuclear EoS and understand the role of strange hyperons in the core collapse.

1. Introduction

The core collapse supernova explosion mechanism is being investigated over the last five decades [1]. Still, the detailed theory of a successful supernova explosion is beyond our reach and is a challenging problem. The inner core of massive stars rebounds as it reaches supra nuclear densities at their final journey, a shock wave is said to form [2]. The shock soon looses its energy and stalls after traversing a few hundred kilometers. The massive star directly ends up in a black hole without an explosion, this falls in the category of failed supernova [3]. However, there are different ideas to revive the shock, which could eventually trigger a delayed supernovae explosion. The second shock could be very important in understanding a successful core collapse supernova explosion.

In recent years existence of strange matter in the high-density core of neutron stars has been the subject of extensive research [4,5]. It may so happen that the phase transition from hadronic to exotic phases might have already occurred in the early post-bounce phase of a core-collapse supernova. This idea of appearance of strange particle has initiated intense interest lately. It is said that a quark-hadron phase transition can trigger the explosion and revive the shock [6], that was stalled due to loss of energy. During the implosion, neutrinos could escape the surface of the star carrying off most of the energy and reach detectors on earth. The neutrino burst ceases once the shock is stalled. However, an additional burst of neutrino is released as the second shock passes through the neutrinosphere. Neutrinos could be a significant observational signature [6].

One of the main inputs in the study of evolution of core-collapse supernovae by numerical simulations, is the equation of state (EoS) for a wide range of density ($10^4 - 10^{15}$g/cc), temperature ($0 - 100$MeV) and composition (proton fraction $0 - 0.6$) [7]. For the supernovae simulations with non-strange particles like neutrons, protons, alpha-particles and nuclei, mainly two sets of EoS are used —Lattimer-Swesty (LS) [8] and Shen, Toki, Oyamatsu and Sumiyoshi

(Shen) EoS [9]. Also there have been supernova EoS based on thermodynamically consistent nuclear statistical equilibrium model [7,10] . The first EoS with non-nucleonic degrees of freedom was presented by Ishizuka et. al. [11]. They studied the emergence of the full baryon octet in the dynamical collapse of a massive static star to a black hole formation [12]. This EoS was recently utilised to study the behaviour of black hole formation and neutrino emission with hyperons and/or pions in [13]. Another set of EoS with hyperon degrees of freedom was constructed for supernova simulations [14] using the relativistic mean field model (RMF).

It is obvious that the inclusion of strange degrees of freedom softens the EoS. A stiffer EoS can sustain more mass against the collapse. A soft EoS on the other hand favors lower maximum masses compared to the stars having nucleonic degrees of freedom only. The recent measurement of the Shapiro delay in the radio pulsar PSR J1614-2230 which yielded a mass of $1.97 \pm 0.04 M_{solar}$, puts an important constraint on the neutron star mass and may rule out most of the soft EoS [15]. However, it is at present not possible to rule out any exotica with this observation as many model calculations including hyperons and/or quark matter could still be compatible with it [14, 16, 17], as many of these approaches are model dependent.

In this paper, we report the effect of hyperons on the black hole formation using the spherically-symmetric General relativistic hydrodynamic code, *GR1D* [18], designed to follow the evolution of stars beginning from the onset of core collapse. We adopt two sets of the Shen EoS–for nucleon(np) [9] and hyperon(npY) degrees of freedom [14]. We comment on the neutrino signal that might be observed as a result of phase transition from nucleonic to hyperonic matter.

The paper is arranged as follows. In Section 2, we briefly describe the EoS, adopted for the simulations. Section 3 is devoted to result discussion. Finally we summarise in Section 4.

2. The equation of state (EoS) and the numerical simulation

We use the nuclear and hyperonic equation of state by Shen et. al [9, 14] for our simulations. Shen nuclear EoS is based on a relativistic mean field model at intermediate and high densities ($\rho > 10^{14.2}$ g/cc). At low temperature (T$\leq$ 14MeV), and $\rho < 10^{14.2}$ g/cc, Thomas Fermi approximation is used. Leptons are treated as uniform non-interacting relativistic particles and their contributions are added separately. Minimisation of free energy is done at low density. The calculation has been done in RMF model with the TM1 parameter set [19], the parameters of the model are obtained by fitting the experimental data for binding energies and charge radii of heavy nuclei. With the TM1 parameter set, the nuclear matter saturation density is $0.145 fm^{-3}$, the binding energy per nucleon is 16.3 MeV, the symmetry energy is 36.9 MeV and the compressibility is 281MeV [19]. Shen et. al. included the Λs in their EoS table [14]. For the parameters of Λ hyperons, they use the experimental mass value $M_\Lambda = 1115.7$ MeV. The coupling constant for hyperon-vector meson interaction is taken based on naive quark-model and that of hyperon-scalar meson interactions is determined by fitting experimental binding-energies data for single-Λ hypernuclei [20]. Λs appear when the threshold condition, $\mu_n = \mu_\Lambda$ is satisfied, where μ_n and μ_Λ are the chemical potentials of the neutron and Λ respectively. Other hyperons, Ξ & Σ are excluded due to their relatively higher threshold and lack of experimental data.

We use the open source code *GR1D* [18] for the supernova simulations. *GR1D* is a spherically-symmetric, general-relativistic Eulerian hydrodynamics code for low and intermediate mass progenitors. It is designed to follow the evolution of stars beginning from the onset of core collapse to black hole formation and makes use of several microphysical EoS.

3. Results & Discussion

We report our simulation results for a $40 M_{solar}$ progenitor model of Woosley et. al [21] using GR1D [18] for Shen EoS- nucleon(np) as well as and hyperon(npY) [9, 14]. We solved the Tolman-Oppenheimer-Volkov equation for zero temperature (T=0) EoS of neutron stars(NS) assuming neutrino-less β equilibrium. The maximum mass of the neutron star for np EoS is

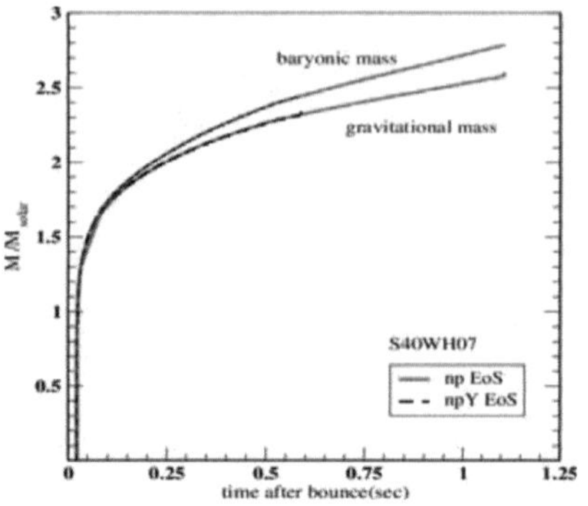

Figure 1. Baryonic mass and gravitational mass vs. time after bounce

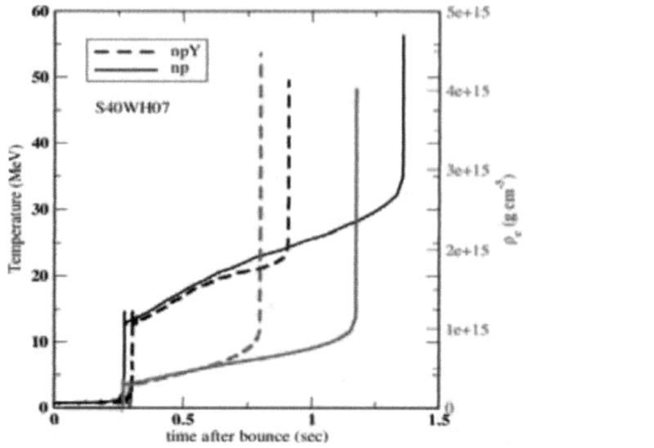

Figure 2. Central density (ρ_c) vs time after bounce

2.18 M_{solar}, whereas for npY EoS, the maximum mass reduces to $1.82 M_{solar}$. The corresponding radii are 12 and 12.5 km respectively.

In Fig. 1 we plot the baryonic and gravitational mass of proto-neutron star(PNS), obtained from simulations. The maximum mass is higher than that of NS. When accretion pushes PNS over its maximum mass, a black hole(BH) is formed. The spike in the gravitational mass correspond to a blow-up and the BH formation. For the np EoS, this happens for a $2.71 M_{solar}$ star at 1.09sec after bounce, whereas for npY EoS (the black dashed lines) this happens much earlier at 0.57sec after bounce for a $2.38 M_{solar}$ star.

Fig. 2 shows the evolution of central density (ρ_c) and temperature (T) for the np (solid lines) and npY EoS (dashed lines). The bounce corresponds to the spikes at real timeline $t_{bounce} = 0.27$sec, which we take as t=0 in the figure. The value of t_{bounce} is same for the np and npY EoS, as the contribution of hyperon is not important at that time as would be discussed later. The onset of BH formation is marked by a sharp rise in the value of ρ_c (the black lines). Similar trend is noticed in the temperature profile (red lines in online version). Owing to the hyperon emergence, the contraction of PNS is accelerated, which leads to quicker rise in temperature and central density. Or in other words, the stiffer EoS leads to larger post-bounce time to BH-formation.

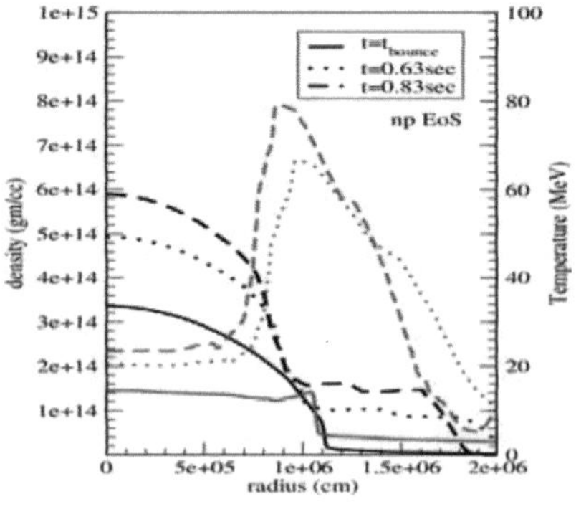

Figure 3. Density and temperature profiles for np EoS

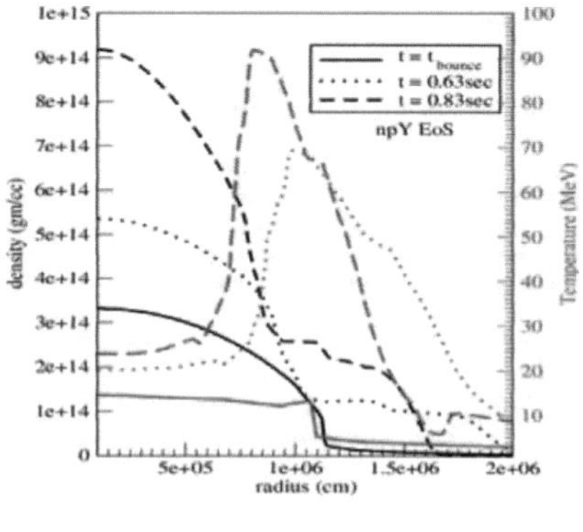

Figure 4. Density and temperature profiles for npY EoS

In Figs. 3 and 4, we compare the density(black lines) and temperature (red lines in online version) profiles for np and npY cases. The density rises from less than ρ_0 ($\rho_0 \simeq$

Journal of Physics: Conference Series **426** (2013) 012004 doi:10.1088/1742-6596/426/1/012004

$2.4 \times 10^{14} gm/cm^{-3}$) at the surface to a few times ρ_0 at the core. The plateau in the density profile could be attributed to strong thermal pressure there. At core bounce, the core density is $1.4\rho_0$. With intense accretion, the central density shoots to $2\rho_0$ at 0.63 sec, and $2.5\rho_0$ at 0.83secs for np EoS. The density and temperature profiles are similar in both the cases, the central density is slightly above that of the np case at t=0.63 sec, as Λ just starts appearing in the system. However, at t=0.83sec, owing to substantial amount of Λ, the central density rises almost $3.8\rho_0$, which is 2.8 times its value at core bounce.

With time, the temperature also attains a peak at the mid-radius region. The peak rises from 66.8Mev at 0.63sec to 79.35Mev at 0.83sec in np case (Fig. 3). This is due to accretion and compression of shock heated material onto the PNS surface. At this region, the thermal pressure support is enough to flatten the density profile. In inner core ($\sim$ 6km) the material is not shock heated, rather is heated by adiabatic compression. The temperature peak is further raised to 91.7MeV at 0.83sec in the presence of Λ hyperons (Fig. 4).

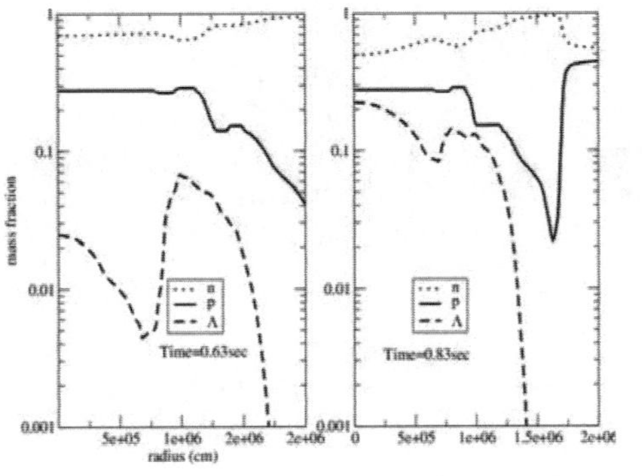

Figure 5. Snapshot of particle fraction as a function of radius at t=0.63sec and 0.83sec

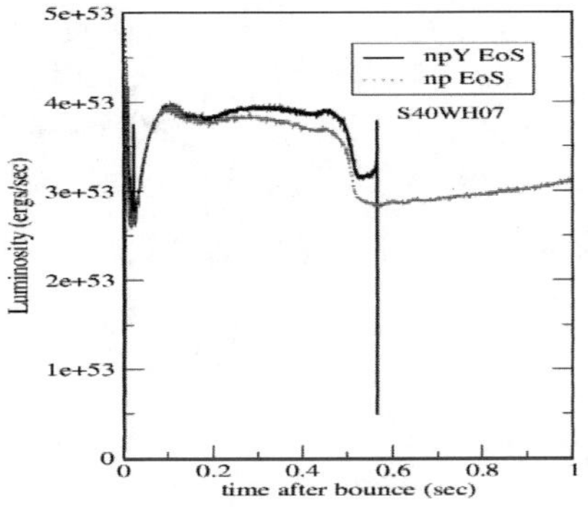

Figure 6. Luminosity of total neutrinos ($\nu_e, \bar{\nu}_e$ and ν_μ) as a function of time after bounce.

Next we compare the compositions of PNS in Fig. 5. We have noticed that initially at core bounce the system consists of neutron and protons only; hyperons appears first in the collapse at 0.16sec after core bounce (assuming 10^{-3} considerable amount of fraction). The central density that was just above normal nuclear matter density (Fig 2) at bounce rises to $3.79 \times 10^{14} gm/cm^{-3}$ at 0.16sec after bounce. We have seen that the appearance of Λ hyperon is delayed until the matter density reaches at least $2\rho_0$, the threshold density shifts to lower density with increasing temperature [14]. The temperature also increases to 16.26 MeV. We display two snapshots at 0.63sec and 0.83 sec after core bounce in the two panels of Fig. 5. It is interesting to note that hyperons appear off-center owing to high temperature, although density is still on the plauteu. At 0.63sec after core bounce, the abundance of Λ becomes significant at R $\simeq$ 10km, as temperature is maximum there (Fig. 5). It even falls sharply after reaching the peak due to fall in temperature, only to rise at the core again owing to high density there. At a later time, the high central density forbids it from dropping too low, once it reaches the peak at mid-radius region. Thus, Λ becomes one of the major components at the core.

In the final figure, the evolution of neutrino-luminosity is plotted for the np and npY EoS. We find a short neutrino burst($\sim$ 1sec) before the PNS, born temporarily in a failed-supernova, terminates in a black hole. The resulting neutrino burst in np and npY cases are quite similar, differ only in earlier termination of burst in the latter. The soft npY EoS lowers the critical mass of PNS, thus accelerates the mass accretion onto it and triggers the gravitational instability at 0.63sec. However, no second neutrino burst is observed as was in quark-hadron phase transition [6]. The quark EoS is stiff while the npY is a soft one. So, npY though triggers black hole formation, fails to generate second shock.

4. Summary

We studied the effect of hadron-hyperon phase transition in core-collapse supernova using general relativistic hydrodynamic simulation GR1D [18]. By following the dynamical collapse of a new-born proto-neutron star from the gravitational collapse of a $40M_{solar}$ star adopting Shen hyperonic EoS table [14], we noticed that hyperons appear just before bounce. It appears off center at first due to high temperature and prevails at the center just before the black hole formation, when the density becomes quite high. Hyperons triggers the black-hole formation, but fails to generate the second shock as the EoS is softened too much with the appearance of hyperons. Hyperon emergence in the collapse produces an intense but short neutrino burst, which terminates at the black hole formation. However, no second neutrino burst is observed as in quark-hadron phase transition.

There are possibilities for other strange degrees of freedom in the form of kaon condensates to appear in the highly dense matter. We have seen such a phase transition can support a maximum mass [5], which is well above the observed mass [15]. It would be intriguing to investigate if a hadron-antikaon condensed matter can revive the second shock. A successful shock revival would have observational consequence in the form of neutrino signatures. Until now, only one supernova, SN1987A, has been detected by its neutrinos. Post SN1987A, more advanced neutrino facilities, such as ice-cube and super-Kamiokande are expected to detect the neutrino signals more efficiently and frequently.

Acknowledgments

The author acknowledges travel support from the organisers of FAIRNESS 2012 and would like to thank Debades Bandyopadhyay, Evan O'connor and Christian Ott for insightful discussions.

References

[1] A. Burrows, arXiv:1210.4921
[2] G. Brown, Sci. American, **252** 60 (1985).
[3] S. E. Woosley, A. Heger, and T. A. Weaver, Rev. Mod. Phys. **74** 1015 (2002).
[4] N.K. Glendenning, Compact stars, (Springer, New York, 1997).
[5] S. Banik and D. Bandyopadhyay, Phys. Rev. **C63** 035802 (2001), Phys. Rev. **C64** 055805 (2001), Phys. Rev. **C66** 065801 (2002), Phys. Rev. **D67** 123003 (2003).
[6] T. Fischer, I. Sagert, G. Pagliara, M. Hempel, J. Schaffner-Bielich, T. Rauscher, F.-K. Thielemann, R. Kppeli, G. Martnez-Pinedo, M. Liebendrfer ApJ. Supp. **194**:39, (2011).
[7] M. Hempel and J. Schaffner-Bielich, Nucl. Phys. A **837** 210 (2010).
[8] J. M. Lattimer and F. D. Swesty, Nucl. Phys. A **535** 331 (1991).
[9] H. Shen, H. Toki, K. Oyamatsu, K. Sumiyoshi, Nucl.Phys. **A637** 435 (1998).
[10] S. Furusawa, S. Yamada, K. Sumiyoshi, H. Suzuki, Astrophys.J. **738** 178 (2011).
[11] C. Ishizuka, A. Ohnishi, K. Tsubakihara, K. Sumiyoshi, S.Yamada, J. Phys. **G35** 085201 (2008).
[12] K. Sumiyoshi, C. Ishizuka, A. Ohnishi, S. Yamada, H. Suzuki, APjL,**690**, L43 (2009).
[13] K. Nakazato, S. Furusawa, K. Sumiyoshi, A. Ohnishi, S. Yamada, H. Suzuki, Astrophys.J.**745** 197 (2012).
[14] H. Shen, H. Toki, K. Oyamatsu, K. Sumiyoshi, ApjS, **197** 20 (2011).
[15] P. B. Demorest, T. Pennucci,S. M. Ransom, M. S. E. Roberts & J. W. T. Hessels, Nature **467** 1081 (2010).
[16] J. M. Lattimer, M. Prakash, arXiv:1012.3208
[17] E. Massot, J. Margueron, and G. Chanfray, EPL **97** 39002 (2012)
[18] E. O'Connor, C. D. Ott, ApJ **730** 70 (2011)
[19] Y. Sugahara and H. Toki, Nucl. Phys. **A579** 557 (1994).
[20] H. Shen,F. Yang, H. Toki, Prog. Theo. Phys., **115** 325 (2006).
[21] S. E. Woosley, A. Heger, and T. A. Weaver, Rev. Mod. Phys. **74** 1015 (2002).

Meson Production in Coherent Antiproton-Nucleus Annihilation

S Lourenço[1], H Lenske[1] and S Wycech[2]

[1] Inst. f. Theoretische Physik, JLU Giessen Heinrich-Buff Ring 16, D-35392 Giessen, Germany
[2] National Centre for Nuclear Studies, Hoza 69, 00-681, Warsaw, Poland

E-mail: stefanie.lourenco@physik.uni-giessen.de

Abstract. The study of a fully quantum mechanical description of coherent meson production in antinucleon annihilation on nuclei is presented. Two different reaction mechanisms are presented including initial and final state interactions, which are both derived from a folding approach. The underlying fundamental antinucleon-nucleon $\overline{N}N$ and pion-nucleon πN interactions enter into the optical potentials, which are folded with Hartree-Fock-Bogoliubov nuclear densities. Existing approaches to pion nucleus interactions have been extended to higher energies beyond the Δ-resonance. Cross sections are shown.

1. Introduction

Meson and hadron production by antinucleon-nucleus annihilation reactions is well suited to explore a broad spectrum of final particle configurations and physics phenomena. Varying the incident energy, the whole spectrum of flavour states can be scanned through. Annihilation of an incoming antiproton on both protons and neutrons of the target nucleus allows to study the dependence of production dynamics on the initial charge state. The upcoming FAIR@GSI facility with its large variety of envisioned experiments will provide ideal experimental conditions for such investigations. Corresponding efforts are being made on the theoretical side. Of highest interest is to identifying physics cases of general relevance. Here, we are investigating meson production in antiproton-nucleus reactions [1], intended as exploratory studies for the PANDA [2] experiment and, if realized at a later stage of FAIR, also for the nuclear structure-oriented use of high energy antiprotons aimed for within the AIC proposal [3]. Meson production in hadronic reactions can be obtained by different reaction scenarios. A well studied case is meson production out of high energy collisions of two heavy ions, see for example [4]. Here, our point of view is different: We consider coherent $\overline{p}A$ reactions, leaving the residual nucleus B intact and leading otherwise to few-body exit channels with a limited number of emerging hadrons. As a concrete and typical example we treat explicitly the case of two pion production. Coherent reactions have the distinct advantage that the full quantum mechanical structure of the production process is accessible, including the details of the interactions of the projectile and the ejectiles with the initial and final nuclei, respectively. Hence, different to heavy ion collisions, determined by reaction probabilities, in coherent reactions one is dealing with quantum mechanical amplitudes, including information on modulus and phase. The reaction scenarios for the coherent meson production are discussed in section 2. For their treatment $\overline{p}A$ initial and mB final state interactions are required. Our approach to these interactions is briefly summarized in section 3 and section 4, respectively. In both cases, we use a description in terms of microscopically determined optical potentials. Both the $\overline{p}A$ and the πA interactions are expressed in terms of state-of-the-art HFB nuclear structure input for ground state densities and binding energies, also far off stability. The reaction amplitudes by themselves had to be extended into energy regions hitherto not studied, neither experimentally nor theoretically. The paper closes with a summary and an outlook in section 5.

2. Reaction Mechanisms and Production Vertices

Two-meson production is inherently leading to a 3-body exit channel. This 3-body problem is not – or cannot be – solved exactly. In order to find reliable approximations, it is worthwhile to consider the reaction in more detail. A closer inspection shows that we can distinguish two meaningful reaction scenarios:

2.1. Peripheral In-Flight Reactions

In this reaction mechanism the antinucleon in flight picks up a nucleon of the target nucleus and forms an intermediate $\overline{N}N$ correlated state, subsequently decaying into mesons in free space. The in-flight scenario resembles in the first step a traditional pick-up reaction as long known in (p, d) or $(p, 2p)$ reactions, but with the presence of a nucleus, there are more nucleons available to take care of the recoil momentum. The therefore low-momentum transfer character of this reaction is reflected in clearly visible nuclear structure effects indicated by the diffraction pattern of the angular distributions shown in in figure 1. As a first step we calculated the production of a hadron cloud. In figure 1 is shown the pick-up of valence neutrons of ^{48}Ni, stable ^{58}Ni and neutron-rich ^{78}Ni target nuclei.

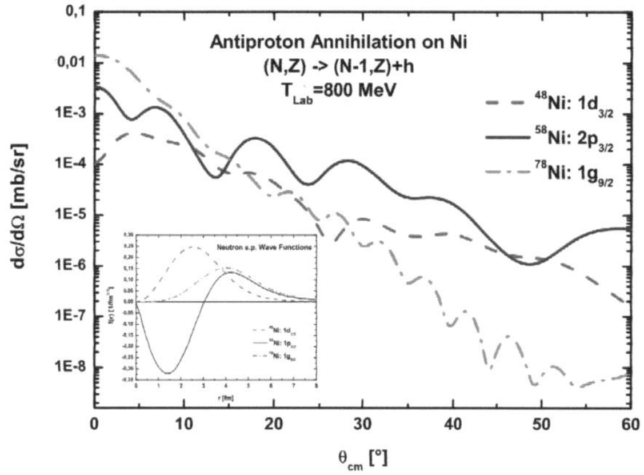

Figure 1. As a first step we calculate the cross section of the production of a hadron cloud, not resolving its inner structure. The differential cross sections are shown for valence neutrons of ^{48}Ni, stable ^{58}Ni and neutron-rich ^{78}Ni and $T_{lab} = 800$ MeV.

2.2. In-Situ Reactions

In contrast to the in-flight reaction mechanism the antinucleon in this reaction is stopped inside the target nucleus, similarly as in former experiments at LEAR. The annihilation and production of mesons takes place in-medium. The nucleus must absorb a large part of the incoming momentum. Hence, short range correlations and nuclear form factor effects will play a dominant role. As a first step we consider the production of a ρ-meson, which then decays into two pions. The relative motion of those pions has been integrated out. In figure 2 the cross

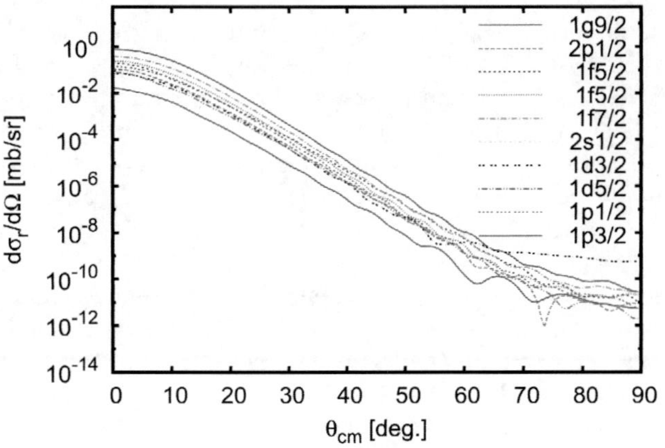

Figure 2. As a first step we calculate the cross section of one meson production, namely ρ^-, which then decays into two pions. The relative motion of the two pions has been integrated out. The differential cross sections are shown for meson production on a ^{78}Ni target and T_{lab} = 800 MeV.

section is shown for the valence neutrons of ^{78}Ni target.

As has been discussed above the two scenarios are different in their energy-momentum balance, but in a general description, both can be expressed in terms of cross section:

$$d^9\sigma_{\alpha\beta} = N_{\alpha\beta}\left(\frac{\hbar c}{2\pi}\right)^9 \frac{d^3k_1}{E_1}\frac{d^3k_2}{E_2}\frac{d^3k_B}{E_B}\left|M_{\alpha\beta}\left(\vec{k_1},\vec{k_2},\vec{k_B};\vec{k_\alpha}\right)\right|^2 \quad (1)$$

$$\times\delta\left(\vec{k_1}+\vec{k_2}+\vec{k_B}\right)\delta\left(E_1+E_2+E_B-\sqrt{s}\right) \quad (2)$$

where $N_{\alpha\beta}$ is a normalization constant and the subscripts 1 and 2 stand for each pion, while B denotes the residual nucleus, α corresponds to the initial state and $\sqrt{s}$ is the total centre of mass energy, while the deltas account for energy and momentum conservation. Studying the production mechanism one finds a separation of scales. The influence of the nuclear matter proceeds on the scale of the Fermi-momentum ($\approx$ 250 MeV/c), while the production involves baryon exchange ($\approx$ 1 GeV/c), this allows to factorize the production amplitude into overlap wave function φ_B and the production amplitude $t_{\overline{N}N\to 2\pi}$ in the matrix element:

$$M_{\alpha\beta} \approx t_{\overline{N}N\to 2\pi}(s)\langle\chi^{(-)}_{1\beta}\chi^{(-)}_{2\beta}|\varphi_B|\chi^{(+)}_\alpha\rangle. \quad (3)$$

The outgoing wave functions $\chi^{(-)}_{1\beta}\chi^{(-)}_{2\beta}$ may be of mesonic or baryonic character, depending on the scenario, discussed above. The antiproton and meson self-energies are entering into the scattering wave functions of the incoming and outgoing particles. The wave functions are calculated in eikonal approximation.

3. Initial State Interaction

The prevailing reaction mechanism of the initial state $\overline{N}A$ interaction is annihilation. Due to the strong absorption the antinucleon penetrates only a little into the nucleus. In principle, the $\overline{N}N$ interaction is modified by in-medium self-energies and Pauli-blocking of the target nucleons, but the importance of the low-density surface region justifies, as an approximation, to separate nuclear properties from the $\overline{N}N$ dynamics. This is done within a folding approach, where the $T^{(A)}_{\overline{N}N} \propto T_{\overline{N}N}$ is folded over the nuclear density ρ_N, taken from self-consistent Hartee-Fock-Bogoliubov calculations. This leads to a good approximation of the optical potential

$$U_{opt}(\mathbf{r}) = V - iW = \sum_{N=p,n}\int\frac{d^3q}{(2\pi)^3}\rho_N(q)t_{\bar{p}N}(T_{Lab},q^2)e^{i\mathbf{q}\cdot\mathbf{r}}. \quad (4)$$

Journal of Physics: Conference Series **426** (2013) 012005 doi:10.1088/1742-6596/426/1/012005

At those energies where theoretical results are available, the $\overline{N}N$ interactions are taken from microscopic models, as for example Paris [5] and Jülich/Bonn [6]. Those models derive the $\overline{N}N$ potential by G-parity transformation of their NN potential. In a meson-exchange picture, that means a change of the sign of the contribution of π, ω and δ mesons. From the analysis of data one finds a a rather shallow real part, indicating that the dispersive contributions are effectively reducing the otherwise very deep static potential obtained by G-parity, e.g. [7]. However, the dominating imaginary part strongly suppresses the incoming antiproton flux already in the surface region, thus inhibiting direct access to the nuclear interior. Considering the Jülich model in more detail, we find as a typical structure that a few annihilation channels are treated explicitly while the majority of states is treated schematically by a purely phenomenological $\overline{N}N$ optical potential. This optical potential is fitted to describe the data up to 300 MeV incident kinetic energy. At higher energies where theoretical results are missing, we are depending presently on a phenomenological approach using a T-matrix where the on-shell strength is derived from the $\overline{N}N$ cross section by means of the optical theorem. Finally, we mention that for central $\overline{p}A$ collisions transport theory is the appropriate approach, as discussed e.g. in [4].

4. Final State Interaction

Similarly to the initial state interaction we describe the final state meson nucleus interaction within a folding approach, leading to an expression similar to eq. 4. However, the leading order part which is linear in the density must be extended because meson interactions with correlated pairs of nucleons are known to be important. Hence, the pion-nucleus contains significant contributions being proportional to the square of the target density:

$$U(r) = 4\pi f_{\pi N}\rho(r) + \delta U(\rho^2) \tag{5}$$

where ρ is the nuclear density and $f_{\pi N}$ the pion-nucleon scattering amplitude. The leading order term, however, is modified by genuine in-medium contributions, denoted by δU. Important effects are the two-nucleon absorption mechanism and $N^{-1}\Delta$ excitations off the medium, both leading to contributions of at least ρ^2, see e.g. [8]. There have been detailed discussions on a microscopic level of the density dependence of pion-nucleus potentials for the low energy s-wave case [8]. Reactions like two nucleon absorption have been studied [9] and are implemented in our work. Because the two produced pions are sharing the momentum of almost 2 M_N we developed a semi-phenomenological description by extending an existing approach of [10]. As a starting point a Kisslinger type potential is used:

$$U_{opt} = U_s + \overrightarrow{\partial}\ U_p\ \overrightarrow{\partial}. \tag{6}$$

The derivatives covering the U_p already indicates, that the p-wave part mainly contributes at the nuclear surface. As discussed in [11] in order to extend and improve the energy dependence of the s-wave part U_s by taking into account higher resonances. By using a Krell-Ericson transformation $\Phi = (1 - U_p)^{-1/2}\psi$ one derives a local potential:

$$U_N(r) = \frac{(\hbar c)^2}{2\omega}\left\{\frac{U_s}{1-U_p} - \frac{k^2 U_p}{1-U_p} - \left[\frac{\frac{1}{2}\vec{\nabla}^2 U_p}{1-U_p} + \left(\frac{\frac{1}{2}\vec{\nabla} U_p}{1-U_p}\right)^2\right]\right\}, \tag{7}$$

where ω is the energy and k the momentum of the pion. A Schrödinger equation is solved in eikonal-approximation, leading to a good description of the data over the whole energy range as shown in figure 3 and figure 4.

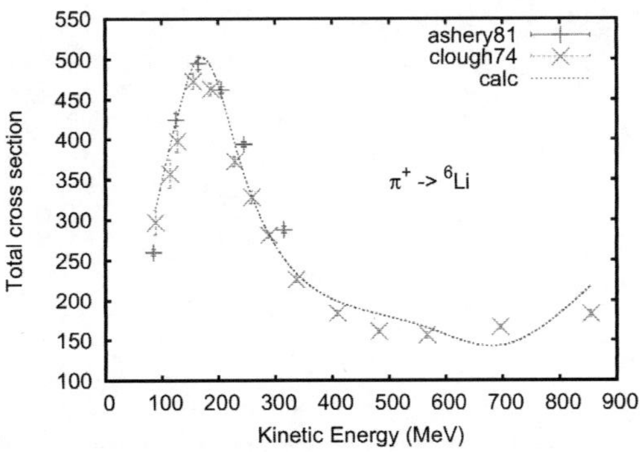

Figure 3. Total cross section for π^+ ^{6}Li interaction. The calculation (dashed line) is compared with data [12] and [13].

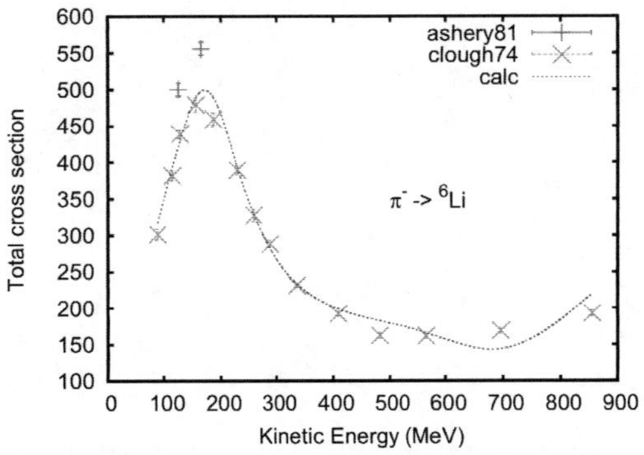

Figure 4. Total cross section for π^- ^{6}Li interaction. The calculation (dashed line) is compared with data [12] and [13].

5. Summary and Outlook

Meson production in coherent, peripheral $\overline{p}A$ reactions has been investigated in a fully quantum mechanical approach. Initial and final state interactions were treated by using realistic nuclear structure input and elementary reaction amplitudes. The different scales defined by the soft-scale nuclear properties on the one hand and by the hard-scale production dynamics on the other hand allowed to use a separation ansatz. The reactions as such take place in the low density region of the nuclear surface. Hence, a $t\rho$ folding approach is an appropriate first approximation. Cross sections for the two types of reaction mechanisms were presented in Distorted Wave Eikonal Approximation (DWEA). The in-flight scenario could be of potential interest also for the search for $\overline{p}p$ or $\overline{p}n$ baryonium bound states which will produce a distinct component in the spectra of the outgoing particles. The in-situ scenario is ideal for searching Pontecorvo-type reactions [14] given by the re-absorption of most of the produced mesons on the target nucleons. The two-meson channels, although being only a minor part of the $\overline{p}A$ total cross section are of particular interest because of their comparatively simple channel configuration and clean kinematics. Extensions to $\overline{K}K$ production is feasible and, in fact, is a work in progress. At higher energies, also $B\overline{B}$ baryon-antibaryon channels will open, e.g. $\Lambda\overline{\Lambda}$ hyperon production. These reactions are in principle measurable at PANDA, thus opening a new research path.

Acknowledgments

Partially supported by DFG grant Le439/8-1

[1] Lourenco S, Lenske H, Wycech S, *Hyp. Int.* **209** (2012) 117
[2] PANDA Proposal `http://www-panda.gsi.de/archive/public/panda_pb.pdf`
[3] Proposal for an Antiproton-Ion Collider at FAIR, (AIC), GSI, Darmstadt (2005)
[4] Gaitanos T, Larionov A B, Lenske H, Mosel U, *Nucl. Phys.* A **881** (2012) 240
[5] El-Bennich B, Lacombe M, Loiseau B, Wycech S, *Phys. Rev.* C **79** (2009) 054001
[6] Mull V, Holinde K, *Phys. Rev.* C **51** (1995) 2360
[7] Mishustin I N, Satarov L M, Bürvenich T J, Stöcker H, Greiner W, Phys. Rev. C **71** (2005) 035201
[8] Oset E, Toki H, Weise W, *Phys. Rep.* **83** (1982) 281
[9] Mutazz Nuseirat, Lodhi M A K, Gibbs W R, *Phys. Rev.* C **58** (1988) 314
[10] Johnson M B, Satchler G R, *Ann. of Phys.* **248** (1996) 134
[11] Lourenco S, Lenske H, Wycech S, *Genshikaku Kenkyu Suppl.* Proc. SNP workshop (2012) in print
[12] Ashery D, Navon I, Azuelos G, Walter H K, Pfeiffer H J, Schlepütz F W, *Phys. Rev.* C **23** (1981) 2173
[13] Clough A S et al., *Nucl. Phys.* B **76** (1974) 15
[14] Pontecorvo B M, *Zh. Eksp. Teor. Fiz.* **30** (1956) 947 *Sov. Phys. JETP 3 (1956) 966*

Journal of Physics: Conference Series **426** (2013) 012006 doi:10.1088/1742-6596/426/1/012006

Monte Carlo simulations of Microdosimetry for Space Research at FAIR

Lucas Burigo[1,2], **Igor Pshenichnov**[1,3], **Igor Mishustin**[1,4] **and Marcus Bleicher**[1,2]

[1] Frankfurt Institute for Advanced Studies, Johann Wolfgang Goethe University, 60438 Frankfurt am Main, Germany

[2] Institut für Theoretische Physik, Johann Wolfgang Goethe University, 60438 Frankfurt am Main, Germany

[3] Institute for Nuclear Research, Russian Academy of Sciences, 117312 Moscow, Russia

[4] Kurchatov Institute, Russian Research Center, 123182 Moscow, Russia

E-mail: `burigo@fias.uni-frankfurt.de`

Abstract. The exposure to high charge and energy (HZE) particles is one of major concerns for humans during their missions in space. As radiation effects essentialy depend on charge, mass and energy of cosmic-ray particles, the radiation quality has to be investigated, e.g. by means of microdosimetry measurements on the board of a spacecraft. We benchmark the electromagnetic models of the Geant4 toolkit with microdosimetry data obtained with a walled Tissue Equivalent Proportional Counter (TEPC) with beams of HZE particles. Our MCHIT model is able to reproduce in general the response functions and microdosimetry variables for nuclear beams from He to Fe with energies of 80–400 MeV per nucleon.

1. Introduction

Future plans of space exploration include the construction of a station on the Moon and traveling to Mars [1].One of the most important issues related to such projects is the health hazard to astronauts caused by exposure to radiation in space. Several research programs are in progress to investigate the physical and biological effects of space radiation and develop mitigating strategies to reduce risk to humans [2].

The radiation environment in deep space (far from the magnetic field of the Earth) is usually classified in two components: the solar particle events (SPE) and the galactic cosmic rays (GCR). The first component mainly consists of protons emitted by the Sun. Strategies to mitigate its effects include the use of proper shielding or the schedule of missions during phases of low solar activity [3]. However, the shielding from GCR represented by various nuclei from hydrogen to iron is much less effective because of their energy spectra extending to high energies. High-charge (Z) and energy (E) nuclei from GCR, known as HZE particles, can not be stopped by any shielding available in space missions. The maximum of the spectrum for specific GCR nuclei is between $100A$ and $1000A$ MeV. The fluxes of $10A$ GeV nuclei is just two orders of magnitude lower [3]. Even though the fluxes of heavy nuclei (e.g. Fe) are much lower than the flux of protons or helium nuclei, the energy deposition of the former escalates with the square of the nuclear charge. As a result, the physical dose from GCR HZE particles is comparable to one from GCR protons. As soon as the biological effectiveness of each particle is considered, the dose equivalent delivered by iron nuclei, for example, becomes even higher than by protons [4]. Therefore, the radiation effects due to HZE particles must be thoroughly investigated.

The experiments which investigated radiation effects in space have employed several types of detectors on the board of the Space Shuttle [5] and at the International Space Station [6]. Due to its compact size the Tissue Equivalent Proportional Counter (TEPC) [7] is frequently employed for microdosimetry measurements in space. Such kind of detectors has also been used in ground-based experiments with HZE particles from accelerators to model irradiation in space. TEPC usually consists of a gas cavity surrounded by a plastic shell made of a tissue equivalent material. The response function of the detector gives the probability distribution of lineal energy, y, which is defined as the energy imparted by particle to the sensitive volume divided by the mean chord length of the cavity.

Future studies for space exploration shall continue to use TEPCs following the recommendations of the National Council on Radiation Protection and Measurements (NCRP) [8]. Several experiments at HIMAC in Japan and AGS in the USA used TEPC detectors for microdosimetry measurements of HZE particles in the energy range of $200A$ to $1\,000A$ MeV [9–11]. Similar measurements, but with heavy nuclei at higher energies are expected to be performed at the BIOMAT facility at FAIR [12]. These experiments will contribute to a better understanding of radiation effects induced by cosmic rays.

2. Simulations with MCHIT

A computational model which is able to simulate the propagation of energetic nuclei in extended media serves as a powerful tool in investigating radiation effects. In particular, the response function of a TEPC which represents the probability distribution of lineal energy can be estimated by the Monte Carlo method. The patterns of energy deposition around the track of a HZE particle can be also studied by Monte Carlo methods [13] in order to understand biological effects of GCR.

The Monte Carlo model for Heavy Ion Therapy (MCHIT) was developed at FIAS to study physics phenomena relevant to heavy-ion therapy of cancer [14]. MCHIT is based on Geant4 which is a rapidly developing toolkit for calculations in nuclear, particle, space and medical physics [15]. It comprises software libraries for detailed simulations of the particle propagation in various media. MCHIT describes a wide set of experimental data related to heavy-ion therapy. This includes depth-dose profiles for protons and carbon nuclei in tissue-like materials, yields of secondary nuclear fragments, energy spectra and angular distributions of secondary neutrons. It was also demonstrated [16] that MCHIT reproduces microdosimetry spectra measured by walled TEPCs for neutron and carbon beams.

In this work the MCHIT model is extended to describe microdosimetry data relevant to space research. The experimental data collected at HIMAC and AGS [9–11] are used to validate the model for calculations with beams of HZE particles. In these experiments a walled TEPC detector was irradiated by ions with energies from $200A$ to $1000A$ MeV. Detectors upstream and downstream of the TEPC were used to measure the impact parameter of the incoming ion and select events only without beam fragmentation. Due to this selection, the response functions of the TEPC measured in these experiments are sensitive only to electromagnetic processes induced by beam nuclei in TEPC, as fragmentation events of beam nuclei are rejected.

In this work MCHIT is build with Geant4 of version 9.5 with patch 1. The electromagnetic processes are simulated by means of G4EmStandard_opt3 and G4EmPenelope [17] Physics Lists which contain sets of electromagnetic models. The Penelope models have a lower energy threshold of 250 eV for production of delta-electrons compared to the standard electromagnetic models (with a 990 eV threshold). Therefore, once the difference in results obtained with these two Physics Lists is found, this can indicate the importance of transporting of low-energy electrons.

Journal of Physics: Conference Series **426** (2013) 012006 doi:10.1088/1742-6596/426/1/012006

The geometry of TEPC is represented in simulations by a cavity of 12.7 mm in diameter surrounded by a 2.54 mm thick plastic shell. No wires and electric potential inside the gas cavity are simulated. The gas pressure is set to the values used in experiments [9–11] for each ion beam corresponding to simulated tissue objects with diameter of 1 and 3 μm. A total of 2×10^7 primary ions are simulated for each beam energy. In each event the energy deposition inside the gas cavity and the impact parameter are scored to calculate the response function of the detector.

3. Results

The energy imparted to TEPC by 210A MeV ^{20}Ne in central collisions with small impact parameters b <0.8 mm is presented in Fig. 1. MCHIT successfully describes this distribution with both sets of electromagnetic models. Figure 2 demonstrates the mean energy deposition as a function of the impact parameter. MCHIT describes well the mean energy deposition at impact parameters smaller than the cavity radius (6.35 mm). At the interface between gas and plastic there is a bump in the mean energy deposition in the experimental data and simulation results. This phenomenon is due to the enhancement of production of delta electrons when a primary ion grazes the gas/wall interface. Such delta electrons enter the gas cavity and elevate the energy deposition in grazing collisions. Both Physics Lists describe this phenomenon well despite of different energy thresholds for production of delta electrons. One can conclude that delta electrons with kinetic energy below 990 eV do not affect simulation results. The green line presents the product of the linear energy transfer (LET) and the chord length at a given impact parameter. It deviates from the experimental results and MCHIT simulations due to the above-described enhancement of energy deposition in grazing collisions and also due the violation of the charged-particle equilibrium [9] for this detector.

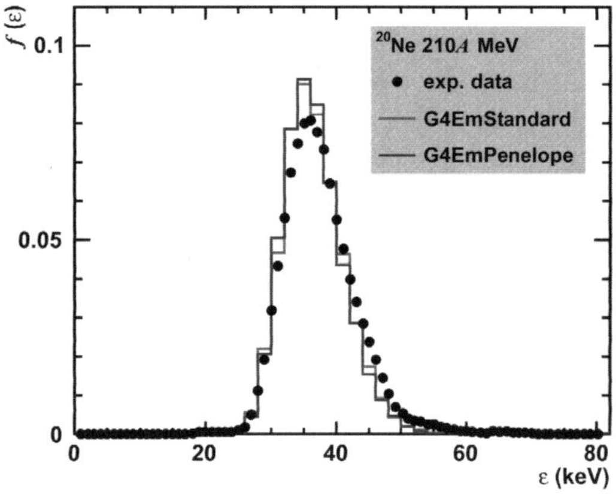

Figure 1. Energy imparted to TEPC by 210A MeV neon ions at impact parameters b <0.8 mm. Experimental data from [10].

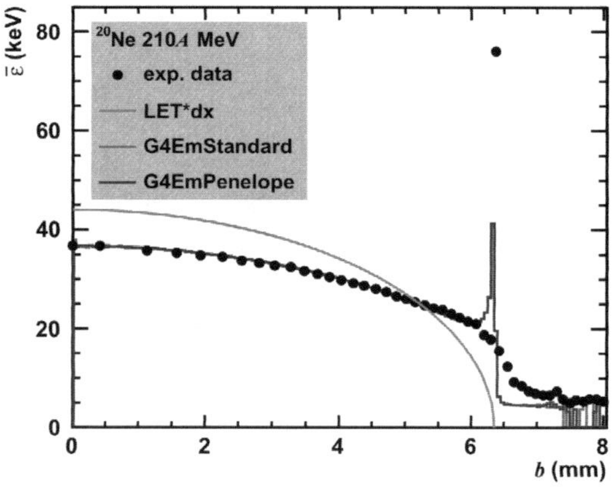

Figure 2. Energy imparted to TEPC by 210A neon ions as a function of impact parameter. Experimental data from [10].

Calculations of the mean energy imparted to TEPC by 360A MeV ^{56}Fe ions as a function of impact parameter with G4EmStandard_opt3 and G4EmPenelope provide again equivalent results with the lowest possible energy thresholds for delta electrons. Figure 3 presents simulation results only with G4EmStandard_opt3 but using various energy thresholds for production of delta electrons. We can see that calculations with the energy threshold of 100 keV can not reproduce the enhancement of energy in grazing collisions, while with the energy threshold of 10 keV this phenomenon is reproduced. The mean energy imparted to TEPC in events with impact parameter smaller than 3 mm is overestimated with both models even with the lowest possible energy thresholds for delta electrons. This overestimation needs further investigations.

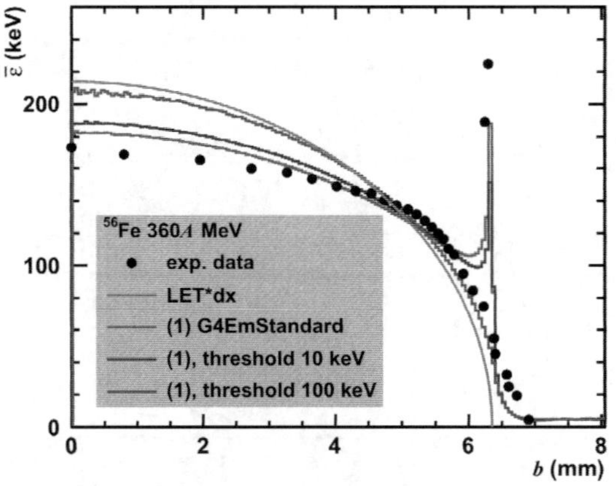

Figure 3. Energy imparted to TEPC by 360A MeV iron ions as a function of impact parameter. Experimental data from [9].

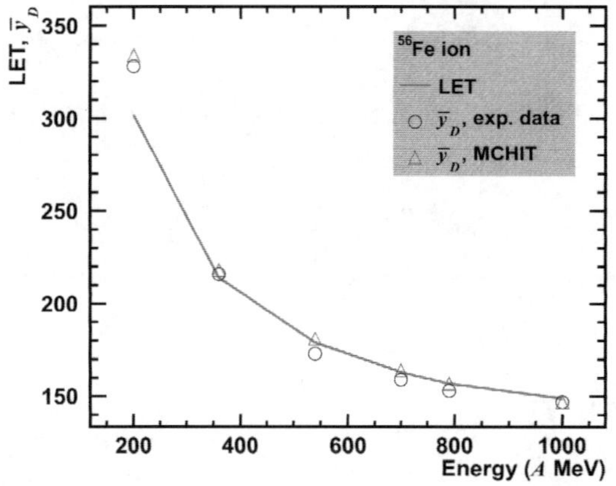

Figure 4. Comparison of LET with $\bar{y}_D$ for iron ions of several beam energies. Experimental data from [9].

MCHIT results obtained with Penelope models for the frequency-average $\bar{y}_F$ and dose-average $\bar{y}_D$ lineal energy for several ions are listed in Table 1. The experimental values [10, 11] and results of the Monte Carlo code FLUKA [18] are shown for comparison. Calculations with Geant4 and FLUKA agree well the experimental values. Although TEPC does not measure Lineal Energy Transfer (LET) directly, $\bar{y}_D$ for heavier ions approximates LET, as already pointed out by Gersey *et. al.* [9]. This demonstrates the usefulness of TEPC data for estimating biological effects of radiation as several biological models describe such effects as a function of LET. The relation between LET and measured and calculated $\bar{y}_D$ is presented in Fig. 4 for iron ions with energy from 200A to 1000A MeV. A good agreement between simulations and data is found. $\bar{y}_D$ and LET agree well especially for high energy iron ions. Therefore, $\bar{y}_D$ can be used as an approximation of LET for iron ions.

4. Conclusions

Geant4/MCHIT describes well the TEPC response function as well as the microdosimetry parameters for a broad selection of ions and beam energies relevant for space research. Both electromagnetic Physics Lists G4EmStandard_opt3 and G4EmPenelope describe reasonably well the experimental data. MCHIT describes the wall effect due to the enhanced production of delta electrons inside the gas cavity when a primary ion grazes the wall/gas interface. The validation of MCHIT for space research should also include experiments which take into account fragmentation reactions for benchmarking of hadronic models. The inclusion of electromagnetic models for low energy processes from the Geant4-DNA project is foreseen as another extension. It will allow to study the patterns of energy deposition at few nanometers scale around the ion track.

Acknowledgments

The presented results were obtained in the framework of NanoBIC-NanoL project. L.B. is grateful to the Beilstein Institute for support. This work was also partially supported by HIC

Table 1. Results for microdosimetry parameters $\bar{y}_F$ and $\bar{y}_D$ simulated with MCHIT. Experimental data from [10, 11]. Simulations results with FLUKA from [18].

Ion	Energy (A MeV)	LET (keV/μm)	$\bar{y}_F$ (keV/μm) exp.	$\bar{y}_F$ FLUKA	$\bar{y}_F$ MCHIT	$\bar{y}_D$ (keV/μm) exp.	$\bar{y}_D$ FLUKA	$\bar{y}_D$ MCHIT
^{4}He	225	1.68	1.56	1.67	1.61	2.58	2.52	2.41
^{12}C	215	15.6	13.4	14.41	14.0	16.1	17.11	16.4
^{12}C	389	11.2	9.93	10.2	9.95	12.4	12.97	12.2
^{14}N	80	43	44	43.8	42.3	47	48.3	46.0
^{16}O	385	19.9	17.9	17.54	17.2	20.8	21.68	20.8
^{20}Ne	210	44	42	43.5	41.5	48	48.1	45.4
^{28}Si	375	61.9	50.4	49.7	49.0	59.8	65	62.4
^{28}Si	780	46	39	41.6	41.3	47	47.7	46.4
^{56}Fe	355	219	184	186.4	183	224	228.3	220

for FAIR within the Hessian LOEWE-Initiative. Our calculations were performed at the Center for Scientific Computing (CSC) of the Goethe University Frankfurt. We are grateful to the staff of the Center for support.

References

[1] NASA The NASA vision for space exploration URL `http://www.nasa.gov/exploration/home/index.html`
ESA The ESA exploration program (Aurora) URL `http://www.esa.int/SPECIALS/Aurora/`
[2] Schimmerling W, Cucinotta F and Wilson J 2003 *Adv. Space Res.* **31** 27–34
Durante M, Kraft G, ONeill P, Reitz G, Sabatier L and Schneider U 2007 *Adv. Space Res.* **39** 1082–6
[3] Durante M and Cucinotta F A 2011 *Rev. Mod. Phys.* **83** 1245–81
[4] Cucinotta F A and Durante M 2006 *Lancet. Oncol.* **7** 431–5
[5] Badhwar G *et al* 1995 *Radiat. Meas.* **24** 283–9
Badhwar G D, Keith J E and Cleghorn T F 2001 *Radiat. Meas.* **33** 235–41
[6] Reitz G, Beaujean R, Benton E, Burmeister S, Dachev T, Deme S, Luszik-Bhadra M and Olko P 2005 *Radiat. Prot. Dosim.* **116** 374–9
[7] ICRU 1983 *Report No. 36, Microdosimetry* (Bethesda, MD: ICRU)
[8] NCRP 2006 *Report No. 153, Information Needed to Make Radiation Protection Recommendations for Space Missions Beyond Low-Earth Orbit* (Bethesda, MD: NCRP)
[9] Gersey B B, Borak T B, Guetersloh S B, Zeitlin C, Miller J, Heilbronn L, Murakami T and Iwata Y 2002 *Radiat. Res.* **157** 350–60
[10] Guetersloh S, Borak T, Taddei P, Zeitlin C, Heilbronn L, Miller J, Murakami T and Iwata Y 2004 *Radiat. Res.* **161** 64–71
[11] Taddei P J, Borak T B, Guetersloh S B, Gersey B B, Zeitlin C, Heilbronn L, Miller J, Murakami T and Iwata Y 2006 *Radiat. Meas.* **41** 1227–34
[12] Durante M 2006 BIOMAT. Biophysics collaboration: applications of relativistic heavy ions in radiobiology and space radiation protection FAIR Newsletter 3 11-13
[13] Friedland W, Dingfelder M, Kundrt P and Jacob P 2011 *Mutation Research/Fundamental and Molecular Mechanisms of Mutagenesis* **711** 28–40
[14] Mishustin I, Pshenichnov I and Greiner W 2010 *Eur. Phys. J.* D **60** 109–114
Pshenichnov I, Botvina A, Mishustin I and Greiner W 2010 *Nucl. Instrum. Methods* B **268** 604–15
[15] Agostinelli S *et al* 2003 *Nucl. Instrum. Methods* A **506** 250–303
Allison J *et al* 2006 *IEEE T. Nucl. Sci.* **53** 270–8
[16] Burigo L, Pshenichnov I, Mishustin I and Bleicher M 2012 Submitted to publication.
[17] Geant4 Collaboration Physics lists for electromagnetic processes URL `http://geant4.web.cern.ch/geant4/collaboration/working_groups/electromagnetic/physlist.shtml`
[18] Böhlen T T, Dosanjh M, Ferrari A, Gudowska I and Mairani A 2011 *Phys. Med. Biol.* **56** 6545–61

Journal of Physics: Conference Series **426** (2013) 012007 doi:10.1088/1742-6596/426/1/012007

The Beam Energy Scan at RHIC: Recent Results from STAR

Alexander Schmah for the STAR collaboration
Lawrence Berkeley National Laboratory, 1 Cyclotron Road, Berkeley, CA 94720, USA
E-mail: aschmah@lbl.gov

Abstract. The first phase of the beam energy scan (BES) program at RHIC was successfully completed in the years 2010 and 2011. The main goals are the search for the QCD critical point and to find signatures for a phase transition between the hadron gas and the QGP phase. For this purpose several observables like higher moments of net-proton distributions, R_{CP} of charged hadrons or the elliptic flow v_2 are studied. Further new observations and results like the centrality dependence of chemical freeze-out parameters (T_{ch} and μ_B) or di-electron spectra in comparison with model calculations are discussed.

1. Introduction

Several heavy-ion measurements from the past decade indicate the existence of a QCD phase with partonic degrees of freedom at high energy A+A collisions. The quark-gluon plasma (QGP) is expected to have fundamentally different characteristics compared to the hadron gas phase which is observed at lower energies. Lattice QCD calculations show a crossover between the hadron gas and the QGP phase at low baryon chemical potentials μ_B [1], whereas a first order phase transition is expected at higher μ_B values. This would result in a QCD critical point. The two main goals of the beam energy scan (BES) at the Relativistic Heavy-Ion Collider (RHIC) facility are to find direct signatures for a transition between the two phases and for the critical point [2]. For this purpose, a series of Au+Au experiments at different energies were accomplished in the years 2010 and 2011 at RHIC.

Within the range of the chosen BES energies it is expected that fireballs are created which exhibit properties around this transition line. For each of these energies multiple observables like azimuthal anisotropies (v_n), p_T spectra, particle yields and ratios, azimuthal HBT, di-lepton spectra, R_{CP} etc. are studied. A rapid change at a certain energy of one or more of these observables could indicate an underlying phase transition. In addition, the disappearance of QGP signatures like number-of-constituent-quark (NCQ) scaling are studied as a function of energy. Comparisons to model predictions are performed as well, to identify an agreement or significant differences with the data. To find signatures of the critical point event-by-event measurements are performed such as the net proton or K/π fluctuation measurements.

2. Results

In the following sections, a selection of BES results from the STAR experiment is highlighted. The first part shows the energy dependence of the chemical freeze-out conditions. In the next two sections di-lepton spectra and the charged hadron nuclear modification factor R_{CP} are discussed

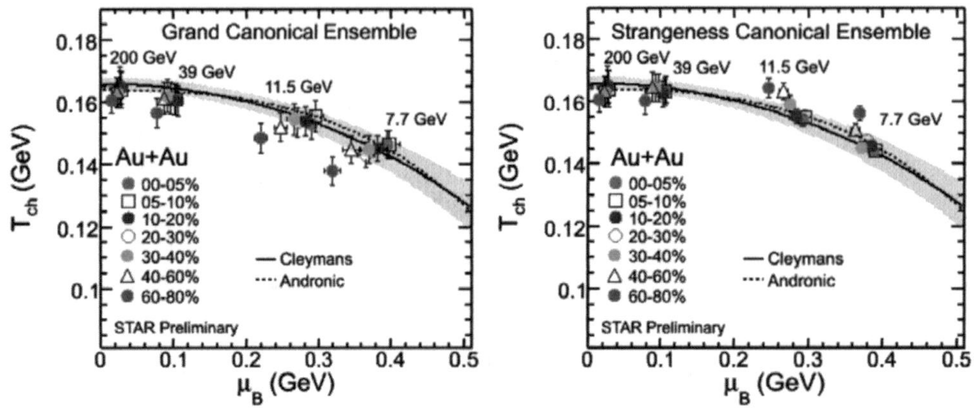

Figure 1. (Color online) Centrality dependence of the chemical freeze-out parameters T_{ch} and μ_B for $\sqrt{s_{NN}}$ = 7.7, 11.5, 39 and 200 GeV for the grand canonical fit (left) and the strangeness canonical fit (right).

in comparison with model calculations. It is followed by recent results from identified particle elliptic flow and higher moments from net-proton distributions.

2.1. Chemical freeze-out conditions

Bulk properties from heavy-ion reactions, such as the chemical and kinetic freeze-out temperatures, can be extracted from particle spectra. Figure 1 shows the chemical freeze-out parameters T_{ch} and μ_B for various energies and centralities [3], extracted from THERMUS [4] fits to particle ratio yields from a grand canonical ensemble fit (left) and a strangeness canonical ensemble fit (right). A centrality dependence of these parameters is observed at the lower energies 39, 11.5 and 7.7 GeV, whereas the parameters seem to be constant at 200 GeV. The temperature dependence between the two ensembles is slightly different. At SIS energies a similar study was performed [5].

2.2. Di-electron spectra

In Fig. 2 the di-electron invariant mass distributions from $\sqrt{s_{NN}}$ = 19.6, 62.4, and 200 GeV 0-80% central Au+Au collisions are shown in comparison with cocktail and model calculations [6]. The combinatorial and correlated background was subtracted from the data and it is corrected for efficiency. Conversion pairs were removed at $\sqrt{s_{NN}}$ = 62.4 and 200 GeV but not at 19.6 GeV. In the vacuum cocktail simulation the ρ^0 contribution was not included since this vector meson has a short live time and decays inside the dense medium which is expected to modify the spectral function. This modification is included in the sum of the cocktail and model calculations [7, 8] shown as dashed lines. In this model a complete evolution of the QGP and thermal dilepton rates in the QGP and hadron-gas (HG) phases are convoluted with an isentropic fireball evolution. The model plus cocktail calculations can describe the low mass part of the di-electron spectra over a wide range of energies.

2.3. Nuclear modification factor R_{CP}

The energy loss of high energetic partons in heavy-ion reactions and the corresponding suppression in the yield is one of the key signatures for the QGP phase [9]. Figure 3 shows the nuclear modification factor R_{CP} of unidentified charged hadrons for various energies from the

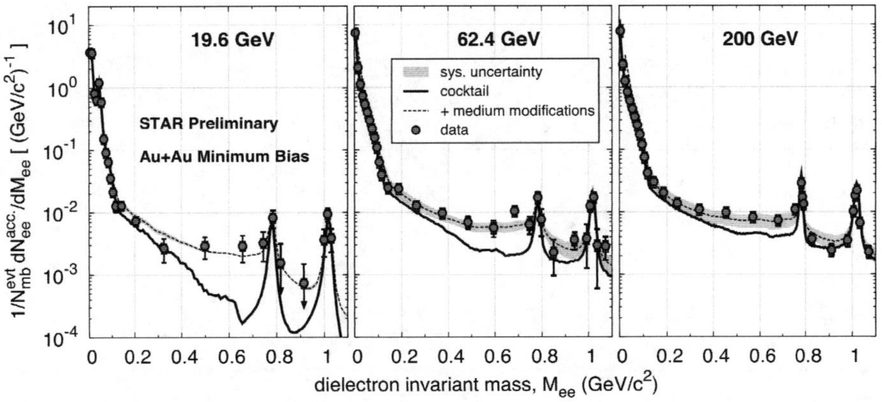

Figure 2. (Color online) Di-electron invariant mass distributions from $\sqrt{s_{NN}}$ = 19.6, 62.4, and 200 GeV 0-80% central Au+Au collisions. The data points contain statistical and systematic error bars. The solid black line shows a cocktail simulation without the ρ^0 contributions. The dashed lined is the sum of the cocktail simulation and a model calculation with an in-medium modified ρ^0.

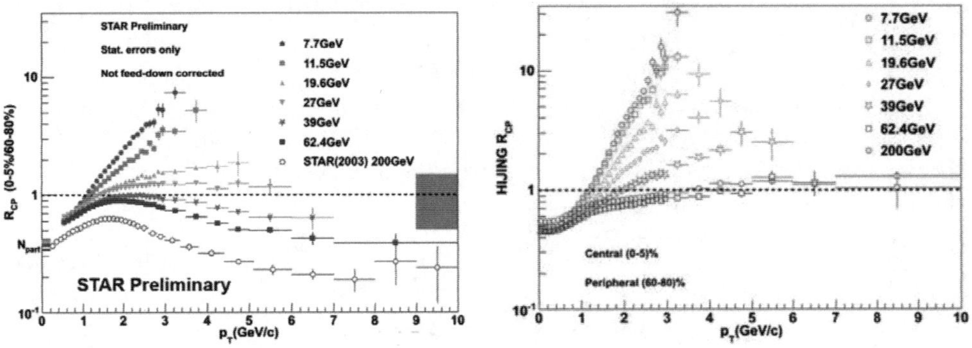

Figure 3. (Color online) The nuclear modification factor $R_{\rm CP}$ for unidentified charged hadrons from different energies of the BES (left) and from HIJING calculations (right).

BES (left) and for HIJING calculations (right). The data is corrected for detector efficiency. The data shows a clear suppression at high transverse momenta at $\sqrt{s_{NN}}$ = 39, 62.4, and 200 GeV, whereas the $R_{\rm CP}$ becomes larger than 1 at lower energies. The HIJING simulations include the Cronin effect through k_T broadening but have in-medium energy loss turned off. The $R_{\rm CP}$ flattens out to 1 at high p_T for $\sqrt{s_{NN}}$ = 62.4 and 200 GeV, as we would expect for jets that do not experience in-medium effects, but at lower collision energies we see a significant enhancement that is qualitatively very similar to what we observe in data. This suggests that for the lower collision energies the Cronin effect becomes more dominant, which makes it more difficult to draw conclusions about the magnitude of partonic energy loss.

2.4. Elliptic flow of identified particles

The elliptic flow v_2 is the second order harmonic of the azimuthal particle distribution relative to the reaction plane. It is generated by the pressure gradient of the overlapping nuclei, mainly during the early stage of the reaction. The number-of-constituent quark (NCQ) scaling of v_2

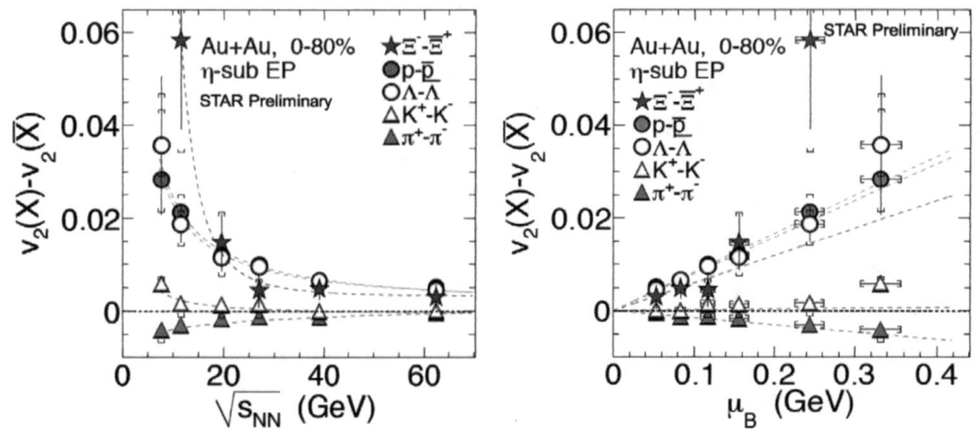

Figure 4. (Color online) The difference in v_2 between particles (X) and their corresponding anti-particles ($\bar{\text{X}}$) (see legend) as a function of $\sqrt{s_{NN}}$ (left) and the baryon chemical potential μ_B (right) for 0-80% central Au+Au reactions.

at top RHIC energies is interpreted as one of the signatures for the QGP phase [10]. On the left hand side of Fig. 4 the energy dependence of the mean difference in v_2 between particles and their corresponding anti-particles is plotted. At higher energies we observe a negligible difference between K^+ and K^-, and between π^+ and π^-. A difference of about 0.005 is found for baryons at 39 and 62.4 GeV. This constant behaviour changes at energies below 27 GeV. Baryons show an increase with a magnitude in the v_2 difference of up to 0.05. Pions show the opposite behaviour compared to kaons with respect to their charge. On the right hand side of Fig. 4 the dependence of the v_2 difference as a function of the baryon chemical potential is shown. The values of μ_B are from the parametrization of Ref. [11]. A linear increase of the difference with μ_B is observed which could point to a direct connection between the two quantities. From the observed difference in v_2 between particles and corresponding anti-particles at lower energies, which also holds for the transverse momentum dependence, one can infer that the NCQ scaling between these particles is broken. The splitting could be an indication that hadronic interactions become more important at lower energies. Also other effects like transported quarks or hadronic potentials could explain qualitatively the observations as shown in [12, 13].

2.5. Higher moments of net-proton distributions

The existence of a QCD critical point would most probably lead to a significant increase of event-by-event fluctuations for $T-\mu_B$ trajectories which come close to the critical area or pass the phase transition. It was suggested to study observables which have a connection to baryon number conservation like net-proton distributions [14]. In Fig. 5 the net-proton moment products $\kappa\sigma^2$ and $S\sigma$, which are connected to baryon number susceptibilities and the correlation length, are shown in comparison with Poisson and UrQMD calculations for four different centrality bins [15]. The $\kappa\sigma^2$ and normalized $S\sigma$ for the most central 0-5% Au+Au reactions are for all energies below the Poisson expectation, whereas the results from more peripheral reactions show values above unity at the lowest energies and below unity for higher energies.

3. Summary and outlook

The first results from the beam energy scan program at STAR contain important new systematics for the search of the QCD phase transition and the critical point. Event-by-event dynamical

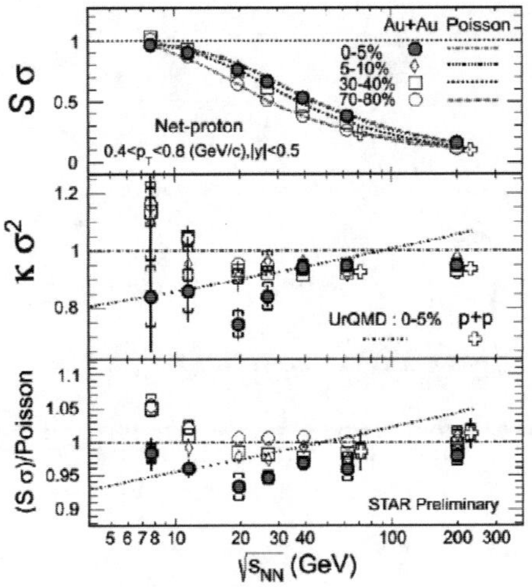

Figure 5. (Color online) Energy dependence of $\kappa\sigma^2$ and $S\sigma$ of net-proton distributions for different centralities of Au+Au reactions. The STAR data is compared with Poisson expectations and UrQMD transport model calculations.

fluctuations, particle spectra and bulk correlations like elliptic and directed flow were studied as a function of energy and centrality. A centrality dependence of the chemical freeze-out parameters was observed which shows a slight difference between the grand canonical and strangeness canonical calculation. A comparison of low mass di-leptons pairs to model calculations show a good agreement for $\sqrt{s_{NN}}$ = 19.6, 62.4, and 200 GeV. Further energies will be studied in the future. The nuclear modification factor $R_{\rm CP}$ shows a suppression at high p_T for energies above $\sqrt{s_{NN}} = 27$ GeV and values above unity for the lower energies. HIJING model calculations show a similar trend. For the first time an energy dependence of particle-antiparticle v_2 difference was measured from which one can infer that the number-of-constituent quark scaling between particles and anti-particles is broken at the lower energies. The net-proton higher moment products $\kappa\sigma^2$ and $S\sigma$ for most central Au+Au collisions are below the Poisson expectation whereas relative values above unity are observed for lower energies and more peripheral collisions.

It is planned to have another phase of BES experiments at RHIC, starting from the year 2015. The CBM experiment at the future FAIR facility will be able to measure the interesting QCD phase space region at maximum baryon densities with high precision.

[1] Y. Aoki, G. Endrodi, Z. Fodor, S. D. Katz, K. K. Szabo, Nature, 443 (2006) 675
[2] M. M. Aggarwal *et al.* (STAR Collaboration), arXiv:1007.2613 [nucl-ex] (2010)
[3] S. Das (STAR Collaboration), arXiv:1210.6099 [nucl-ex], (2012)
[4] S. Wheaton, J. Cleymans, M. Hauer, Computer Physics Communications, 180 (2009)
[5] J. Cleymans, H. Oeschler and K. Redlich, Phys. Rev. C **59**, 1663 (1999)
[6] F. Geurts (STAR Collaboration), arXiv:1210.5549 [nucl-ex], (2012)
[7] R. Rapp, J. Wambach, H. van Hees, arXiv:0901.3289v1 [hep-ph]
[8] R. Rapp, Phys. Rev. C 63, 054907; R. Rapp, private communication (2001)
[9] J. Adams *et al.* (STAR Collaboration), Phys. Rev. Lett. 91 172302 (2003)
[10] J. Adams *et al.* (STAR Collaboration), Phys. Rev. Lett. 92, 052302 (2004)
[11] S. K. Tiwari, P. K. Srivastava and C. P. Singh, Phys. Rev. C **85**, 014908 (2012)
[12] J. C. Dunlop, M. A. Lisa, and P. Sorensen, Phys. Rev. C **84**, 044914 (2011)
[13] J. Xu, L. -W. Chen, C. M. Ko, and Z. -W. Lin, Phys. Rev. C **85**, 041901 (2012)
[14] V. Koch, A. Majumder and J. Randrup, Phys. Rev. Lett. 95, 182301 (2005)
[15] X. Luo (STAR Collaboration), arXiv:1210.5573 [nucl-ex], (2012)

Journal of Physics: Conference Series **426** (2013) 012008 doi:10.1088/1742-6596/426/1/012008

Signals for the QCD phase transition and critical point in a Langevin dynamical model

Christoph Herold[1,2], Yu-Peng Yan[3,4], Marcus Bleicher[1,2]

[1] Institut für Theoretische Physik, Goethe-Universität, Max-von-Laue-Str. 1, 60438 Frankfurt am Main, Germany

[2] Frankfurt Institute for Advanced Studies (FIAS), Ruth-Moufang-Str. 1, 60438 Frankfurt am Main, Germany

[3] School of Physics, Institute of Science, Suranaree University of Technology, Nakhon Ratchasima 30000, Thailand

[4] Thailand Center of Excellence in Physics (ThEP), Commission on Higher Education, Bangkok 10400, Thailand

E-mail: `herold@th.physik.uni-frankfurt.de`

Abstract. The search for the critical point is one of the central issues that will be investigated in the upcoming FAIR project. For a profound theoretical understanding of the expected signals we go beyond thermodynamic studies and present a fully dynamical model for the chiral and deconfinement phase transition in heavy ion collisions. The corresponding order parameters are propagated by Langevin equations of motions on a thermal background provided by a fluid dynamically expanding plasma of quarks. By that we are able to describe nonequilibrium effects occurring during the rapid expansion of a hot fireball. For an evolution through the phase transition the formation of a supercooled phase and its subsequent decay crucially influence the trajectories in the phase diagram and lead to a significant reheating of the quark medium at highest baryon densities. Furthermore, we find inhomogeneous structures with high density domains along the first order transition line within single events.

1. Introduction

What do we know about the phase diagram of QCD and especially the transition between the confined and deconfined or chirally broken and restored phases? For vanishing baryochemical potential μ_B, lattice QCD data confirm a smooth crossover [1]. Studies of effective models indicate a first order phase transition at large values of μ_B [2, 3], which necessarily leads to the existence of a critical end point (CEP). In a thermalized system, the correlation length at the CEP diverges, which lead to the suggested search for an enhancement of event-by-event fluctuations in heavy-ion collisions [4, 5]. Nevertheless, the finite size of such systems together with the divergent relaxation times near the CEP will at least weaken the expected signal [6]. On the other hand, nonequilibrium effects like spinodal instabilities [7] may enhance fluctuations at a dynamical fist order phase transition.

We investigate these effects within the setup of nonequilibrium chiral fluid dynamics [8, 9, 10, 11, 12] with explicit propagation of the Polyakov loop, taking into account effects of both the chiral and the (de-)confinement phase transition. We present results of fluid dynamic expansions with finite baryon number densities through the crossover, the CEP and the first order phase transition.

2. Nonequilibrium chiral fluid dynamics with propagation of the Polyakov loop

The starting point for our studies is the Lagrangian of the Polyakov-quark-meson model [13]

$$\mathcal{L} = \bar{q}\left[i\left(\gamma^{\mu}\partial_{\mu} - ig_{s}\gamma^{0}A_{0}\right) - g\left(\sigma + i\gamma_{5}\vec{\tau}\cdot\vec{\pi}\right)\right]q + \frac{1}{2}\left(\partial_{\mu}\sigma\right)^{2} + \frac{1}{2}\left(\partial_{\mu}\vec{\pi}\right)^{2} - U\left(\sigma,\vec{\pi}\right) - \mathcal{U}(\ell,\bar{\ell}) \quad (1)$$

with the mesonic potential and a polynomial form of the Polyakov loop potential [14]

$$U\left(\sigma\right) = \frac{\lambda^{2}}{4}\left(\sigma^{2} - \nu^{2}\right)^{2} - h_{q}\sigma - U_{0}\ , \quad (2)$$

$$\frac{\mathcal{U}}{T^{4}}\left(\ell,\bar{\ell}\right) = -\frac{b_{2}(T)}{4}\left(|\ell|^{2} + |\bar{\ell}|^{2}\right) - \frac{b_{3}}{6}\left(\ell^{3} + \bar{\ell}^{3}\right) + \frac{b_{4}}{16}\left(|\ell|^{2} + |\bar{\ell}|^{2}\right)^{2}\ . \quad (3)$$

We set the pion fields to zero, so from now on $\pi = \langle\pi\rangle = 0$. The relevant degrees of freedom are the lights quarks $q = (u, d)$, the sigma field σ the Polyakov loop ℓ as the order parameters of the chiral and deconfinement phase transition respectively. The relation between ℓ and the temporal component of the color gauge field A_0 is

$$\ell = \frac{1}{N_c}\left\langle tr_{c}P\exp\left(ig_{s}\int_{0}^{\beta}\mathrm{d}\tau A_{0}\right)\right\rangle_{\beta}\ , \quad (4)$$

with P denoting the path ordering operator. We assume fast local equilibration of the quarks and antiquarks, treating them as an ideal fluid. Their local thermodynamic properties are given by the grand canonical potential which in mean field approximation reads

$$\Omega_{q\bar{q}} = -2N_{f}T\int\frac{\mathrm{d}^{3}p}{(2\pi)^{3}} \quad (5)$$

$$\left\{(\ln\left[1 + 3(\ell + \bar{\ell}\mathrm{e}^{-\beta(E-\mu)})e^{-\beta(E-\mu)} + \mathrm{e}^{-3\beta(E-\mu)}\right] \quad (6)$$

$$+ \ln\left[1 + 3(\bar{\ell} + \ell\mathrm{e}^{-\beta(E+\mu)})e^{-\beta(E+\mu)} + \mathrm{e}^{-3\beta(E+\mu)}\right]\right\}\ . \quad (7)$$

Here $E = \sqrt{p^2 + g^2\sigma^2}$ is the quark energy and $\mu = \mu_B/3$ the quark chemical potential. We have the thermodynamic relations

$$p(\sigma,\ell,T) = -\Omega_{q\bar{q}}(\sigma,\ell,T,\mu)\ , \quad (8)$$

$$e(\sigma,\ell,T) = T\frac{\partial p(\sigma,\ell,T,\mu)}{\partial T} + \mu\frac{\partial p(\sigma,\ell,T,\mu)}{\partial\mu} - p(\sigma,\ell,T,\mu)\ , \quad (9)$$

$$n(\sigma,\ell,T) = \frac{\partial p(\sigma,\ell,T,\mu)}{\partial\mu}\ . \quad (10)$$

together with the conservation laws for energy-momentum and baryon number

$$\partial_{\mu}\left(T_{q}^{\mu\nu} + T_{\sigma}^{\mu\nu} + T_{\ell}^{\mu\nu}\right) = 0 \quad (11)$$

$$\partial_{\mu}n^{\mu} = 0\ . \quad (12)$$

which determine the dynamics of the quark fluid. The interaction of this locally thermalized background with the order parameter fields is incorporated into Langevin equations of motion. For the sigma field this could be derived self-consistently within the two-particle irreducible effective action [10],

$$\partial_{\mu}\partial^{\mu}\sigma + \eta_{\sigma}(T)\partial_{t}\sigma + \frac{\partial V_{\mathrm{eff}}}{\partial\sigma} = \xi_{\sigma}\ , \quad (13)$$

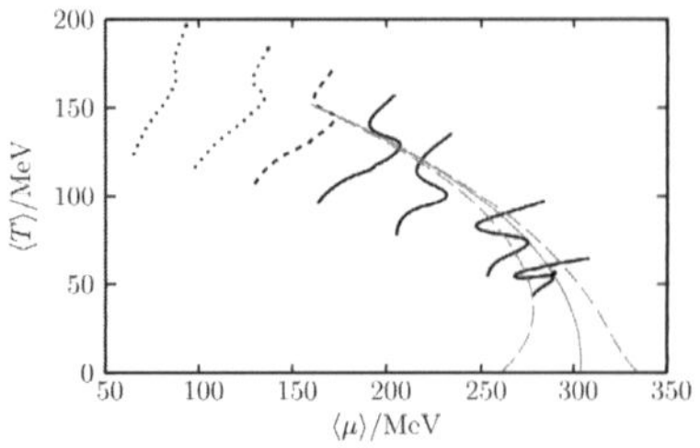

Figure 1. (Color online) Trajectories of the hydrodynamic simulation in the phase diagram. Supercooling delays the transition until the system leaves the spinodal region, indicated by the light dashed lines.

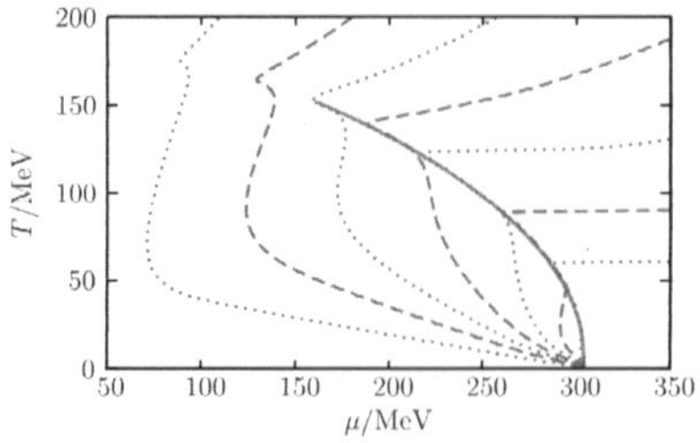

Figure 2. (Color online) First order phase transition line and adiabats for the Polyakov-quark-meson model. Curves correspond from left to right to values of entropy per baryon $S/A = \{24, 16, 12.5, 10, 8, 6, 4\}$.

where the damping coefficient η_σ is temperature dependent and vanishes only around the CEP allowing for critical fluctuations. For the Polyakov loop we deploy the idea that the dynamics of the order parameter in the vicinity of the phase transition can be described by a relaxation equation

$$\eta_\ell \partial_t \ell T^2 + \frac{\partial V_{\rm eff}}{\partial \ell} = \xi_\ell \,. \tag{14}$$

Note that the Polyakov loop ℓ is originally defined as a static object in Euclidean space which makes this ansatz purely phenomenological, similar to work done in [15, 16]. The stochastic noise fields ξ_σ and ξ_ℓ are connected to the respective damping coefficients via dissipation-fluctuation relations that ensure relaxation to the proper equilibrium state.

$$\langle \xi_\sigma(t,\vec{x})\xi_\sigma(t',\vec{x}')\rangle = \frac{1}{V}\delta(t-t')\delta(\vec{x}-\vec{x}')m_\sigma\eta_\sigma \coth\left(\frac{m_\sigma}{2T}\right)\,, \tag{15}$$

$$\langle \xi_\ell(t,\vec{x})\xi_\ell(t',\vec{x}')\rangle T^2 = \frac{1}{V}\delta(t-t')\delta(\vec{x}-\vec{x}')2\eta_\ell T\,. \tag{16}$$

3. Nonequilibrium trajectories

We initialize a spherical droplet with the temperature $T_{\rm ini}$ and quark chemical potential $\mu_{\rm ini}$, the edges smoothed by a Woods-Saxon function

$$T(t=0,\vec{x}) = \frac{T_{\rm ini}}{1+\exp[(|\vec{x}|-r)/a]}\,,\quad \mu(t=0,\vec{x}) = \frac{\mu_{\rm ini}}{1+\exp[(|\vec{x}|-r)/a]}\,. \tag{17}$$

The system is propagated by full (3+1)d fluid dynamics. This resembles the situation after the collision of two heavy nuclei. We choose different initial conditions and extract the temperature and chemical potential averaged over a central volume during the evolution. The resulting trajectories in the T-μ-plane are shown in fig. 1 and compared to the isentropes of the Polyakov-quark-meson model, fig. 2. We see that the trajectories follow the tendency of the isentropes to bend in the direction of positive chemical potential when they hit the crossover or phase transition line. This is a result of the rapid change of the dynamically generated quark mass $m_{\rm q} = g\sigma$ at the phase boundary making this effect more pronounced with growing strength of the transition at larger μ.

Furthermore we observe that the trajectories overshoot the first order phase transition line before this bending happens. This is a result of the nonequilibrium evolution that leads to

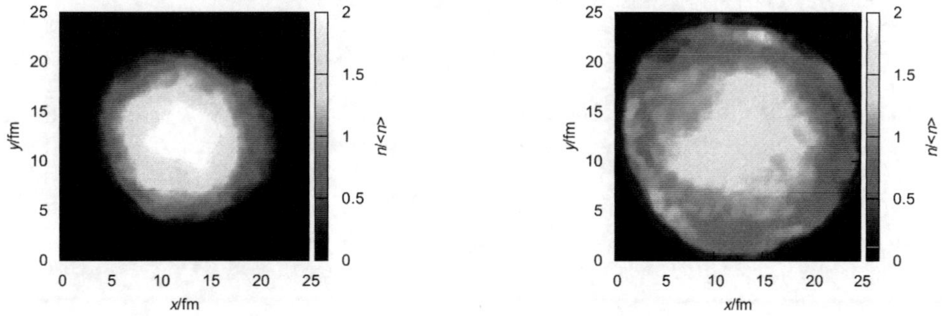

Figure 3. (Color online) Quark number density at $z = 0$ after $t = 6$ fm (left) and $t = 12$ fm (right) for an evolution through the CEP. The system evolves homogeneously, no strong fluctuations occur.

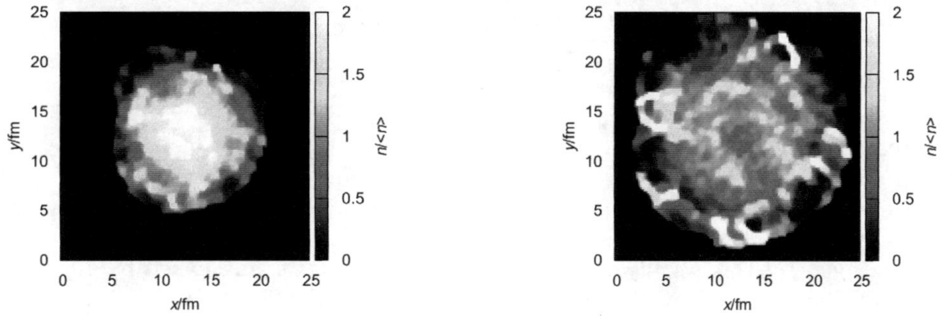

Figure 4. (Color online) Quark number density at $z = 0$ after $t = 6$ fm (left) $t = 12$ fm (right) for a first order transition. High density clusters develop.

supercooling during the evolution. Below T_c the system gets trapped in the metastable high temperature phase until the spinodal line is reached and the barrier separating the minima vanishes. At values of highest μ, where this effect is strongest, it leads to a reheating of the quark fluid.

4. Density fluctuations

To gain better insight into the processes during the evolution it is instructive to study the structure of fluctuations in a single event. The spatial delta functions in eq. 15 and 16 have the effect of averaging the noise as explained in [12]. We now allow for the formation of structures and domains by correlating the noise fields over a spatial volume of 1fm^3 in each time step. The relative baryon number density $n/\langle n\rangle$ in the $z = 0$ plane is shown in fig. 3 for the evolution through the CEP and in fig. 4 for an evolution through the first order transition at high values of μ. For each transition we show an intermediate state at $t = 6$ fm and a late one at $t = 12$ fm. We define $\langle n\rangle$ as the volume average of the baryon number density over all cells with n bigger than a cutoff value.

We observe significant differences between the two transitions. Through the CEP it evolves rather homogeneously, with no strong fluctuations in the quark number density preserving a smooth shape. On the other hand, we observe the formation of irregular structures when the system evolves through the first order phase transition. Here, clusters of high density are embedded in regions with lower density. This is presumably a result of the nonequilibrium evolution through the first order phase transition. The system does not trespass the phase boundary smoothly but by forming domains of the high- and low-temperature phase that coexist for a certain amount of time until all domains of the chirally symmetric and deconfined phase have vanished. Such a behavior leads to large pressure gradients and further on to inhomogeneous structures in the fluid dynamically propagated quantities. In experiment, this should manifest itself e. g. in non-statistical fluctuations of hadron multiplicities [17].

5. Conclusions

We presented a model that allows a dynamical description of the QCD phase transition. A fluid dynamically expanding heat bath of quarks is coupled to the order parameters of the chiral and deconfinement transition. We saw that the nonequilibrium evolution strongly influences the behavior of an expanding system at the phase boundary. Supercooling can be observed in the spinodal region, followed by a reheating at highest baryochemical potential. Within a single event we observe that clusters of high baryon density build during the evolution through the first order phase transition. This is in contrast to the evolution through the CEP where the density remains uniformly distributed. These results can help to improve our understanding of signatures of the phase transition or the CEP at FAIR energies.

Acknowledgments

The authors thank Carsten Greiner for fruitful discussions and Dirk Rischke for providing the SHASTA code with which the fluid dynamic simulations were performed. C. H. gratefully acknowledges the hospitality of the School of Physics at the Suranaree University of Technology where essential parts of this work were developed. This work was supported by GSI and the Hessian LOEWE initiative Helmholtz International Center for FAIR.

References

[1] Aoki Y, Endrodi G, Fodor Z, Katz S D and Szabo K K, *Nature* **443** (2006) 675-78.
[2] Scavenius O, Mocsy A, Mishustin I N and Rischke D H, *Phys. Rev.* C **64** (2001) 045202.
[3] Schaefer B J and Wambach J, *Nucl. Phys.* A **757** (2005) 479
[4] Stephanov M A, Rajagopal K and Shuryak E V, *Phys. Rev. Lett.* **81** (1998) 4816
[5] Stephanov M A, Rajagopal K and Shuryak E V, *Phys. Rev.* D **60** (1999) 114028
[6] Berdnikov B and Rajagopal K, *Phys. Rev.* D **61**, 105017 (2000)
[7] Sasaki C, Friman B and Redlich K, *Phys. Rev.* D **77** (2008) 034024
[8] Mishustin I N and Scavenius O, *Phys. Rev. Lett.* **83** (1999) 3134
[9] Paech K, Stoecker H and Dumitru A, *Phys. Rev.* C **68** (2003) 044907
[10] Nahrgang M, Leupold S, Herold C and Bleicher M, *Phys. Rev.* C **84** (2011) 024912.
[11] Nahrgang M, Leupold S and Bleicher M, *Phys. Lett.* B **711** (2012) 109
[12] Nahrgang M, Herold C, Leupold S, Mishustin I N and Bleicher M, arXiv:1105.1962 [nucl-th].
[13] Schaefer B J, Pawlowski J M and Wambach J, *Phys. Rev.* D **76** (2007) 074023
[14] Ratti C, Thaler M A and Weise W, *Phys. Rev.* D **73** (2006) 014019.
[15] Dumitru A and Pisarski R D, *Phys. Lett.* B **504** (2001) 282
[16] Dumitru A and Pisarski R D, *Nucl. Phys.* A **698** (2002) 444
[17] I. N. Mishustin, *PoS CPOD* **07** (2007) 010

Journal of Physics: Conference Series **426** (2013) 012009 doi:10.1088/1742-6596/426/1/012009

Time-Dependent Hartree-Fock Approach to Nuclear Pasta at Finite Temperature

B. Schuetrumpf[1], **M. A. Klatt**[2], **K. Iida**[3], **J. A. Maruhn**[1], **K. Mecke**[2], **P.-G. Reinhard**[2]

[1]Institut für Theoretische Physik, Universität Frankfurt, D-60438 Frankfurt, Germany
[2]Institut für Theoretische Physik, Universität Erlangen-Nürnberg, D-91058 Erlangen, Germany
[3]Department of Natural Science, Kochi University, 2-5-1 Akebono-cho, Kochi 780-8520, Japan

Abstract. We present simulations of neutron-rich matter at subnuclear densities, like supernova matter, with the time-dependent Hartree-Fock approximation at temperatures of several MeV. The initial state consists of α particles randomly distributed in space that have a Maxwell-Boltzmann distribution in momentum space. Adding a neutron background initialized with Fermi distributed plane waves the calculations reflect a reasonable approximation of astrophysical matter. This matter evolves into spherical, rod-like, and slab-like shapes and mixtures thereof. The simulations employ a full Skyrme interaction in a periodic three-dimensional grid. By an improved morphological analysis based on Minkowski functionals, all eight pasta shapes can be uniquely identified by the sign of only two valuations, namely the Euler characteristic and the integral mean curvature.

1. Introduction

In systems at densities just below normal nuclear density, the spherical shape for nuclear matter is no longer favorable and the system tends to lower the total surface area by elongating the nuclei and forming nuclear rods [1, 2]. Seminal liquid-drop calculations [3, 4] predicted that with further increase in density, the shapes of inhomogeneous nuclear matter change from rods to slabs, tubes, and bubbles until the system melts into uniform matter. For a more realistic description of these so called pasta nuclei, multi-dimensional calculations were performed in various models (e.g. [5, 6, 7]) which showed more complex structures.

Recently, a time-dependent Hartree-Fock (TDHF) approach has been utilized to describe pasta nuclei in zero-temperature supernova matter [8, 9]. In this work we perform such calculations at finite temperature. The TDHF simulations produce a great variety of three dimensional structures. A very general and powerful means to quantify such complex structures is provided by the Minkowski functionals, from integral geometry [10, 11, 12]. They were already introduced for structure analysis of galaxy clusters in Ref. [13] and later used for cosmological problems (see e.g. [14]). They provide an optimal technique to classify the different pasta shapes as already exploited in [15, 16, 17].

2. TDHF description

In this framework, we want to describe the pasta structure with the time-dependent Hartree-Fock (TDHF) approximation proposed by Dirac [18]. In general, the TDHF equations are

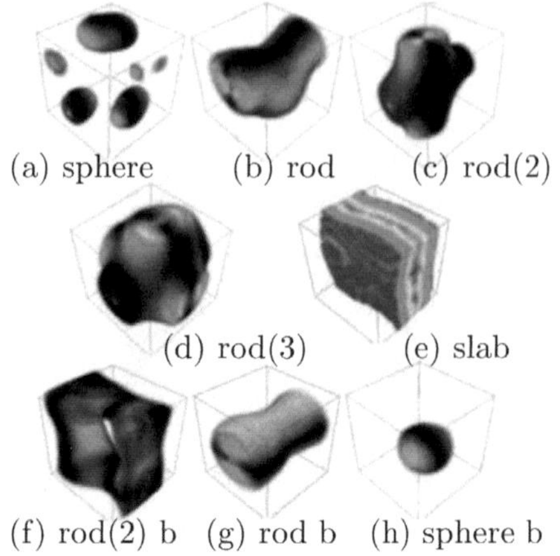

Figure 1. Typical pasta shapes at $T = 7$ MeV. Bubble shape illustrations show gas phase indicated by the color-scale.

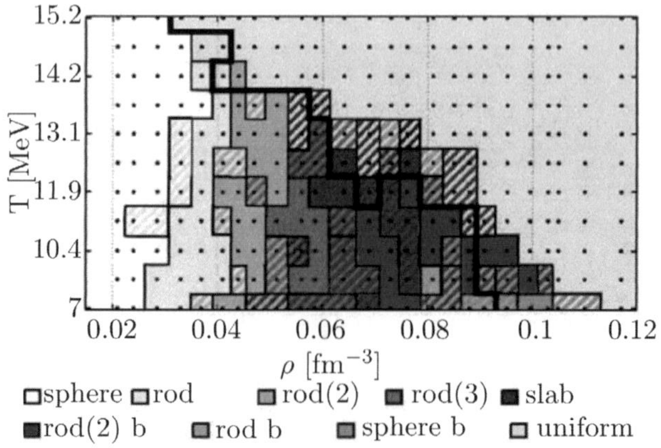

Figure 2. Map of pasta shapes achieved in TDHF calculations. Each dot represents two calculations. The solid black line shows a phase separation line discussed in Sec. 4.

derived from a time-dependent variational principle with the variational space restricted to one single, time dependent Slater determinant. In nuclear applications, TDHF usually employs an effective interaction as, e.g., the Skyrme force, for a review see [19]. With the advance of computational power, it is now possible to perform full 3D calculations without any symmetry restrictions (e.g. [20, 21]).

The TDHF calculations are performed on an 3D grid in coordinate space with 16 grid points in each direction with a grid spacing of 1 fm. We simulate infinite matter using periodic boundary conditions. The Coulomb problem is solved in the jellium approximation. The kinetic energy operator is evaluated in Fourier space and time evolution is performed using a Taylor expansion of the mean-field propagator. For the present calculations we choose Skyrme parametrization SLy6 which was developed with an emphasis on describing neutron rich matter [22].

3. Map of regimes

We initialize the calculations by distributing a number N_α of α particles randomly over the grid and in momentum space following a Maxwell-Boltzmann distribution with a given initialization temperature T_{init}. To that end, we use ground-state wave functions for the α particles from a stationary HF calculation. Further neutrons are introduced in plane wave states using a Fermi distribution with the same temperature to get a proton fraction of 1/3. Spin up and spin down states are occupied simultaneously to avoid a large spin excitation. The number N_α is varied to produce systems with different densities.

The excitation state of matter is characterized by its actual temperature T. Because it is difficult to measure temperature in the quantum simulation of a micro-canonical ensemble we introduce as a rough measure T estimated by the excitation energy E^* through the relation $E^* = (\pi^2 A)(2\varepsilon_F)T^2$ where A is the total number of nucleons and ε_F the Fermi energy of the system [23]. This excitation temperature T is to be distinguished from the "initial temperature" T_{init} which we use to boost stochastically the initial ensemble of α particles and background neutrons. Calculating the excitation temperature T for the different values of the initial temperature T_{init}, we find empirically $T \approx 7\text{ MeV} + \sqrt{T_{\text{init}}(T_{\text{init}} + 100\text{ MeV})}/6$. This is taken henceforth for a rough calibration of T.

Each setup is evolved in time for 1500 fm/c. After that time, the system goes over into a pasta

Table 1. Signs of the integral mean curvature and Euler characteristic for each observed shape.

shape	sph	rod	rod(2)	rod(3)	slab	rod(2) b	rod b	sph b
W_2	>0	>0	>0	- to +	≈ 0	<0	<0	<0
W_3	>0	$=0$	<0	<0	$=0$	<0	$=0$	>0

state where some type of equilibrium is achieved and shapes are satisfyingly stable. Depending on the temperature differently strong fluctuations can be observed.

In Fig. 1 many different pasta shapes are classified. Among the structures found are rod and slab structures discovered, e.g., by QMD calculations [17]. Rod(2) corresponds to rods forming a two-dimensional layer. The shape rod(3) describes three rods in x-, y- and z-direction crossing in one point. Similar shapes were seen in [8, 24] and the rod(3) structure was also found in [6].

Fig. 2 shows for which density and temperature the different shapes appear. Note that two calculations for each point in the map were performed. As we have random initial conditions, the final shapes may differ although the same values for temperature and mean density are assumed. The hatched areas with a mixture of colors indicate the different final states reached in these cases. At higher densities, bubble structures can be seen (the gas phase has a pasta like shape). Many shapes can coexist here in one point of the map.

For slab and rod(3) structures, the final state seems to highly depend on the initial condition. The similarity of these two structures is, that liquid and gas phase have the same shapes, symmetrically complementing each other and filling out almost the same volume.

4. Minkowski Functionals and structure classification

There are four Minkowski functionals W_ν defined for a spatial domain K in three dimensions: they are proportional to its volume $W_0 \propto V$, its surface-area $W_1 \propto A$, the integral mean curvature $W_2 \propto \int_{\partial K} \mathrm{d}A\,(\kappa_1 + \kappa_2)/2$, and the topological Euler-Poincaré characteristic $W_3 \propto \chi$, equal to the integrated Gaussian curvature $\int_{\partial K} \mathrm{d}A\,\kappa_1 \cdot \kappa_2$. Here, κ_1 and κ_2 are the principle curvatures on ∂K, the bounding surface of K. The Euler characteristic is a topological constant. These shape indices form a complete basis of all functionals defined on unions of convex sets which are translational and rotational invariant, additive, and at least continuous on convex sets [25], providing a complete scalar morphological characterization of the pasta shapes.

The TDHF results for pasta matter are given as gray-scale data with a density ρ_i assigned to each voxel i (= grid point). In order to compute the Minkowski functionals of a domain K, the density field is turned into binary data via thresholding at the threshold density ρ_{th}. The Minkowski functionals of the union of all black voxels (each voxel i with $\rho_i > \rho_{th}$), interpreted as a polygon K, can quantify the shape of the pasta matter in dependence of ρ_{th}. Beforehand the marching cube algorithm [26] is applied, in order to reduce voxelization errors. The Minkowski functionals of the polygon are evaluated, e.g., using the linear-time algorithm [27, 28] [1].

The threshold density can take on values $\rho_{th} \in [0, \rho_{\max}]$. But for small thresholds $\rho_{th} \approx 0$ and for high thresholds $\rho_{th} \approx \rho_{\max}$, the Minkowski functionals show erratic behavior; thus, the structure values for these thresholds are ignored [27, 29].

Calculating the values of the Minkowski scalars for all shapes as a function of threshold density (Fig. 3), we see that we can uniquely classify the different shapes in simple fashion: only the signs of integral mean curvature and Euler characteristic are needed (see TAB. 1).

The central regime with solid lines in Fig. 3 indicates the physical values for the threshold density which yield reliably stable values of W_2 and W_3 for lowest value of temperature are

[1] Implementations can be found at `www.theorie1.physik.fau.de`

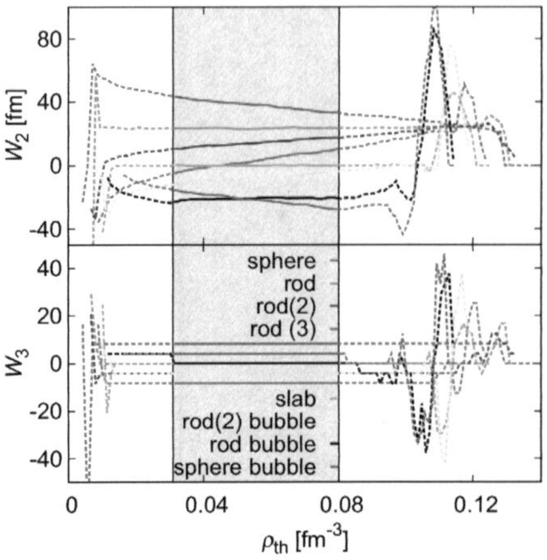

Figure 3. W_2, proportional to the integral mean curvature, and W_3, proportional to the Euler characteristic as function of threshold density for the samples shown in Fig.1.

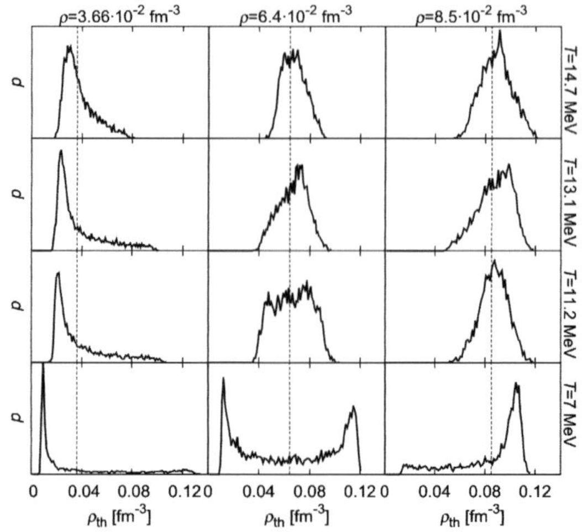

Figure 4. Threshold density profiles for calculations varying in density and temperature. Each plot shows the fraction of grid points p in arbitrary units corresponding to a certain density value. The dashed lines show the mean density.

$\rho_{th} \in [0.03\,\mathrm{fm}^{-3}, 0.08\,\mathrm{fm}^{-3}]$. For low ρ_{th} only few dots are below ρ_{th} due to quantum fluctuations in the gas phase, and for high ρ_{th} there are fluctuations in the liquid phase respectively.

Rod(2), rod(2) bubble, and rod(3) consist in channels and tunnels; thus, the domain has a negative Euler characteristic $\chi < 0$ [30]. The bubble shapes are the complements of the according sphere, rod, or rod(2) shapes. As the sign of both main curvatures κ_1 and κ_2 changes, W_2 also changes sign but the Euler characteristic remain constant.

For the slab and rod(3) shapes, which are symmetric in gas and liquid phase, the integral mean curvature as a function of the threshold ρ_{th} is point symmetric w.r.t. a mid value for the threshold density, trivially true for the slab with $W_2 \approx 0$, but holds also for the rod(3) shape.

Especially at higher temperatures and mean densities mixtures of phases can appear, in a sense that at different threshold densities the pasta matter can take on different shapes (cf. [9]). With increasing threshold densities the shapes can change to shapes which appear usually at lower mean densities. For these cases we take the pasta phase with the bigger range in ρ_{th}.

To show the fraction of liquid and gas phase and the transition to uniform matter, we plotted the threshold density profiles for various calculations in Fig. 4 as an additional observable. For $T = 7\,\mathrm{MeV}$ and low mean density, there is a big peak at small densities and a long tail to high densities. With increasing mean density a peak at high density develops. At the densities where rod(3) and slab shapes appear the peaks for low and high densities have nearly the same width and height. At higher mean densities the peak for low densities disappears and only a tail remains. With increasing temperature the double peak structure vanishes and finally a single peak around the mean density is forming which moves successively to central density with ever higher T.

As the pasta structures vanish with higher density, the variance from the mean value of the calculation of these plots decreases. Going through the systematics of the results, we found as a reasonable value for a limit of the variance to observe pasta structures $10^{-4}\,\mathrm{fm}^{-6}$. The phase separation line computed with this observable is displayed in Fig. 2 as a bold line.

5. Conclusion

In this work we have studied the nuclear composition of supernova matter at sub-nuclear densities using time-dependent Hartree-Fock with the Skyrme energy functional. The initial state is prepared by placing N_α α particles stochastically over the simulation box with periodic boundary conditions and adding $2N_\alpha$ neutrons in plane waves. We have varied the excitation state by initializing the α particles and the neutrons with a certain amount of kinetic energy. As a measure for the actual excitation, we introduce a rough estimate for the resulting temperature based on the excitation energy. The thus given initial states evolve to a topologically stable state of pasta matter within about 1000 fm/c. This allows us to deduce a map of pasta shapes in the plane of temperature and density. Those TDHF results agree qualitatively with the phase diagram calculated in QMD [17]. In realm of "rod" structure, we include further sub-structures, coined rod(2) and rod(3), which appeared in earlier Hartree-Fock calculations [6, 8].

To characterize these structures, we use Minkowski functionals which allows us to characterize the various shapes in terms of a few key numbers. By taking the variance of the density profile as a further observable, we can define a phase border between uniform matter and pasta shapes.

Acknowledgments

This work was supported by the BMBF under contract numbers 06ER9063 and 06FY9086, the DFG under grant ME1361/11 and by Grants-in-Aid for Scientific Research on Innovative Areas through No. 24105008 provided by MEXT. The calculations for this work were performed at the Center for Scientific Computing of Goethe-Universität Frankfurt. B.S., K.I., and J.A.M. acknowledge the hospitality of the Yukawa Institute for Theoretical Physics, where this work was initiated.

[1] Pethick C J and Ravenhall D G 1995 *Ann. Rev. Nucl. Part. Sci* **45** 429–484
[2] Chamel N and Haensel P 2008 *Living Rev. Relativ.* **11**
[3] Ravenhall D G, Pethick C J and Wilson J R 1983 *Phys. Rev. Lett.* **50**(26) 2066–2069
[4] Hashimoto M, Seki H and Yamada M 1984 *Prog. Theor. Phys.* **71** 320–326
[5] Williams R and Koonin S 1985 *Nucl. Phys.* **435** 844 – 858 ISSN 0375-9474
[6] Newton W G and Stone J R 2009 *Phys. Rev. C* **79**(5) 055801
[7] Sonoda H *et al.* 2008 *Phys. Rev. C* **77**(3) 035806
[8] Sébille F, Figerou S and de la Mota V 2009 *Nucl. Phys. A* **822** 51 – 73 ISSN 0375-9474
[9] Sébille F, de la Mota V and Figerou S 2011 *Phys. Rev. C* **84**(5) 055801
[10] Santaló L A 1976 *Integral Geometry and Geometric Probability* (Addison-Wesley)
[11] Schneider R and Weil W 2008 *Stochastic and Integral Geometry (Probability and Its Applications)* (Springer)
[12] Mecke K and Stoyan D 2000 *Statistical Physics and Spatial Statistics - The Art of Analyzing and Modeling Spatial Structures and Pattern Formation* 1st ed (*Lecture Notes in Physics* vol 554) (Springer)
[13] Mecke K, Buchert T and Wagner H 1994 *Astron. Astrophys.* **288** 697
[14] Kerscher M, Mecke K and Schücker P 2001 *Astron. Astrophys.* **377** 1
[15] Watanabe G *et al.* 2003 *Phys. Rev. C* **68**(3) 035806
[16] Watanabe G *et al.* 2004 *Phys. Rev. C* **69**(5) 055805
[17] Sonoda H *et al.* 2008 *Phys. Rev. C* **77**(3) 035806
[18] Dirac P A M 1930 *Math. Proc. Cambridge* **26** 376–385
[19] Bender M, Heenen P H and Reinhard P G 2003 *Rev. Mod. Phys.* **75**(1) 121–180
[20] Maruhn J *et al.* 2005 *Phys. Rev. C* **71** 064328
[21] Umar A S and Oberacker V E 2006 *Phys. Rev. C* **74** 021601
[22] Chabanat E *et al.* 1998 *Nucl. Phys. A* **635** 231 – 256 ISSN 0375-9474
[23] Ashcroft N W and Mermin N D 1976 *Solid State Physics* (Brooks Cole)
[24] Pais H and Stone J R 2012 *Phys. Rev. Lett.* **109**(15) 151101
[25] Hadwiger H 1957 *Vorlesungen über Inhalt, Oberfläche und Isoperimetrie* (Springer)
[26] Lorensen W E and Cline H E 1987 *SIGGRAPH Comput. Graph.* **21** 163–169 ISSN 0097-8930
[27] Schröder-Turk G E *et al.* 2011 *Adv. Mater.* **23** 2535–2553 ISSN 1521-4095
[28] Schröder-Turk G E *et al.* 2010 *ArXiv e-prints* (*Preprint* 1009.2340)
[29] Mickel W *et al.* 2008 *Biophys. J.* **95** 6072–6080 ISSN 0006-3495
[30] Hyde S T 1989 *Z. Kristallogr.* **187** 165–185 ISSN 2194-4946

Journal of Physics: Conference Series **426** (2013) 012010 doi:10.1088/1742-6596/426/1/012010

General relativistic effects on accretion disk neutrinos

O. L. Caballero[1,2]

[1] Institut für Kernphysik, Technische Universität Darmstadt, 64289 Darmstadt, Germany
[2] ExtreMe Matter Institute EMMI, GSI Helmholtzzentrum für Schwerionenforschung GmbH, 64291 Darmstadt, Germany

E-mail: `lcaballe@theorie.ikp.physik.tu-darmstadt.de`

Abstract. The strong gravitational field of a black hole changes the neutrino fluxes emitted from matter accreting around it. We present effects of general relativity on the neutrino fluxes emerging from black hole accretion disks. The changes on the fluxes have implications on the nucleosynthesis resulting from the interaction of the emitted neutrinos and matter outflowing such disks.

1. Introduction

Black hole-neutron star, neutron-neutron star mergers and core collapse supernovae can result in a disk rapidly rotating around a black hole [1, 2]. These disks are interesting for several reasons. As core collapse supernovae they emit a large amount of neutrinos that can be detected at several facilities [3, 4]. The energy released during the merger has been proposed as a source for the generation of gamma ray bursts [5, 6, 7]. Also they have been suggested as the possible site for the production of heavy elements via rapid-neutron captures (r-process) [8, 9, 10, 11, 12, 13].

The presence of a black hole changes the properties of space-time around it affecting the spectra of radiation emerging from the matter located in its vicinity. Since black hole accretion disks are an important source of neutrinos, it is relevant to consider the effects of general relativity on their energies and trajectories. Several studies have been carried out for photons emerging from accretions disks [14, 15, 16]. These studies considered that the radiation is observed at a distant point form the source and that the disk has a smooth flat surface. They showed a strong impact of the general relativistic effects on the emerging photon spectra. To date, studies of the effects of general relativity on neutrinos emitted from accretion disks have focused on neutrino pair annihilation rates and the production of gamma ray bursts [17, 18, 19]. In this work we discuss general relativistic effects such as energy shifts and bending of trajectories when the observer (or absorption point) of the neutrino is in the vicinity of the disk. Besides studying a flat surface we also consider the neutrinos being emitted from the last point of scattering which corresponds to a puffy uneven surface. The results have strong implication in the synthesis of heavy nuclei as discussed in [20].

2. Formalism

In order to study the effects of general relativity on neutrino fluxes it is necessary to calculate energy shifts and bending of trajectories of neutrinos that leave the neutrino surface and arrive at some observation point. The last one could be a point on the trajectory of hot matter that has been ejected from the disk and that can interact with the neutrinos producing new elements.

2.1. Neutrino Spectra

The effective neutrino flux observed at some distance r_{ob} from the black hole is

$$\phi^{eff} = \frac{1}{4\pi}\int d\Omega_{ob} \times \phi_{ob}(E_{ob}), \tag{1}$$

where $d\Omega_{ob}$ is the solid angle that the source subtends as seen by the observer and

$$\phi_{ob}(E_{ob}) = \frac{g_\nu c}{2\pi^2(\hbar c)^3}\frac{E_{ob}^2}{\exp(E_{ob}/T_{ob})+1}, \tag{2}$$

is the observed flux with $g_\nu = 1$. Equation 1 can be re-writen in terms of the emitted neutrino temperature T_{em}, which corresponds to the temperature at the surface of last scattering (neutrino surface) as

$$\phi^{eff} \propto \frac{1}{4\pi}\int d\Omega_{ob} \times \frac{E_{ob}^2}{\exp(E_{ob}(1+z)/T_{em})+1}, \tag{3}$$

where we have used the fact that the energy transforms as $E_{ob} = (1+z)E_{em}$, with z the redshift. According to Equation 1, to obtain the observed fluxes we need to determine the redshifts and the solid angle that the source describes on the observer's sky. Bellow we discuss the steps we follow to accomplish these tasks.

2.2. Neutrino trajectories

The solid angle $d\Omega_{ob}$ that the disk describes on the observer's sky can be written as

$$d\Omega_{ob} = \sin\xi d\xi \times d\alpha, \tag{4}$$

where ξ is the angle that a neutrino forms with the line that joins the center of the black hole and the observation point r_{ob} (radial direction), and the angle α describes the point where the neutrino hits the plane perpendicular to the radial direction (on the observer's sky). These angles are equivalent to θ and φ (in spherical coordinates) of a frame centered at the observer. However, to avoid confusion with the spherical coordinates used to described the disk we have used different greek letters. The details of how to find the angles ξ and α for a non-rotating black hole are given in [20]. Here we summarize the general steps.

Following the neutrino trajectories allows one to find ξ and α (see Figure 1). Because of the black hole presence neutrino trajectories are given by null geodesics. In the Schwarzchild metric null geodesics are described by the equation of motion (in a frame centered at the black hole) [21],

$$\int_{\varphi_{em}}^{\varphi_{ob}} d\varphi = \pm \int_{r_{em}}^{r_{ob}} \frac{dr}{r\sqrt{\frac{r^2}{b^2} - (1-\frac{r_s}{r})}}, \tag{5}$$

where the position of the observer(emitter) is $(r_{ob}, \varphi_{ob})((r_{em}, \varphi_{em}))$. r and φ are the spherical coordinates of these positions. The sign of equation 5 depends on whether the neutrino is approaching or leaving the center of the black hole. In the above equation b, the impact parameter at infinity, is the ratio between the neutrino energy E and its angular momentum L, $b = L/E$. The impact parameter b is a constant along the neutrino trajectory. Finally $r_s = 2M$ with M the black hole mass.

It is possible to solve Equation 5 for b when the initial and final points (r_{em}, φ_{em}) and (r_{ob}, φ_{ob}) are known. Once b is determined the angle ξ between the neutrino trajectory and the radial direction at any point r can be obtained from

$$b = \frac{r\sin\xi}{\sqrt{1-r_s/r}}. \tag{6}$$

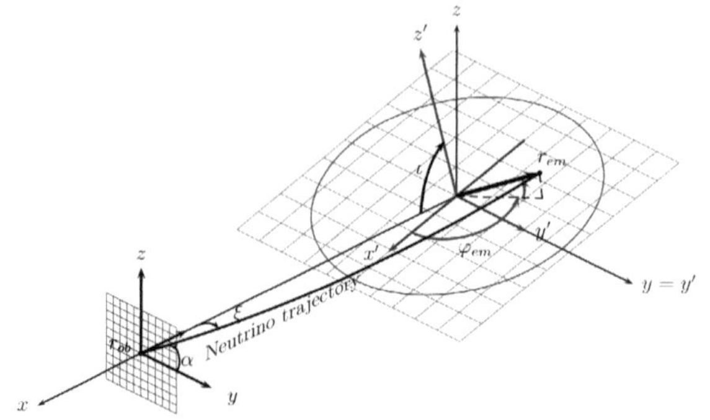

Figure 1. A neutrino is emitted from $(r_{em}, \varphi_{em}$ and observed at (r_{ob}, φ_{ob}). The z-axes of the observer and emitter coordinate frames form an angle ι. On the observer's sky the neutrino forms the angles α andξ.

The angle α in Equation 4 can be found from relationships between spherical triangles formed by the equatorial plane of the disk, the xy plane in the observer's sky, and the plane described by the neutrino trajectory and the line joining the center of the black hole and the observer.

2.3. *Energy shifts*

The red-shift is given by [22],

$$1 + z = \frac{\left(p_t u^t + p_r u^r + p_\theta u^\theta + p_\varphi u^\varphi\right)_{em}}{\left(p_t u^t + p_r u^r + p_\theta u^\theta + p_\varphi u^\varphi\right)_{ob}}, \tag{7}$$

where the subindexes em and ob indicate that the quantities above should be computed with the observer or emitter coordinates accordingly. p is the projection of the neutrino 4-momentum on the 4-velocity u of the emitter/observer, $p_\beta u^\beta$. We consider both the observer and the emitter rotating around the black hole in stationary orbits with angular velocities Ω_{em} and Ω_{ob} respectively. If the black hole has a spin a then the redshifts are given in the Kerr metric by

$$1 + z = \frac{\left[-g_{tt} - 2g_{t\varphi}\Omega_{ob} - g_{\varphi\varphi}\Omega_{ob}^2\right]_{ob}^{1/2}}{\left[-g_{tt} - 2g_{t\varphi}\Omega_{em} - g_{\varphi\varphi}\Omega_{em}^2\right]_{em}^{1/2}} \times \frac{(1 + \Omega_{em} b \cos\eta)}{(1 + \Omega_{ob} b \cos\eta)}, \tag{8}$$

with $g_{\mu\nu}$ the components of the metric tensor [22]. η is the angle that the neutrino trajectory makes with the plane of the disk, which can be found using the law of cosines for spherical triangles, and is given by $\cos\eta = \cos\alpha \sin\iota$, where ι is the angle that the line joining the observer and the center of the black hole forms with the equatorial plane. Finally the angular velocity (calculated for the observer or emitter accordingly) is

$$\Omega = \frac{M^{1/2}}{r^{3/2} + aM^{1/2}}, \tag{9}$$

where a is the black hole spin, which is a input parameter in the disk models.

3. Results

We applied the above formalism to a disk formed during a black hole-neutron star merger. The disk model corresponds to a 3D hydrodynamical simulation [24, 23]. We use as an input for our calculations a snapshot of such simulation. We post-process the data to find neutrino surfaces and therefore neutrino temperatures as described in Ref [4]. The neutrino surfaces found are not smooth. They present sharp variations in temperature, density and height. We adopt the points on these surfaces to be the emission points of the neutrinos r_{em} and for this work we locate the observer at coordinates $r_{ob} = (x, z) = (40, 48)$ km above the neutrino surfaces. The r_{ob} coordinates correspond to the initial point (above the disk surface) of the trajectory of matter ejected from the disk used in the nucleosynthesis studies of Ref. [20].

Figure 2 compares results for effective neutrino fluxes when general relativistic effects are included and when they are ignored (Newtonian fluxes). The red lines are the Newtonian fluxes for electron neutrinos and antineutrinos. The black lines take into account redshifts and bending of trajectories. It can be seen that relativistic effects strongly impact the fluxes specially at high energies.

Since general relativistic effects depend on the position of the emitter and observer it is important to consider the geometry of the neutrino surfaces. To study this effect we consider two scenarios. The first one is the "flat disk" approximation. In this approximation, while the neutrino temperatures are determined by the surface of last scattering, the neutrino trajectories are started from the midplane (z=0) of the disk. The second one is the "puffy disk" where the neutrino trajectories begin at the neutrino surface.

Figure 3 shows the difference between the effective fluxes for electron neutrinos and electron antineutrinos when the flat disk approximation is used versus considering the three dimensional neutrino surfaces. In both cases general relativistic effects have been included. In the case of a puffy surface the neutrinos are emitted farther from the black hole and thus redshift effects are smaller. However, the hottest neutrinos in the puffy disk are emitted from a surface that tilts towards the black hole, making the geometrical factors more important and resulting in smaller fluxes when compared to the flat disk fluxes.

4. Conclusions

We have studied the effects that bending of trajectories and energy shifts have on the neutrino fluxes emitted from an accreting disk around a black hole. We found that these general relativistic effects have a strong impact on the effective neutrino fluxes in particular to the hottest neutrinos which are emitted closer to the black hole. We also studied the consequences that the neutrino surface geometry has on the resulting fluxes. We found that the geometry has an important impact and that a flat disk approximation results in larger neutrino fluxes compared with a puffy disk scenario. When neutrinos are emitted they can interact with outflowing matter from the disk. Both general relativistic effects and geometry considerations influence the ratio of protons to neutrons having important consequences for the synthesis of heavy elements [20]. General relativistic effects should be taking into account in nucleosynthesis studies of black hole accretion disks.

Acknowledgments

O. L. C. thanks G. C. McLaughlin and R. Surman for useful discussions.

References

[1] Rosswog S, Speith R & Wynn G A 2004 *MNRAS* **351**, 1121
[2] Taniguchi K, Baungarte T W, Faber J A & Shapiro S L 2005 *Phys. Rev.* D **72** 044008
[3] McLaughlin G C, & Surman R 2007 *Phys. Rev.* D 75 023005
[4] Caballero O L, McLaughlin G C, & R Surman 2009 *Phys. Rev.* D **80** 123004

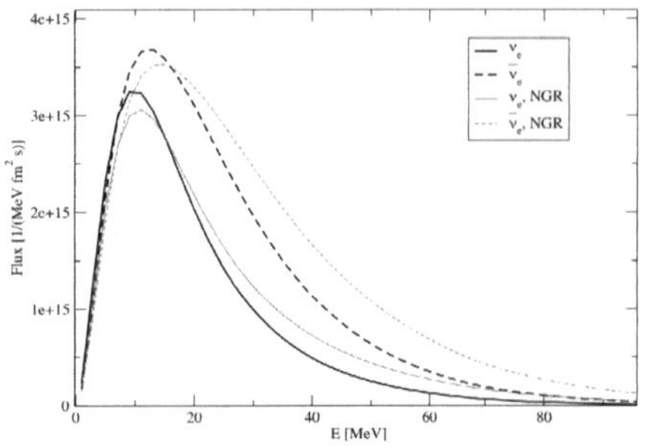

Figure 2. Electron neutrino and antineutrino effective fluxes as seen at a point above the neutrino surfaces located at $(x, y) = (40, 48)$ km. The black thick lines show the fluxes when bending of trajectories and energy shifts have been included. The red thin lines (NGR) correspond to the fluxes when general relativistic effects are ignored.

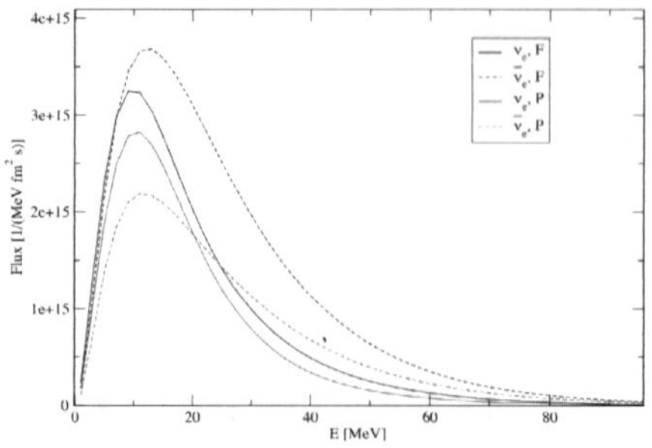

Figure 3. Effective neutrino fluxes obtained at a point above the neutrino surfaces located at $(x, y) = (40, 48)$ km. General relativistic effects are included. The black lines (F) correspond to a flat thin disk. The blue lines (P) show the fluxes when the uneven geometry of the neutrino surface is used.

[5] Popham R, Woosley S E, & Fryer C 1999 *ApJ* **518** 356
[6] Ruffert M & Janka H-Th 1999 *A&A* **344** 573
[7] Setiawan S, Ruffert M & Janka H-Th 2006 *A&A* **458** 553
[8] McLaughlin G C & Surman R 2005 *Nucl. Phys.* A **758** 189
[9] Oechslin R, Janka H-Th, and Marek A 2007 *A&A* **467** 395
[10] Surman R, McLaughlin G C, Ruffert M, Janka H-Th, & Hix W H 2008 *ApJ* **679** L117
[11] Metzger B D, Piro A L, & Quataert E 2009 *MNRAS* **396** 304
[12] Wanajo S, & Janka, H-Th 2010 arXiv:1006.2277
[13] Kajino T, Harikae S, Yoshida T, Nakamura K, and Aoki W 2010 *AIP Conf. Series* **1235** 341
[14] Luminet J-P 1979 *A&A* **75** 228
[15] Cunningham C T 1975 *ApJ* **202** 788
[16] Lorenzin A & Zampieri L 2009 *MNRAS* **394** 1588
[17] Asano K & Fukuyama T 2001, ApJ, **546**, 1019
[18] Birkl R, Aloy M A, Janka H-Th & Muller E 2007 *A&A* **463** 51
[19] Zalamea I & Belobodorov A M 2010 *MNRAS* **410** 2302
[20] Caballero O L, McLaughlin G C, and Surman R, 2012 *ApJ* **745** 170
[21] Muller T 2009 *Gen. Relativ. Gravit.* **41** 541
[22] Misner C W, Thorne K S, & Wheeler J A 1973 *Gravitation* (New York:W. H. Freeman and Company)
[23] Ruffert M & Janka H-Th 2001 *A&A* **380** 544
[24] Setiawan S, Ruffert M & Janka H-Th 2004 *MNRAS* **352** 753

Polarization in low-density neutrons

Alexandros Gezerlis

[1]EMMI, GSI Helmholtzzentrum fr Schwerionenforschung GmbH, 64291 Darmstadt, Germany
[2]Institut fr Kernphysik, Technische Universitt Darmstadt, 64289 Darmstadt, Germany

Abstract. We examine selected aspects of strongly interacting neutrons. We first go over the basics of Green's Function Monte Carlo and then provide some specifics on the variational wave functions used. We then turn to results for the ground-state energies of polarized neutron matter, finding that phase separation is energetically favored with respect to a homogeneous polarized superfluid. Polarized neutron matter is of possible relevance to the physics of highly magnetized neutron stars as well as to terrestrial experiments with ultracold fermionic atoms. Furthermore, the simulation results we show can function as microscopic constraints for phenomenological energy-density functional theories of nuclei. We make this connection more explicit by also discussing the "neutron polaron".

1. Introduction

Neutron matter is a system that has a long history in nuclear physics [1]. While finite nuclei are bound systems composed of neutrons and protons, neutron matter is pure (it contains no protons) and infinite (unbound). These two aspects lead to simplifications, both physically and algorithmically, that make it possible to apply Quantum Monte Carlo (QMC) methods. The lack of protons makes the many-body problem simpler, as now there are only 2 components to deal with and no long-range Coulomb forces. Neutron matter has direct physical relevance to the physics of neutron stars: it impacts their crusts (which contain a sea of dripped neutrons) [2, 3, 4] as well as their cores. Furthermore, the physics of neutron matter is related to properties of neutron-rich nuclei, such as neutron skins, which will be explored at the Facility for Antiproton and Ion Research (FAIR) in Darmstadt and the Facility for Rare Isotope Beams (FRIB) in Michigan. Neutron matter equation of state and pairing gap results can constrain phenomenological treatments of neutron-rich nuclei [5, 6, 7, 8, 9, 10, 11, 12, 13, 14, 15, 16, 17, 18, 19]. The general case of imbalanced neutrons (at larger density) has also been studied by many groups [20, 21, 22, 23, 24, 25].

An appealing aspect of low-density neutron matter is that its physics is especially clean: at intermediate densities only the scattering length and the effective range are relevant, and at even lower density only the former is important. Thus, low-density neutron matter is directly related also to the physics of strongly interacting ultracold atomic gases. This field has experienced an explosion of interest in the last decade, attracting both condensed-matter and nuclear physicists [26, 27]. The main reason for this has to do with the experimentalists' ability to not only cool and trap ^{6}Li and ^{40}K atoms, but also to directly tune the interatomic S-wave scattering length by means of Feshbach resonances. As a result of this, cold atomic systems now routinely function as laboratories of strong interaction theory.

Journal of Physics: Conference Series **426** (2013) 012011 doi:10.1088/1742-6596/426/1/012011

2. Methodology

2.1. Green's Function Monte Carlo

When interested in the ground-state energy of a strongly interacting system, we often start out with a Variational Monte Carlo calculation which is computationally straightforward, but is contaminated with contributions coming from excited states. The results of this variational calculation are then used as input to a Green's Function Monte Carlo (GFMC) simulation. GFMC works by projecting out the lowest-energy eigenstate Ψ_0 from a variational wave function Ψ_V. The ground-state energy is usually computed using a combination of the variational wave function as well as the "evolved" one, giving what is known as a "mixed estimate":

$$\begin{aligned}\langle H\rangle_M &= \frac{\langle\Psi_V|H|\Psi(\tau)\rangle}{\langle\Psi_V|\Psi(\tau)\rangle} = \frac{\langle\Psi_V|He^{-(H-E_T)\tau}|\Psi_V\rangle}{\langle\Psi_V|e^{-(H-E_T)\tau}|\Psi_V\rangle} \\ &= \frac{\langle\Psi_V|e^{-(H-E_T)\tau/2}He^{-(H-E_T)\tau/2}|\Psi_V\rangle}{\langle\Psi_V|e^{-(H-E_T)\tau/2}e^{-(H-E_T)\tau/2}|\Psi_V\rangle} \\ &= \frac{\langle\Psi(\tau/2)|H|\Psi(\tau/2)\rangle}{\langle\Psi(\tau/2)|\Psi(\tau/2)\rangle} = E\left(\tau/2\right) \ , \qquad (1)\end{aligned}$$

where $E\left(\tau/2\right)$ is the expectation value of H in the state $\Psi(\tau/2)$, a quantity which approaches E_0 from above as $\tau \to \infty$. Let us note parenthetically that expectation values of quantities that do not commute with the Hamiltonian can be calculated using a combination of the mixed and variational estimate:

$$\langle\Phi|\hat{S}|\Phi\rangle \approx 2\langle\Phi|\hat{S}|\Psi_V\rangle - \langle\Psi_V|\hat{S}|\Psi_V\rangle \ , \qquad (2)$$

where $\hat{S}$ is the operator corresponding to the relevant physical quantity, and the error in this expression is of second order in $\Phi - \Psi_V$. Such a combination of estimates is often called the "extrapolated estimate".

2.2. Wave function

The choice of the wave function used in these numerical calculations, despite the fact that GFMC is in principle an exact method whose result should not depend on it, is a significant one because this wave function is used in three distinct ways: i) as the initial/trial wave function, ii) in calculating the importance-sampling function, and iii) as the fixed-node constraint.

We have chosen to write the variational wave function in the following form, which is in principle capable of educing pairing in the system under consideration:

$$\Psi_V(\mathbf{R}) = \prod_{i\neq j} f_P(r_{ij}) \prod_{i'\neq j'} f_P(r_{i'j'}) \prod_{i,j'} f(r_{ij'})\Phi(\mathbf{R}) \ . \qquad (3)$$

The f_P and f are called "pair correlation functions" (also known as "Jastrow functions"). In principle they correlate both same-spin pairs and opposite-spin ones. The term $\Phi(\mathbf{R})$ describes the entire collection of N particles and in these studies we have taken it to be either of the "Slater" or of the "BCS" form.

The simplest approximation that can be used to calculate $\Phi(\mathbf{R})$ in Eq. (3) is to describe the particles as being in a free Fermi gas (i.e., let all the correlations lie within the Jastrow functions). Solving this elementary exercise gives the following form for $\Phi(\mathbf{R})$:

$$\Phi_S(\mathbf{R}) = \mathcal{A}[\phi_n(r_1)\phi_n(r_2)\ldots\phi_n(r_{\frac{N}{2}})] \ \mathcal{A}[\phi_n(r_{1'})\phi_n(r_{2'})\ldots\phi_n(r_{\frac{N}{2}'})] \ , \qquad (4)$$

where $\mathcal{A}$ is the antisymmetrizer, and $\phi_n(r_k) = e^{i\mathbf{k_n}\cdot\mathbf{r}_k}/L^{3/2}$. Each of the two terms in Eq. (4) can be written as a Slater determinant (one for spin-up particles and one for spin-down ones).

A choice for $\Phi(\mathbf{R})$ that can also describe pairing is the well-known BCS wave function $\Phi_{BCS}(\mathbf{R})$. In the canonical ensemble, it can also be written as an antisymmetrized product of the pairing functions $\phi(r)$:

$$\Phi_{BCS}(\mathbf{R}) = \mathcal{A}[\phi(r_{11'})\phi(r_{22'})\ldots\phi(r_{\frac{N}{2}\frac{N}{2}'})] \,, \tag{5}$$

or, equivalently, in the form of a determinant. It is possible to generalize such a determinant so as to describe systems that contain n pairs of particles, u unpaired spin-up particles and d unpaired spin-down particles. In this formulation the unpaired particles are in $\psi_{i\uparrow}$ and $\psi_{j'\uparrow}$ single-particle states. This leads to an $(n+u+d)\times(n+u+d)$ determinant. For example, when $u=2$ and $d=3$, one obtains the following determinant:

$$\begin{vmatrix} \phi(r_{11'}) & \phi(r_{12'}) & \ldots & \phi(r_{1(n+d)'}) & \psi_{1\uparrow}(\mathbf{r}_1) & \psi_{2\uparrow}(\mathbf{r}_1) \\ \phi(r_{21'}) & \phi(r_{22'}) & \ldots & \phi(r_{2(n+d)'}) & \psi_{1\uparrow}(\mathbf{r}_2) & \psi_{2\uparrow}(\mathbf{r}_2) \\ \vdots & \vdots & \vdots\vdots\vdots & \vdots & \vdots & \vdots \\ \phi(r_{(n+u)1'}) & \phi(r_{(n+u)2'}) & \ldots & \phi(r_{(n+u)(n+d)'}) & \psi_{1\uparrow}(\mathbf{r}_{n+u}) & \psi_{2\uparrow}(\mathbf{r}_{n+u}) \\ \psi_{1'\downarrow}(\mathbf{r}_{1'}) & \psi_{1'\downarrow}(\mathbf{r}_{2'}) & \ldots & \psi_{1'\downarrow}(\mathbf{r}_{(n+d)'}) & 0 & 0 \\ \psi_{2'\downarrow}(\mathbf{r}_{1'}) & \psi_{2'\downarrow}(\mathbf{r}_{2'}) & \ldots & \psi_{2'\downarrow}(\mathbf{r}_{(n+d)'}) & 0 & 0 \\ \psi_{3'\downarrow}(\mathbf{r}_{1'}) & \psi_{3'\downarrow}(\mathbf{r}_{2'}) & \ldots & \psi_{3'\downarrow}(\mathbf{r}_{(n+d)'}) & 0 & 0 \end{vmatrix} \,. \tag{6}$$

To summarize the situation with the wave functions: for an unpolarized superfluid we use a determinant containing only pair-wave functions; for a normal gas (whether polarized or unpolarized) a product of two single-particle state-determinants (one for spin-up particles, one for spin-down); for a homogeneous superfluid the form of Eq. (6) where $d=0$ and u varies.

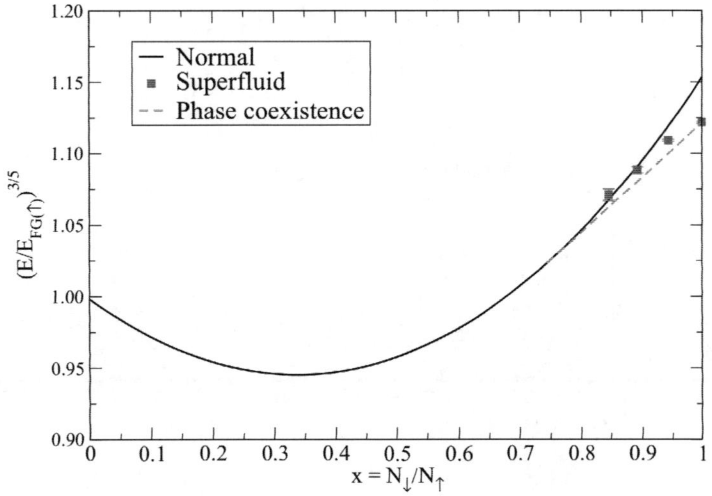

Figure 1. We show the energy of a normal neutron gas (the black solid line shows a simple fit to the Quantum Monte Carlo results), along with homogeneous superfluid results (red square points), and a tangent construction showing the phase coexistence curve (green dashed line). For large values of the relative fraction, the phase separated state is favored with respect to a homogeneous polarized superfluid.

3. Results

We begin by showing ground-state energy results for a fixed low total density of neutrons, in Fig. 1. The polarized normal neutron gas results are the black solid line from Ref. [28]. We also show as red square points the unpolarized superfluid result from Ref. [29] (at $x = 1$) and the homogeneous polarized superfluid results of Ref. [30] (at smaller x). In addition, we show a coexistence curve which is arrived at by a tangent construction from the unpolarized superfluid to the normal polarized curve. This coexistence line meets the normal curve at $x_c = 0.74$. As is obvious, the coexistence line lies below the homogeneous polarized superfluid Quantum Monte Carlo results, implying that these are not energetically favored. This is similar to the behavior of ultracold fermionic gases at unitarity, where $x_c = 0.44$ [31, 32, 27].

We have also focused our attention on the Fermi polaron in nuclear physics. The polaron is a single spin-down impurity embedded in a Fermi sea of spin-up particles, which has been studied by Quantum Monte Carlo methods for ultracold atomic gases [31, 32]. We show the binding energy of the neutron polaron [33] in Fig. 2. Diffusion Monte Carlo calculations for a system of the order of 100 particles provide a tight upper bound, which we use to constrain phenomenological theories (like the Skyrme functionals shown in the figure) that describe neutron-rich nuclei. Such approaches are fit to a number of empirical properties, e.g., the energy of nuclear matter. Skyrme functionals in their present form provide very different predictions for the binding energy of the polaron, none of which obey the expected behavior at weak coupling. Our results thus provide direct input for nuclear density functionals at the extreme physics of the neutron polaron.

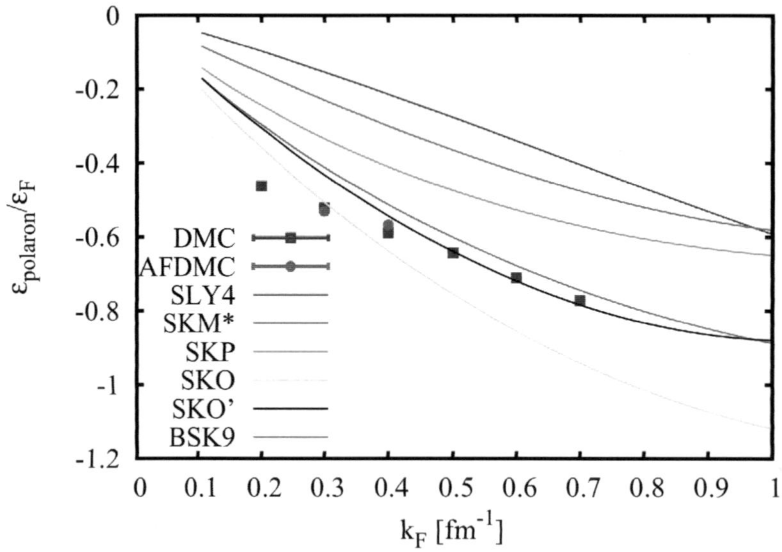

Figure 2. Binding energy of neutron polaron in GFMC and AFDMC compared with Skyrme energy-density functionals. The GFMC results shown include only s-wave interactions, while AFDMC also uses higher partial waves, as well as tensor, and spin-orbit interactions. The microscopic results serve to exclude the behavior exhibited by a large class of phenomenological functionals.

4. Summary

Using insights and experimental guidance from the field of ultracold atomic gases, we have studied polarized low-density neutron matter in the general case and in the extreme case of the polaron. Quantum Monte Carlo provides microscopic benchmarks that can be of use to phenomenological theories of nuclear physics. This line of reasoning can be readily extended, to the case of external potentials, 3 or 4 species, lower-dimensional systems and more. These are all relatively clean systems that have a lot to teach us on the physics of neutron-rich nuclei.

Acknowledgments

I'd like to thank the organizers of FAIRNESS 2012 for their work, as well as Stefano Gandolfi and Rishi Sharma for a series of illuminating discussions. This work is supported by the Helmholtz Alliance Program of the Helmholtz Association, contract HA216/EMMI "Extremes of Density and Temperature: Cosmic Matter in the Laboratory". Computations were performed at the NERSC computing facility and the Juelich Supercomputing Center.

References

[1] C. J. Pethick and D. G. Ravenhall, Annu. Rev. Nucl. Part. Sci. **45**, 429 (1995).
[2] D. Page, J. M. Lattimer, M. Prakash, and A. W. Steiner, Astrophys. J. **707**, 1131 (2009).
[3] D. N. Aguilera, V. Cirigliano, J. A. Pons, S. Reddy, and R. Sharma, Phys. Rev. Lett. **102**, 091101 (2009).
[4] F. J. Fattoyev and J. Piekarewicz, Phys. Rev. C. **82**, 025810 (2010).
[5] B. Friedman and V. R. Pandharipande, Nucl. Phys. **A361**, 502 (1981).
[6] A. Akmal, V. R. Pandharipande, and D. G. Ravenhall, Phys. Rev. C **58**, 1804 (1998).
[7] J. Carlson, J. Morales, Jr., V. R. Pandharipande, and D. G. Ravenhall, Phys. Rev. C **68**, 025802 (2003).
[8] A. Schwenk and C. J. Pethick, Phys. Rev. Lett. **95**, 160401 (2005).
[9] J. Margueron, E. van Dalen and C. Fuchs, Phys. Rev. C **76**, 034309 (2007).
[10] M. Baldo and C. Maieron, Phys. Rev. C **77**, 015801 (2008).
[11] E. Epelbaum, H. Krebs, D. Lee, and U. -G. Meissner, Eur. Phys. J. A **40**, 199 (2009).
[12] E. Epelbaum, H.-W. Hammer, and U. -G. Meissner, Rev. Mod. Phys. **81**, 1773 (2009).
[13] T. Abe and R. Seki, Phys. Rev. C **79**, 054002 (2009).
[14] S. Gandolfi, A. Yu. Illarionov, F. Pederiva, K. E. Schmidt, and S. Fantoni, Phys. Rev. C **80**, 045802 (2009).
[15] A. Rios, A. Polls, and I. Vidaña, Phys. Rev. C **79**, 025802 (2009).
[16] K. Hebeler and A. Schwenk, Phys. Rev. C **82**, 014314 (2010).
[17] N. Chamel, S. Goriely, and J.M. Pearson, Nucl. Phys. **A812**, 72 (2008).
[18] N. Kaiser and W. Weise, Phys. Lett. B **671**, 25 (2009).
[19] N. Kaiser, arXiv:1210.0783.
[20] S. Fantoni, A. Sarsa, and K. E. Schmidt, Phys. Rev. Lett. **87**, 181101 (2001).
[21] I. Vidaña, A. Polls, and A. Ramos, Phys. Rev. C **65**, 035804 (2002).
[22] J. Margueron, J. Navarro, and N. Van Giai, Phys. Rev. C **66**, 014303 (2002).
[23] A. Rios, A. Polls, and I.Vidaña, Phys. Rev. C **71**, 055802 (2005).
[24] F. Sammarruca and P. G. Krastev, Phys. Rev. C **75**, 034315 (2007).
[25] M. Ángeles Pérez-García, J. Navarro, and A. Polls, Phys. Rev. C **80**, 025802 (2009).
[26] J. Carlson, S. Y. Chang, V. R. Pandharipande, and K. E. Schmidt, Phys. Rev. Lett. **91**, 050401 (2003).
[27] S. Giorgini, L. P. Pitaevskii and S. Stringari, Rev. Mod. Phys. **80**, 1215 (2008).
[28] A. Gezerlis and R. Sharma, Phys. Rev. C **85**, 015806 (2012).
[29] A. Gezerlis and J. Carlson, Phys. Rev. C **81**, 025803 (2010).
[30] A. Gezerlis, Phys. Rev. C **83**, 065801 (2011).
[31] C. Lobo, A. Recati, S. Giorgini, and S. Stringari, Phys. Rev. Lett. **97**, 200403 (2006).
[32] S. Pilati and S. Giorgini, Phys. Rev. Lett. **100**, 030401 (2008).
[33] M. M. Forbes, A. Gezerlis, K. Hebeler, T. Lesinski, and A. Schwenk, under preparation (2012).

Journal of Physics: Conference Series **426** (2013) 012012 doi:10.1088/1742-6596/426/1/012012

Hadronic resonances at FAIR energies

Sascha Vogel

Frankfurt Institute for Advanced Studies, Ruth-Moufang-Str. 1, 60438 Frankfurt, Germany
SUBATECH, Laboratoire de Physique Subatomique et des Technologies Associées
University of Nantes - IN2P3/CNRS - Ecole des Mines de Nantes
4 rue Alfred Kastler, F-44072 Nantes Cedex 03, France

E-mail: `svogel@th.physik.uni-frankfurt.de`

Abstract. These proceedings cover the analysis of hadronic resonances in heavy ion collisions. The model used for these studies is the Ultra-relativistic Quantum Molecular Dynamics (UrQMD) model. The model will be briefly explained, resonance observables will be highlighted and various kinematical issues will be investigated. Special emphasis will be put on the FAIR energy regime, especially highlighting the Compressed Baryonic Matter (CBM) program.

1. Introduction

With the advent of the Facility for Anti-Proton and Ion Research (FAIR) new detailed studies of the phase diagram of nuclear matter will be possible. For the first time it will be feasible to study the region of high baryo-chemical potential with a high luminosity experiment, namely the Compressed Baryonic Matter (CBM) experiment.

The investigation of hot and dense matter created in heavy ion collision is one of the most complex challenges in present day physics. Various uncertainties go into the understanding of heavy ion collisions, with the additional complexity that a lot of input is energy-dependent. The initial state is unknown, although speculations at high energies exist [1, 2, 3, 4]. The proper theoretical treatment of interactions is still heavily debated (see e.g. [5, 6]) and the transition from a possibly deconfined phase to a hadronic phase is quite poorly understood. While Lattice QCD studies give some input at vanishing baryochemical potential μ_B, the extrapolations into the high μ_B region are poorly understood (for a recent pedagogical review see e.g. [7].) In order to understand the hot and dense matter created in such collisions resonances provide an unique approach to learn about the hot and dense phase which is produced in heavy ion collisions [8, 9, 10, 11, 12, 13, 14, 15, 16, 17, 18, 19, 20, 21, 22, 23, 24, 25, 26]. Strongly decaying resonances (e.g. ρ or K^* mesons) decay on time scales of several fm/c. That means they decay inside the hot and dense medium and thus can potentially reveal the properties of the matter they decayed in. This is of course dependent on the kinematics of the particles and will be discussed later in detail.

These proceedings are structured as follows: After a brief introduction to the theoretical model in Section 2 the effects of baryons will be presented and explained in Section 3. The possibility to explore the dense phase of heavy ion collisions using hadronic decay channels of resonances will be discussed in Section 4. The manuscript will end with conclusions.

2. Model description

For the calculations discussed in these proceedings the Ultra-relativistic Quantum Molecular Dynamics (UrQMD) model has been used. UrQMD is a non-equilibrium transport approach, which relies on the covariant Boltzmann equation. Cross sections are calculated by the principle of detailed balance and the additive quark model or are fitted to available data. Note that due to the geometrical collision criterion detailed balance is only possible up to the $2 \to 2$ level. UrQMD does not include any explicit in-medium modifications for vector mesons or effects to describe the restoration of chiral symmetry. The model allows to study the full space time evolution of all hadrons, resonances and their decay products in hadron-hadron or nucleus-nucleus collisions. This permits to explore the emission patterns of resonances in detail and to gain insight into their origins and decay channels. For previous studies of resonances within this model see [15, 16, 17, 18, 19, 20, 21, 22, 23, 24]. For further details about the UrQMD model the reader is referred to [27, 28]. Recent development concerning the inclusion of a hydrodynamic phase are described in [29].

Experimentally, the reconstruction of resonances is challenging. One often applied technique is to reconstruct the invariant mass spectrum for single events. Then, an invariant mass distribution of mixed events is generated (here, the particle pairs are uncorrelated by definition). The mixed event distribution is subtracted from the invariant mass spectrum of the single (correlated) events. As a result one obtains the mass distributions and yields (after all experimental corrections) of the resonances by fitting the resulting distribution with a suitable function (usually a Breit-Wigner function peaked around the pole mass of the respective resonance).

If the resonance spectral function changes in the hadronic medium this is in principle visible in the difference spectrum between true and mixed events.

However, if a daughter particle (re-)scatters before reaching the detector the signal for the experimental reconstruction is blurred or even lost. Especially for strongly interacting decay products this effect can be sizeable. It is therefore difficult to judge whether a deviation from an expected Breit-Wigner distribution is due to an initial deformation or an increase of the initial width or due to the momentum dependence of the rescattering cross section of the daughter particles.

What makes this analysis even tougher is the fact that the resonances decay over a wide range of densities and therefore only an average value is measured. If this average value is dominated by resonance decays at low density the information from the high density phase is blurred and may offer only a limited view on the high density phase of the heavy ion collision.

UrQMD offers a different technique for the extraction of resonances which we apply here. We follow the individual decay products of each decaying resonance (the daughter particles). If the daughter particles do not rescatter in the further evolution of the system, the resonance is counted as 'reconstructable'. The advantage of this method is that it allows to trace back the origin of each individual resonance to study their spatial and temporal emission pattern. Because UrQMD follows the space time evolution of all particles it is possible to link production and decay point of each individual resonance. This method also allows to explore the reconstruction efficiency in different decay branches.

In order to calculate at which density the resonance decays we have to determine the baryonic density. The baryon density is calculated locally at the position of the resonance in the rest frame of the baryon current (Eckart frame) as $\rho_B = j^0$ with $j^\mu = (\rho_B, \vec{0})$. Details on the calculation of the baryon density are discussed in [21]. In all figures we present the density in units of ground state density, where a value of 0.16 $1/\text{fm}^3$ is assumed.

3. Baryon kinematics

In heavy ion collisions at FAIR energies the baryon density is extremely high. Thus effects, which couple to baryons are expected to be very prominent in this energy regime. One important fac-

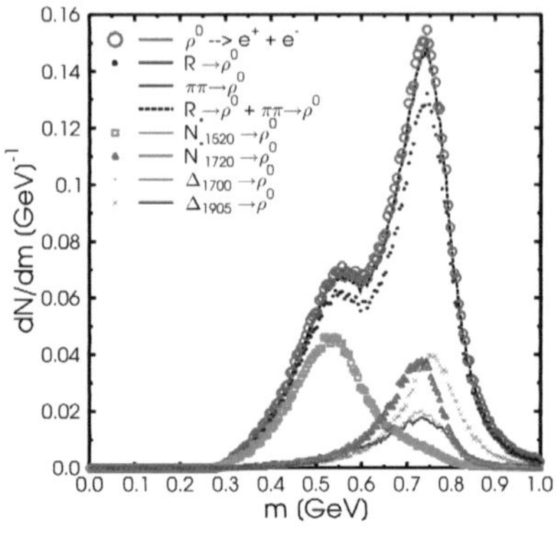

Figure 1. (Color online) Mass spectrum for ρ mesons. Shown is a detailed analysis of the various production channels, with a prominent second peak from the decay of N^*_{1520} baryon resonance decays. For details regarding the different production channels, please refer to the legend.

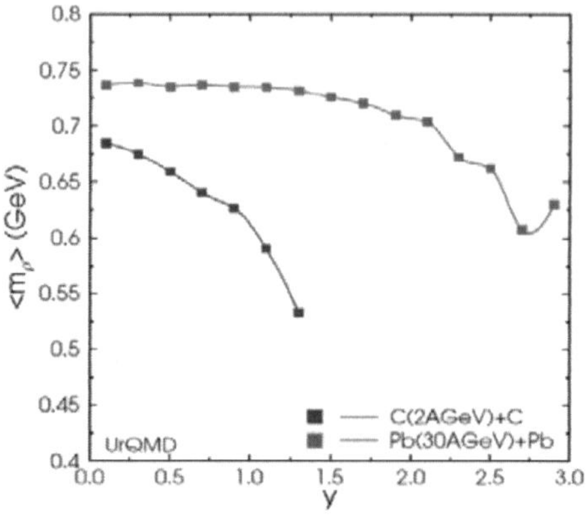

Figure 2. (Color online) Average ρ mass as a function of rapidity. Shown are calculations for Pb+Pb collisions at E_{lab}=30 AGeV (upper line) and C+C collisions at E_{lab}=2 AGeV (lower line). One observes a clear drop when going to higher rapidities. For more details, please refer to the main text.

tor one should keep in mind though is the kinematics of baryons, which can alter the collision dynamics by a great extent. This can be for example seen in Fig. 1. Shown is the ρ meson mass spectrum as a function of the mass for C+C collisions at 2 AGeV. The thick dots depict the total mass spectrum (which can in principle be observed in the dileptonic decay channels), while the curves below depict contributions from individual production channels. For a detailed legend please refer to the figure. One observes a clear double peak structure with one peak at the ρ meson pole mass of roughly 770 MeV and another peak at roughly 500 MeV. The origin of the low mass peak is explained by the contribution of the N^*_{1520} baryon resonance, which decays to roughly 20% via the $N^*_{1520} \rightarrow N + \rho$ channel. This is a purely kinematical effect and should be taken into consideration when analysing more sophisticated effects of the hot and dense medium created in such highly energetic collisions. One should note that this effects vanishes at higher energies, since the mesonic contributions (i.e. the $\pi\pi \rightarrow \rho$ production channel) become more important and dominate the whole mass spectrum. Thus the more baryons are in the system (compared to anti-baryons and mesons) the more prominent the low mass peak becomes.

Another insightful observable apart from the mass spectrum is the rapidity dependence of the average ρ meson mass. Depicted in Fig. 2 is the average ρ meson mass as a function of rapidity for two different collision systems and energies. One observes that it drops when going to higher rapidity, which can be explained with the same line of argument as used before. Since the baryon to meson ratio increases with increasing rapidity the aforementioned effect is quite prominent at higher rapidities. This is in line with the observation, the higher the rapidity (and thus the baryon to meson ratio) is, the lower the average ρ meson mass becomes. This should be kept in mind at FAIR energies, since the baryonic contributions are quite large.

4. Density distribution of decayed resonances

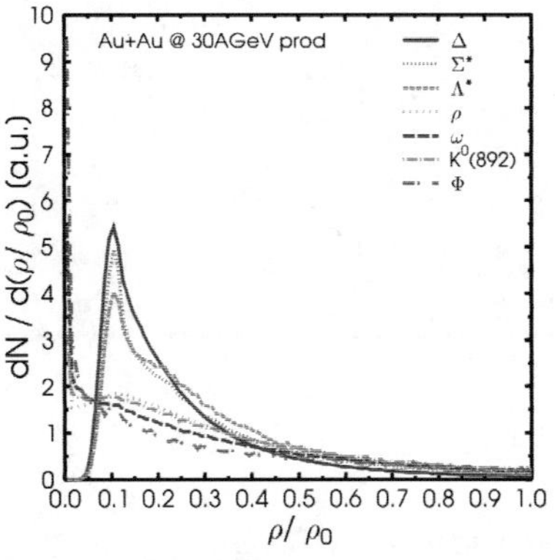

Figure 3. (Color online) Probability distribution of baryon density at the production vertex for various reconstructable resonances in central (b $\leq$ 3.4 fm) Au+Au collisions at 30 AGeV as a function of baryon density. One observes that most resonances which can be reconstructed in the hadronic decay channel originate from low baryon density.

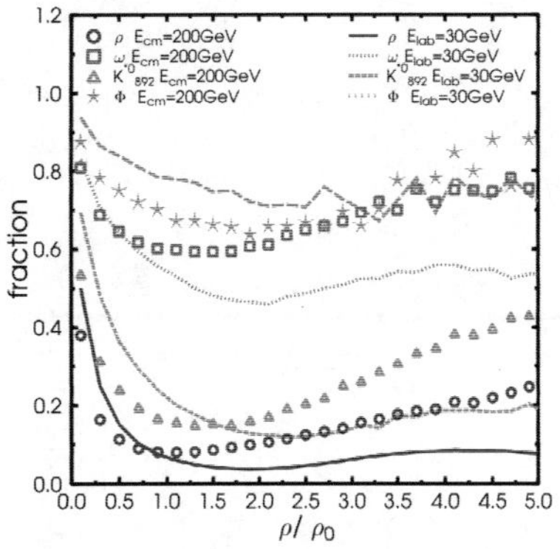

Figure 4. (Color online) Fraction of reconstructable meson resonances as a function of baryon density at the point of production. Shown are various resonances produced in Au+Au collisions at E_{cm} = 200 AGeV (symbols) and E_{lab} = 30 AGeV (lines). Baryon resonances are not shown, however exhibit the same qualitative behaviour.

One central question when analysing resonances is the question where and when they decay and are produced during the collisions and whether we can reconstruct them experimentally or not. Depicted in Fig. 3 is the probability distribution of baryon density at the production vertex for various reconstructable resonances in central (b$\leq$ 3.4 fm) Au+Au collisions at 30 AGeV as a function of baryon density. Most reconstructable resonances originate from very low baryon density. This is of course expected and despite the prominent peaks at very low densities it does not mean that all resonances originate from those low densities.

A different way to look at this is to check the percentage of reconstructable resonances (or the probability to reconstruct the resonance) as a function of baryon density. This is depicted in Fig. 4 on a scale of 0 to 1 for various mesonic resonance species (namely ρ, ω, $K^*(892)$, ϕ). One should note that baryons exhibit exactly the same behaviour, but are not shown here for space reasons. As naively expected the probability is largest at very low densities and decreases when going to higher densities. At roughly 2 times ground state density a surprising effect sets in. The decreasing trend stops and the probability to reconstruct resonances increases when going to even higher baryon density. This effect can be explained when looking at the average transverse momentum of the produced resonances. At higher baryon density resonances have a higher average p_T, which leads to the fact that they escape the collision zone before re-interacting or decaying. Even if they did not decay they are close enough to the edge of the interaction region and the decay products can then leave that region undisturbed. For a more detailed explanation please refer to [23].

5. Conclusions

Understanding the dynamics and kinematics of hadronic resonances proves to be a difficult task both experimentally and theoretically. The FAIR energy regime offers unique opportunities to explore the high density regime of the phase diagram of Quantum-Chromodynamics. Here especially the CBM experiment with its high rates and excellent detectors will play a key role in the experimental program. It will be possible to measure hadronic resonances with great detail, although one should keep in mind that kinematic effects might blur the result. As discussed in these proceedings especially baryon resonance kinematic and the effect of rescattering have a direct influence on observable spectra, particle yields and masses. Only by taking these effects into account one might be able to construct a coherent picture of hadronic resonance physics at FAIR.

References

[1] E. Iancu, A. Leonidov and L. D. McLerran, Nucl. Phys. A **692** (2001) 583
[2] E. Ferreiro, E. Iancu, A. Leonidov and L. McLerran, Nucl. Phys. A **703** (2002) 489
[3] M. Gyulassy and L. McLerran, Nucl. Phys. A **750** (2005) 30
[4] A. Dumitru, F. Gelis, L. McLerran and R. Venugopalan, Nucl. Phys. A **810** (2008) 91
[5] W. A. Horowitz and M. Gyulassy, Phys. Lett. B **666** (2008) 320
[6] W. A. Horowitz, AIP Conf. Proc. **1441** (2012) 889
[7] O. Philipsen, arXiv:1009.4089 [hep-lat].
[8] G. Agakichiev *et al.* [HADES], Phys. Rev. Lett. **98**, 052302 (2007)
[9] X. Lopez *et al.*, Phys. Rev. C **76**, 052203 (2007).
[10] S. V. Afanasev *et al.* [NA49], J. Phys. G **27**, 367 (2001).
[11] D. Adamova *et al.* [CERES], Phys. Rev. Lett. **91**, 042301 (2003)
[12] J. Adams *et al.* [STAR], Phys. Rev. Lett. **97**, 132301 (2006)
[13] B. I. Abelev *et al.* [STAR], Phys. Rev. C **78**, 044906 (2008)
[14] P. Fachini, J. Phys. G **35**, 044032 (2008).
[15] M. Bleicher and J. Aichelin, Phys. Lett. B **530** (2002) 81
[16] M. Bleicher, Nucl. Phys. A **715** (2003) 85
[17] M. Bleicher and H. Stoecker, J. Phys. G G **30** (2004) S111
[18] S. Vogel and M. Bleicher, Phys. Rev. C **74** (2006) 014902
[19] D. Schumacher, S. Vogel and M. Bleicher, Acta Phys. Hung. A **27** (2006) 451
[20] S. Vogel and M. Bleicher, Phys. Rev. C **78** (2008) 064910
[21] S. Vogel, H. Petersen, K. Schmidt, E. Santini, C. Sturm, J. Aichelin and M. Bleicher, Phys. Rev. C **78** (2008) 044909
[22] K. Schmidt, E. Santini, S. Vogel, C. Sturm, M. Bleicher and H. Stocker, Phys. Rev. C **79** (2009) 064908
[23] S. Vogel, J. Aichelin and M. Bleicher, Phys. Rev. C **82** (2010) 014907
[24] S. Vogel, J. Aichelin and M. Bleicher, J. Phys. G G **37** (2010) 094046
[25] G. Torrieri, S. Steinke, W. Broniowski, W. Florkowski, J. Letessier and J. Rafelski, Comput. Phys. Commun. **167** (2005) 229
[26] G. Torrieri, S. Jeon, J. Letessier and J. Rafelski, Comput. Phys. Commun. **175** (2006) 635
[27] S. A. Bass, M. Belkacem, M. Bleicher, M. Brandstetter, L. Bravina, C. Ernst, L. Gerland and M. Hofmann *et al.*, Prog. Part. Nucl. Phys. **41** (1998) 255 [Prog. Part. Nucl. Phys. **41** (1998) 225]
[28] M. Bleicher, E. Zabrodin, C. Spieles, S. A. Bass, C. Ernst, S. Soff, L. Bravina and M. Belkacem *et al.*, J. Phys. G G **25** (1999) 1859
[29] H. Petersen, J. Steinheimer, G. Burau, M. Bleicher and H. Stocker, Phys. Rev. C **78** (2008) 044901
[30] C. Markert, R. Bellwied and I. Vitev, Phys. Lett. B **669** (2008) 92
[31] R. Rapp and J. Wambach, Adv. Nucl. Phys. **25** (2000) 1
[32] H. van Hees and R. Rapp, Phys. Rev. Lett. **97** (2006) 102301
[33] H. van Hees and R. Rapp, Nucl. Phys. A **806** (2008) 339
[34] C. Amsler *et al.* [Particle Data Group Collaboration], Phys. Lett. B **667** (2008) 1.

Journal of Physics: Conference Series **426** (2013) 012013 doi:10.1088/1742-6596/426/1/012013

Soft Photons from transport and hydrodynamics at FAIR energies

Andreas Grimm[1,2], Bjørn Bäuchle[1]

[1] Frankfurt Institute for Advanced Studies, D-60438 Frankfurt am Main, Germany
[2] Institut für Theoretische Physik, Goethe-Universität, D-60438 Frankfurt am Main, Germany

E-mail: grimm@fias.uni-frankfurt.de, baeuchle@th.physik.uni-frankfurt.de

Abstract. Direct photon spectra from uranium-uranium collisions at FAIR energies (E_{lab} = 35 AGeV) are calculated within the hadronic Ultra-relativistic Quantum Molecular Dynamics transport model. In this microscopic model, one can optionally include a macroscopic intermediate hydrodynamic phase. The hot and dense stage of the collision is then modeled by a hydrodynamical calculation. Photon emission from transport-hydro hybrid calculations is examined for purely hadronic matter and matter that has a cross-over phase transition and a critical end point to deconfined and chirally restored matter at high temperatures. We find the photon spectra in both scenarios to be dominated by Bremsstrahlung. Comparing flow of photons in both cases suggests a way to distinguish these two scenarios.

1. Motivation

One of the major goals in high energy nuclear physics was and is the examination of a new strongly interacting deconfined phase of matter, the Quark-Gluon-Plasma (QGP) [1, 2], the state of matter which is believed to have filled the universe shortly after the Big Bang. Many hints for the existence of this phase have been found at the Relativistic Heavy Ion Collider (RHIC). The relevant observables are, e.g., jet quenching and high elliptic flow [3, 4, 5, 6]. The CERN-SPS program has also found evidence for this new state of matter. One has to mention the enhanced K/π ratio (horn) and the step in the mean transverse mass excitation function for pions, kaons and protons [7]. More recently, LHC has confirmed these results at much higher energies [8]. The ALICE Collaboration has shown that the QGP remains a nearly ideal liquid even at much higher energies compared to RHIC [9].

One of the biggest challenges facing heavy-ion physicists is to get information from the hot and dense collision area, the so-called fireball. The small space-time volume of the fireball makes it impossible to probe it directly, so all information has to be obtained from its decay products. Single particles carry only information about their last collision, which usually happens in the cold phase at the end of the collision. Hence, the QGP can only be probed indirectly by most of the possible observables. One of the probes suggested for searching the QGP were photons and leptons. The small interaction cross-section of electromagnetic probes leads to a mean free path which is large compared to the interaction zone. Thus, most photons and electrons leave the interaction zone undisturbed and carry information of the matter present when they were created through the entire collision. Possible electromagnetic probes, which are used for measurements, are single- and dileptons and photons. One further distinguishes between direct photons that are created in elementary collisions, the so-called direct photons, and decay photons. This leads to the biggest challenge in measuring photons in heavy-ion collisions: Most of the photons created in the collision are decay photons from the late stages of the reaction, namely from π^0 and η decays. The dominant process is $\pi^0 \rightarrow \gamma\gamma$. To gain direct photon data, one has to subtract the huge background of decay photons. Some experiments have measured direct photon spectra. Explicit data points at low transverse momenta have been published by the WA98 Collaboration (CERN-SPS) [10] and the PHENIX Collaboration (BNL RHIC) [11, 12]. Theoretical studies have been performed earlier, e.g., with the Ultra-relativistic Quantum Molecular Dynamics (UrQMD) model (see section 2) by Dumitru *et al.* [13] and Bäuchle *et al.* [14] and with hadron-string dynamics (HSD) by Bratkovskaya *et al.* [15]. Hydrodynamic approaches for direct photon calculations have been used for example in: [16, 17, 18, 19, 20, 21]. A comparison between PHENIX data [11, 12] and the model established in [14] can be found in [22].

In the present work, we investigate the composition of direct photon production and elliptic flow v_2 in U+U-collisions at $E_{lab} = 35$ AGeV using the model described in [14]. We compare calculations with hadronic degrees of freedom to calculations with a first order phase transition to a QGP. In Section 2 and 3, we briefly present the underlying model used for calculations and the photon sources, and in Section 4 we analyze our calculations for direct photon emission and the elliptic flow v_2.

2. The Model

This work is based on the microscopic Ultra-relativistic Quantum Molecular Dynamics (UrQMD) model [23, 24, 25], which is originally a hadronic transport model with hadronic and string degrees of freedom. In UrQMD, all non-charmed baryons and mesons with masses up to 2.2 GeV are included. For resonances with high masses, the widths increase and the particles overlap. Therefore, UrQMD replaces resonances with continuous string excitations for higher energies. For hard scatterings, i.e., scatterings with momentum transfer higher than $Q^2 = 1.5$ (GeV)^2 and a minimal center of mass energy $\sqrt{s_{min}} = 10$ GeV, PYTHIA [26] is used. The cross-sections in UrQMD are tabulated from experimental data, parametrized or, if no such data exists, extracted via detailed balance or the additive quark model (AQM). Microscopic models on the basis of an effective solution of the Boltzman equation can be used for non-equilibrium systems and provide all information about the particle trajectory at any time of the collision. On the other hand, these kinds of models are in most cases restricted to binary collisions ($2 \rightarrow n$). This implies large mean free paths of the particles, which may not be fulfilled in the high-density intermediate stage of the collision. One possible description for the hot and dense stage of the collision is ideal hydrodynamics. High elliptic flow observed at RHIC experiments seem to be consistent with ideal hydrodynamic calculations and points to the applicability of this approach [27, 28, 29]. In UrQMD version 3.3, a hybrid approach with a micro- and macroscopic description was incorporated into the UrQMD model [30, 31, 32, 33]. It is now possible to switch from the transport to the hydrodynamic description and back. At the beginning, the particles are propagated on straight trajectories in the so-called cascade mode without potentials. After the first collisions have taken place, fluidization happens and the intermediate hydrodynamic stage starts. The time for this is chosen to be when the nuclei have passed through each other, i.e., $t = 2R\sqrt{2m_N/E_{lab}}$ with R being the radius of the nucleus, m_N the nucleon mass and E_{lab} the kinetic beam energy. Spectators are not fluidized. The particlization, i.e., the transition from the fluid-based to the particle-based description, starts when the energy density falls below a certain value in all propagated cells at the same longitudinal position. Particles are created using the Cooper-Frye formula [34]. All further scatterings and decays are again performed in the microscopic UrQMD model. For the hydrodynamic calculations, different equations of state (EoS) with different degrees of freedom are used. The Hadron Gas EoS (HG-EoS) [35] with only hadronic degrees of freedom can be applied for baseline calculations to compare with cascade

Journal of Physics: Conference Series **426** (2013) 012013 doi:10.1088/1742-6596/426/1/012013

calculations. The Deconfinement EoS (DE-EoS) [36] includes a first-order phase-transition to a chirally restored and deconfined phase of matter at high baryo-chemical potentials, a critical end point and a cross-over at low baryo-chemical potentials and therefore can be applied for photon emission from QGP.

3. Photon emission sources

Photons are not implemented natively in the UrQMD model. Within our model, photons are calculated perturbatively without changing the underlying evolution of the fireball. This treatment is justified, since the electromagnetic cross-section, responsible for photon production, is negligibly small compared to the strong cross-sections dealing with the evolution of the medium.

For the emission of photons from the hydrodynamic stage of the evolution the parametrization by Turbide *et al.* [18] is used. The cross-sections for the emission from the transport evolution description are from Kapusta *et al.* [37]. The two channels implemented in both descriptions and yielding most of the photons are namely $\pi\pi \to \gamma\rho$ and $\pi\rho \to \gamma\pi$. The $\pi\rho \to \gamma\pi$ rate from [18] also includes an a_1-meson as an intermediate stage. Emission from QGP and Bremsstrahlung processes ($\pi\pi \to \pi\pi\gamma$) [38] are only taken into account in the hydrodynamic description. A comprehensive discussion of other hadronic channels can be found in [14]. Photons from primordial pQCD scatterings can be neglected at FAIR energies.

4. Results

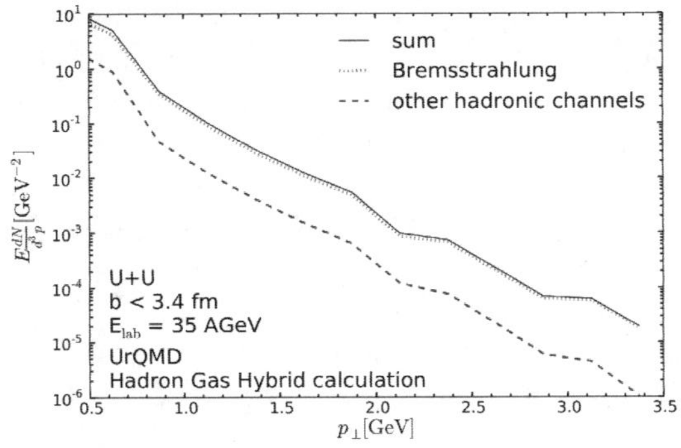

Figure 1. (Color online.) Photon emission from HG-EoS hybrid calculations of central U+U-collisions at $E_{\rm lab} = 35$ AGeV. We show the total spectra (black solid line), emission from Bremsstrahlung processes (red dotted line) and other hadronic channels (blue dashed line).

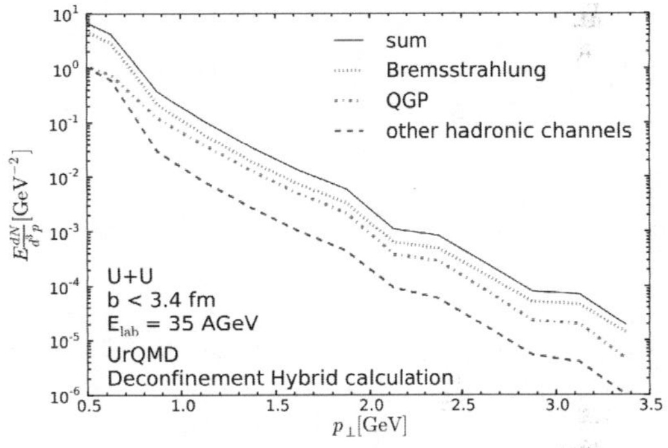

Figure 2. (Color online.) Photon emission from DE-EoS hybrid calculations of central U+U-collisions at $E_{\rm lab} = 35$ AGeV. We show the total spectra (black solid line), emission from Bremsstrahlung processes (red dotted line), other hadronic channels (blue dashed line) and emission from QGP (magenta dash-dotted line).

Figures 1 and 2 show the yield of direct photons from transport and hydrodynamic calculations with the Hadron Gas EoS (HG-EoS) and the Deconfinement EoS (DE-EoS), respectively. We show the spectra for photons from Bremsstrahlung (red dotted lines), QGP (magenta dash-dotted line, only Figure 2), all other channels (blue dashed lines) and the sum of all channels (black solid lines). Emission from Bremsstrahlung is the largest contribution to the total spectra in both cases. In calculations with a phase transition to deconfined matter (see

Figure 2), the amount of photons from Bremsstrahlung is reduced, since a sizeable portion of the system is in the deconfined state and therefore does not emit photons from hadronic processes. Emission from the QGP, however, makes up for the loss almost exactly, so that the total direct photon spectra from both calculations are very similar.

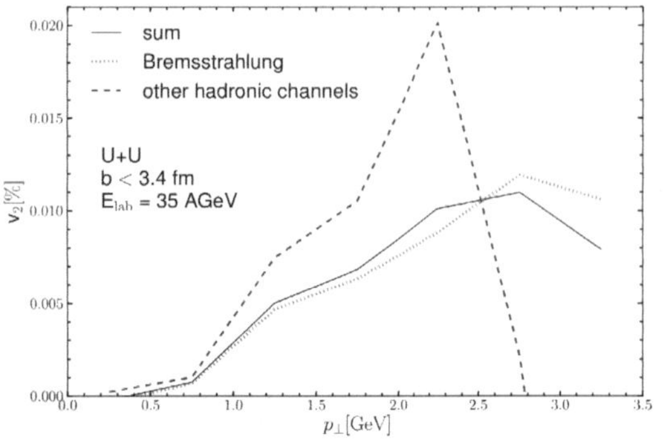

Figure 3. (Color online.) Direct photon elliptic flow v_2 for central U+U-collisions at $E_{\rm lab}$ = 35 AGeV in Hadron Gas EoS Hybrid calculations. We show the flow of all photons (black solid line), flow of Bremsstrahlung photons (red dotted line) and flow from other hadronic channels (blue dashed line).

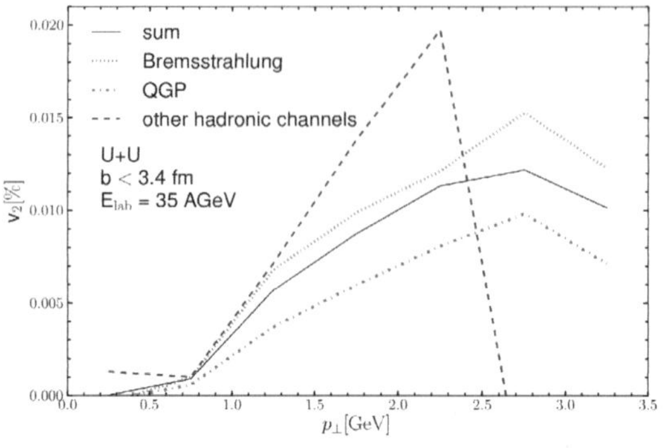

Figure 4. (Color online.) Direct photon elliptic flow v_2 for central U+U-collisions at $E_{\rm lab}$ = 35 AGeV in Deconfinement EoS Hybrid calculations. We show the flow of all photons (black solid line), flow of Bremsstrahlung photons (red dotted line), flow from other hadronic channels (blue dashed line) and flow of photons from QGP (magenta dash-dotted line).

In Figures 3 and 4, we show results for direct photon elliptic flow HG-EoS and DE-EoS hybrid calculations, respectively. We show the elliptic flow v_2 for photons from Bremsstrahlung (red dotted lines), QGP (magenta dash-dotted line, only Figure 2), all other channels (blue dashed lines) and the sum of all channels (black solid lines). The elliptic flow from other hadronic channels is large compared to the flow from Bremsstrahlung and QGP in both scenarios. In both scenarios, elliptic flow is dominated by the largest contributions. The magnitude of flow from these contributions shows the space of time in which emission from these channels happens: QGP emission (DE-EoS calculations only) happens in the early stages of the collision, in which the underlying hadronic medium has not yet built up strong elliptic flow. Thus, photons emitted from this stage carry only little elliptic flow themselves.

Photons from processes that dominate in the late part of the fireball evolution, however, have a large hadronic flow present at their emission time imprinted in their elliptic flow pattern. Comparing flow results from both scenarios, we can re-assess the conclusion we drew for direct photon spectra in Figures 1 and 2. There, we concluded that QGP emission in the Deconfinement EoS calculations basically substitutes Bremsstrahlung emission from Hadron Gas calculations in the early stages. Indeed, we see a larger elliptic flow from Bremsstrahlung in the Deconfinement EoS calculations than in the Hadron Gas EoS calculations, pointing to a later average emission time.

Unlike in the case of the spectra, the remaining difference between both calculations remains quite sizeable, with direct photon elliptic flow peaking about 20 % higher where a phase transition is present.

5. Summary

We have shown calculations for direct photon emission from central Uranium+Uranium-collisions at $E_{\rm lab}$ = 35 AGeV and compared the emission patterns from matter with a first order phase transition to chirally restored and deconfined matter and critical end point (DE-EoS) to that of purely hadronic matter without phase transition. We find the direct photon spectra to be essentially unchanged, while we predict direct photon elliptic flow to be sensitive to the choice of matter. High-quality experiments, as they are being built at the FAIR facility, will be able to distinguish both scenarios.

6. Acknowledgments

This work has been supported by the Frankfurt Center for Scientific Computing (CSC), the GSI and the BMBF. We would also like to thank for the support of the Hessian LOEWE initiative through the Helmholtz International Center for FAIR.

References

[1] J. W. Harris and B. Muller, Ann. Rev. Nucl. Part. Sci. **46**, 71 (1996)
[2] S. A. Bass, M. Gyulassy, H. Stoecker and W. Greiner, J. Phys. G **25**, R1 (1999)
[3] J. Adams *et al.* [STAR Collaboration], Nucl. Phys. A **757**, 102 (2005)
[4] B. B. Back *et al.*, Nucl. Phys. A **757**, 28 (2005)
[5] I. Arsene *et al.* [BRAHMS Collaboration], Nucl. Phys. A **757**, 1 (2005)
[6] K. Adcox *et al.* [PHENIX Collaboration], Nucl. Phys. A **757**, 184 (2005)
[7] C. Alt *et al.* [NA49 Collaboration], Phys. Rev. C 77, 024903 (2008)
[8] E. Shuryak, Physics **3**, 105 (2010)
[9] K. Aamodt *et al.* [ALICE Collaboration], Phys. Rev. Lett. **105**, 252302 (2010)
[10] M. M. Aggarwal *et al.* [WA98 Collaboration], arXiv:nucl-ex/0006007
[11] S. S. Adler *et al.* [PHENIX Collaboration], Phys. Rev. Lett. **94**, 232301 (2005)
[12] A. Adare *et al.* [PHENIX Collaboration], arXiv:0804.4168 [nucl-ex]
[13] A. Dumitru, M. Bleicher, S. A. Bass, C. Spieles, L. Neise, H. Stocker, and W. Greiner, Phys. Rev. C **57**, 3271 (1998)
[14] B. Bäuchle and M. Bleicher, Phys. Rev. C **81**, 044904 (2010)
[15] E. L. Bratkovskaya, S. M. Kiselev, and G. B. Sharkov, Phys. Rev. C **78**, 034905 (2008)
[16] J. I. Kapusta, P. Lichard, and D. Seibert, Phys. Rev. D **44**, 2774 (1991); 47, 4171(E) (1993)
[17] D. K. Srivastava and B. Sinha, Phys. Rev. C **64**, 034902 (2001)
[18] S. Turbide, R. Rapp, and C. Gale, Phys. Rev. C **69**, 014903 (2004)
[19] S. Turbide, C. Gale, S. Jeon, and G. D. Moore, Phys. Rev. C **72**, 014906 (2005)
[20] F. M. Liu, T. Hirano, K. Werner, and Y. Zhu, J. Phys. G **36**, 064072 (2009)
[21] F. M. Liu, T. Hirano, K. Werner, and Y. Zhu, Phys. Rev. C **80**, 034905 (2009)
[22] B. Bäuchle and M. Bleicher, Phys. Rev. C **82**, 064901 (2010)
[23] S. A. Bass *et al.*, Prog. Part. Nucl. Phys. **41**, 255 (1998)
[24] M. Bleicher *et al.*, J. Phys. G **25**, 1859 (1999)
[25] H. Petersen, M. Bleicher, S. A. Bass, and H. Stocker, arXiv:0805.0567 [hep-ph]
[26] T. Sjostrand, S. Mrenna, and P. Z. Skands, J. High Energy Phys. 05 (2006) 026
[27] P. F. Kolb, P. Huovinen, U. W. Heinz and H. Heiselberg, Phys. Lett. B **500**, 232 (2001)
[28] P. F. Kolb, U. W. Heinz, P. Huovinen, K. J. Eskola and K. Tuominen, Nucl. Phys. A **696**, 197 (2001)
[29] P. Huovinen, P. F. Kolb, U. W. Heinz, P. V. Ruuskanen and S. A. Voloshin, Phys. Lett. B **503**, 58 (2001)
[30] D. H. Rischke, S. Bernard, and J. A. Maruhn, Nucl. Phys. A **595**, 346 (1995)
[31] D. H. Rischke, Y. Pursun, and J. A. Maruhn, Nucl. Phys. A **595**, 383 (1995); **596**, 717(E) (1996)
[32] H. Petersen, J. Steinheimer, G. Burau, M. Bleicher, and H. Stocker, Phys. Rev. C **78**, 044901 (2008)
[33] J. Steinheimer, V. Dexheimer, M. Bleicher, H. Petersen, S. Schramm, and H. Stocker, Phys. Rev. C **81**, 044913 (2010)
[34] F. Cooper and G. Frye, Phys. Rev. D **10**, 186 (1974)
[35] D. Zschiesche, S. Schramm, J. Schaffner-Bielich, H. Stoecker and W. Greiner, Phys. Lett. B **547** (2002) 7 [arXiv:nucl-th/0209022]
[36] J. Steinheimer, S. Schramm and H. Stocker, arXiv:0909.4421 [hep-ph]
[37] J. I. Kapusta, P. Lichard and D. Seibert, Phys. Rev. D **44** (1991) 2774 [Erratum-ibid. D **47** (1993) 4171]
[38] K. Dusling, NPA 839 (2010) 70

Journal of Physics: Conference Series **426** (2013) 012014 doi:10.1088/1742-6596/426/1/012014

Hadron phenomenology in the Dyson-Schwinger approach

Gernot Eichmann

Institut für Physik, Karl-Franzens-Universität Graz, 8010 Graz, Austria

E-mail: gernot.eichmann@uni-graz.at

Abstract. I present a brief overview of the Dyson-Schwinger approach and its application to hadron properties. Recent results for baryon and tetraquark masses as well as nucleon and Δ elastic and transition form factors are discussed.

1. Introduction

Probing hadrons with external currents reveals their basic structure properties and provides a connection with the underlying quark and gluon dynamics in Quantum Chromodynamics (QCD). Precision electron scattering and electroproduction measurements at JLab and other facilities have provided groundbreaking insight in the nucleon's structure via elastic and transition form factors, polarizabilities and generalized parton distributions [1, 2]. They have stimulated the development of tools to address questions related to quark orbital angular-momentum correlations, the transition between perturbative and non-perturbative regions, or pion-cloud rescattering effects in the chiral and low-momentum region. Future measurements at PANDA/FAIR via $p\bar{p}$ collisions will be instrumental in extending these studies towards a comprehensive understanding of hadron spectroscopy and hadron structure.

A promising tool for studying hadron phenomenology is the Dyson-Schwinger approach. The properties of QCD and hadrons are encoded in QCD's Green functions which can be determined from QCD's quantum integral equations of motion, the Dyson-Schwinger equations (DSEs) [3–5]. In combination with covariant bound-state equations (Bethe-Salpeter and Faddeev equations), they constitute an ab-initio framework for computing hadron properties. The approach is Poincaré-covariant throughout every step and provides access to all momentum scales and all quark masses without the need for extrapolations. Since one operates directly with QCD's degrees of freedom, observable phenomena at the hadron level can be systematically traced back to their microscopic origin in terms of convolutions of Green functions.

Of course, the approach comes with its own limitations: Green functions depend on the gauge, and the existence of infinitely many of them makes a truncation of the system unavoidable. This has consequences for the calculated observables: for example, at the present level of feasibility in the baryon sector, pseudoscalar-meson cloud effects and dynamical decay widths in the low quark-mass region are not yet implemented. That shortcoming can also serve as an advantage as it allows insight into the 'quark core' of a baryon. It will be imperative for future studies to introduce such effects, either from a phenomenological perspective or directly at the quark-gluon level.

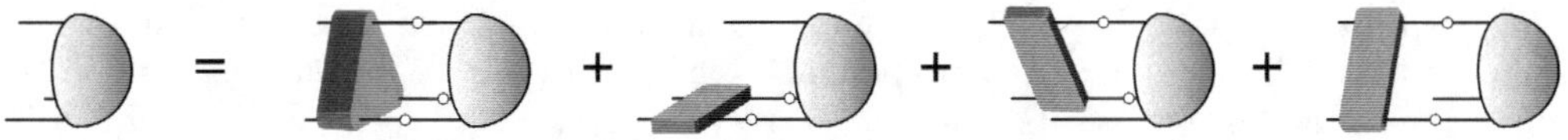

Figure 1. Covariant three-body equation for a baryon's bound-state amplitude, Eq. (1).

2. Covariant bound-state equations

The nonperturbative substructure of hadrons can be resolved through the combination of DSEs and bound-state equations. For baryons, the essential relations are derived from the properties of the quark six-point function G. Baryons correspond to poles in G, and a baryon's bound-state amplitude Ψ satisfies a self-consistent integral equation which is illustrated in Fig. 1:

$$\Psi = \mathbf{K}\, G_0\, \Psi\,, \qquad \text{with} \quad G_0 = S \otimes S \otimes S \quad \text{and} \quad \mathbf{K} = K_{(3)} + \Big[S^{-1} \otimes K_{(2)} \Big]_{\text{perm}}. \tag{1}$$

The subscript 'perm' here and in the equations below indicates three permutations with respect to the quark lines. The equation can be solved once the dressed quark propagator S and the two- and three-quark irreducible kernels $K_{(2)}$ and $K_{(3)}$, which encode the interactions at the quark-gluon level, are determined. This is also the point where a truncation becomes necessary. Omitting the term $K_{(3)}$ yields the covariant Faddeev equation that traces the binding of three quarks in a baryon to its quark-quark correlations. The simplest ansatz for $K_{(2)}$ is the rainbow-ladder kernel; it connects two quarks through a dressed gluon propagator and tree-level quark-gluon vertices. The dressed quark propagator S is then obtained as solution of the quark DSE in the same truncation, and its implementation in the Faddeev equation yields, by iteration, all dressed-gluon ladder exchanges between quark pairs.

The coupling of the baryon to external $q\bar{q}$ currents is encoded in the current matrix element J^μ and the scattering amplitude $J^{\mu\nu}$. They are given by [6]

$$J^\mu = \overline{\Psi}_f\, G_0\, \mathbf{\Lambda}^\mu\, G_0 \Psi_i\,, \qquad J^{\mu\nu} = \overline{\Psi}_f\, G_0 \Big[\mathbf{\Lambda}^{\{\mu} G\, \mathbf{\Lambda}^{\nu\}} - \mathbf{\Lambda}^{\mu\nu} \Big]\, G_0 \Psi_i\,, \tag{2}$$

where Ψ_i and Ψ_f correspond to the incoming and outgoing baryons and

$$\begin{aligned} \mathbf{\Lambda}^\mu &= \Big[\Gamma^\mu \otimes S^{-1} \otimes S^{-1} - \Gamma^\mu \otimes K_{(2)} \Big]_{\text{perm}}, \\ \mathbf{\Lambda}^{\mu\nu} &= \Big[\Gamma^{\mu\nu} \otimes S^{-1} \otimes S^{-1} + \Gamma^{\{\mu} \otimes \Gamma^{\nu\}} \otimes S^{-1} - \Gamma^{\mu\nu} \otimes K_{(2)} \Big]_{\text{perm}}. \end{aligned} \tag{3}$$

In addition to S and $K_{(2)}$, these equations contain the quark-antiquark vertices Γ^μ and $\Gamma^{\mu\nu}$ with one or two external current legs. Their self-consistent calculation from inhomogeneous Bethe-Salpeter equations generates meson poles in the respective J^{PC} channels which dictate the timelike structure of form factors and reproduce features that are familiar from vector-meson dominance models. We have already assumed a rainbow-ladder truncation in Eq. (3); in the general case additional terms would appear in $\mathbf{\Lambda}^\mu$ and $\mathbf{\Lambda}^{\mu\nu}$. The rainbow-ladder truncation ensures that all ingredients, if calculated consistently, satisfy vector and axialvector Ward-Takahashi identities. Electromagnetic gauge invariance is therefore respected at the hadron level, and the Gell-Mann-Oakes-Renner relation for the pion mass and the Goldberger-Treiman relation between the axial and pion-nucleon coupling follow automatically.

Solving Eq. (1) yields a baryon's mass, and from Eq. (2) one can extract baryon elastic and transition form factors as well as, in principle, the nucleon's Compton scattering or pion electroproduction amplitude. The relations for mesons are simpler but obtained along the same lines. Even in rainbow-ladder, solving these equations is complicated by their nature as multi-loop integrals and the fact that their ingredients exhibit a rich momentum, Dirac-Lorentz and color-flavor structure. Nevertheless, once the kernel $K_{(2)}$ is specified, it permits a self-consistent calculation of a wide range of hadron observables without further approximations. Virtually all pseudoscalar-meson, vector-meson, nucleon and Δ-baryon properties calculated so far exhibit only a small sensitivity to the momentum dependence of the quark-gluon coupling in $K_{(2)}$. In that way, these results effectively depend only on a single input scale.

3. Hadron spectroscopy

Considerable progress has been recently made with the covariant Faddeev equation for ground-state baryons. With a rainbow-ladder interaction $K_{(2)}$ that is fixed only to the experimental pion decay constant, the Faddeev equation yields nucleon and Δ masses of $M_N = 0.94$ GeV and $M_\Delta = 1.26$ GeV at the u/d-quark mass, and their current-mass evolution agrees well with lattice data [7–9]. These results complement a range of meson observables that are also well described within a rainbow-ladder truncation, see [10] for a review. Attempts to go beyond rainbow-ladder are underway but so far they have been limited to the meson sector. It is interesting that both N and Δ have a pronounced quark-diquark structure: under the assumption of diquark dominance, the Faddeev equation can be simplified to a quark-diquark two-body equation that implements scalar and axialvector diquarks [11]. The corresponding results in the rainbow-ladder setup agree within $5 - 10\%$ with those from the three-body equation [7].

So far there have been only few studies of meson and baryon excitations in the Dyson-Schwinger approach [12–14]. The open questions in this respect concern the internal structure of baryon resonances, the origin of the observed level ordering, and the connection between the dynamical generation of resonances at the hadron level, where a single 'core' state can generate multiple resonances via meson-baryon interactions [15], versus the generation of these resonances within QCD itself. Results from a simple quark-diquark model suggest that the observed mass ordering of positive- and negative-parity nucleon excitations and the nature of the $N(1440)$ Roper resonance might be caused by diquark clustering in baryons [14]. This has long been suspected as the origin of missing nucleon resonances; nevertheless, such features have not been seen in recent lattice calculations [16].

The idea of diquark clustering has also been the guiding assumption for a recent tetraquark study in the rainbow-ladder truncation [17]. The calculation yields a light scalar four-quark state with a mass of ~ 400 MeV that is predominantly a pion-pion molecule with diquark-antidiquark admixture. It suggests an identification with the elusive $\sigma/f_0(500)$ and thereby provides further support for the identification of the light scalar-meson nonet in terms of tetraquarks.

4. Form factors

Extending these studies to baryon form factors, the rainbow-ladder setup of Eqs. (1–3) has been applied to a number of processes over the past years: nucleon electromagnetic [8], axial and pseudoscalar form factors [18] and Δ electromagnetic form factors [19] in the three-body approach, and N and Δ electromagnetic [20, 21], $N \to \Delta\gamma$ [22] and $N \to \Delta\pi$ [23] transition form factors in the quark-diquark simplification. These works build upon earlier quark-diquark studies with modeled ingredients [11,24]. The calculations were performed in the low– to mid–Q^2 region up to a few GeV^2 and in a current-mass range roughly up to the strange-quark mass.

Without going into details, I will exemplify a few common observations that pertain to all these studies. The overall agreement with experimental data, and also with lattice results at larger quark masses, is generally quite good. Deviations are visible in the low-momentum and small quark-mass region; they are attributable to missing pion-cloud corrections which are not implemented at the present stage. A typical example is the $N \to \Delta\gamma$ magnetic dipole transition form factor $G_M^*(Q^2)$ in Fig. 2 which is comparable to quark model results and to the EBAC

Journal of Physics: Conference Series **426** (2013) 012014 doi:10.1088/1742-6596/426/1/012014

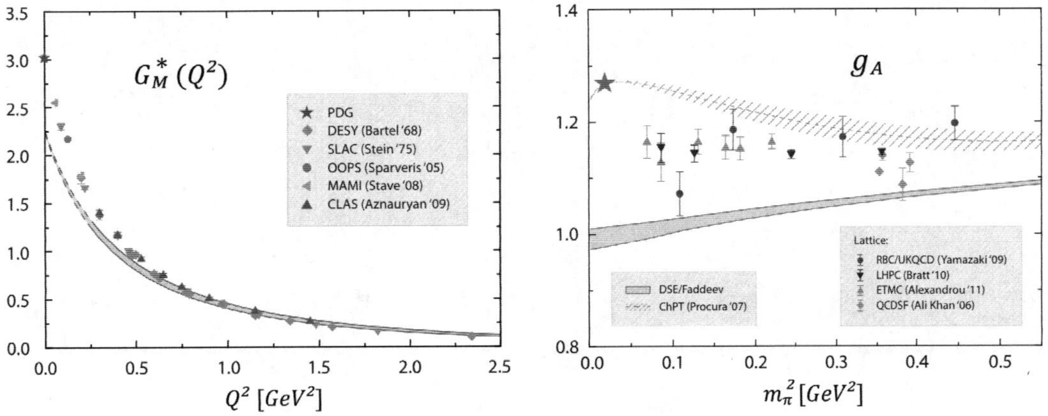

Figure 2. *Left panel:* Q^2-dependence of the $N \to \Delta\gamma$ magnetic dipole transition form factor $G^\star_M(Q^2)$ compared to experiment [22]. *Right panel:* Quark-mass dependence of the nucleon's axial charge g_A, compared to lattice results and a chiral expansion [18].

analysis if meson-baryon rescattering effects are switched off [25]. The nucleon 'quark core' is also visible for various charge radii, which underestimate the experimental values, or in the nucleon axial coupling g_A in Fig. 2. An intriguing example is the nucleon's isoscalar anomalous magnetic moment where leading-order chiral corrections cancel: there, the calculated value agrees quite precisely with the experimental number $\kappa_s = -0.12$ [8].

Via Eq. (3), the form factors inherit the properties of the quark-antiquark vertices Γ^μ. The bulk structure of electromagnetic form factors is determined by the Ball-Chiu vertex, which is the minimal construction that satisfies electromagnetic gauge invariance and depends only on the dressed quark propagator [26]. The second essential ingredient are timelike (π, ρ, a_1) meson poles in the t channel: they are selfconsistently generated in the vertices and important for the timelike and low-momentum structure of the (pseudoscalar, vector, axialvector) form factors. The properties of the quark core are mostly current-mass independent: expressed in terms of dimensionless quantities, no significant change happens for any of the calculated form factors. This means for example that, in the absence of chiral non-analyticities, the quark core of the $\Omega(1672)$ baryon is essentially identical to that of the $\Delta(1232)$ [21].

Finally, the results underline the importance of quark orbital angular momentum in the Q^2-evolution of the form factors. Poincaré covariance entails the appearance of p waves in the ($s-$wave dominated) N and Δ bound-state amplitudes which are absent in non-relativistic quark models. They are, for example, responsible for the small negative value of the $N \to \Delta\gamma$ quadrupole transition ratio $R_{EM}(Q^2)$ and the falloff of the electric proton form factor $G^p_E(Q^2)$ which points toward a possible zero crossing, cf. Fig. 3.

5. Conclusions and outlook

It is imperative to develop versatile theoretical tools to accompany the experimental efforts that are currently made. Existing calculations of nucleon and Δ elastic and transition form factors in the Dyson-Schwinger approach have already offered interesting insights. These studies must be complemented by technical improvements in the future, most importantly the implementation of pion-cloud corrections and hadronic decay channels via truncations beyond rainbow-ladder. Future investigations will also aim at more sophisticated systems and processes such as baryon excitations, nucleon-to-resonance transition form factors, Compton scattering, pion electroproduction, or timelike $p\bar{p}$ annihilation processes.

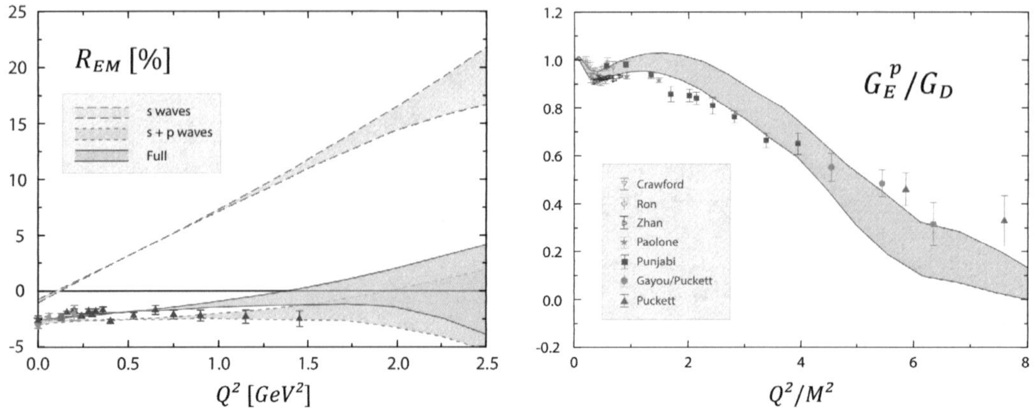

Figure 3. *Left panel:* Quark orbital angular-momentum decomposition of the $N \to \Delta\gamma$ electric quadrupole transition ratio $R_{EM}(Q^2)$ [22]. *Right panel:* Electric proton form factor $G_E^p(Q^2)$, normalized by the dipole and compared to experimental data [8].

Acknowledgments
I would like to thank the organizers of the FAIRNESS 2012 workshop for an inspiring meeting in a beautiful environment. I am indebted to the people who collaborated with me on the various topics that I have presented in this talk. This work was supported by the Austrian Science Fund FWF under Erwin-Schrödinger-Stipendium No. J3039.

References
[1] I. G. Aznauryan and V. D. Burkert, Prog. Part. Nucl. Phys. **67**, 1 (2012).
[2] J. Dudek *et al.*, arXiv:1208.1244 [hep-ex].
[3] C. D. Roberts and A. G. Williams, Prog. Part. Nucl. Phys. **33**, 477 (1994).
[4] R. Alkofer and L. von Smekal, Phys. Rept. **353**, 281 (2001).
[5] C. S. Fischer, J. Phys. G **32**, R253 (2006).
[6] G. Eichmann and C. S. Fischer, Phys. Rev. D **85**, 034015 (2012).
[7] G. Eichmann, R. Alkofer, A. Krassnigg and D. Nicmorus, Phys. Rev. Lett. **104**, 201601 (2010).
[8] G. Eichmann, Phys. Rev. D **84**, 014014 (2011).
[9] H. Sanchis-Alepuz, G. Eichmann, S. Villalba-Chavez and R. Alkofer, Phys. Rev. D **84**, 096003 (2011).
[10] A. Bashir *et al.*, Commun. Theor. Phys. **58**, 79 (2012).
[11] M. Oettel, G. Hellstern, R. Alkofer and H. Reinhardt, Phys. Rev. C **58**, 2459 (1998).
[12] M. Blank and A. Krassnigg, Phys. Rev. D **84**, 096014 (2011).
[13] A. Holl, A. Krassnigg and C. D. Roberts, Phys. Rev. C **70**, 042203 (2004).
[14] C. Chen, L. Chang, C. D. Roberts, S. Wan and D. J. Wilson, Few Body Syst. **53**, 293 (2012).
[15] N. Suzuki *et al.*, Phys. Rev. Lett. **104**, 042302 (2010).
[16] R. G. Edwards, J. J. Dudek, D. G. Richards and S. J. Wallace, Phys. Rev. D **84**, 074508 (2011).
[17] W. Heupel, G. Eichmann and C. S. Fischer, arXiv:1206.5129 [hep-ph], to be published in Phys. Lett. B.
[18] G. Eichmann and C. S. Fischer, Eur. Phys. J. A **48**, 9 (2012).
[19] H. Sanchis-Alepuz, R. Alkofer and R. Williams, PoS QNP **2012**, 112 (2012).
[20] G. Eichmann, *PhD Thesis*, University of Graz, arXiv:0909.0703 [hep-ph].
[21] D. Nicmorus, G. Eichmann and R. Alkofer, Phys. Rev. D **82**, 114017 (2010).
[22] G. Eichmann and D. Nicmorus, Phys. Rev. D **85**, 093004 (2012).
[23] V. Mader, G. Eichmann, M. Blank and A. Krassnigg, Phys. Rev. D **84**, 034012 (2011).
[24] I. C. Cloet, G. Eichmann, B. El-Bennich, T. Klahn and C. D. Roberts, Few Body Syst. **46**, 1 (2009).
[25] B. Julia-Diaz, T.-S. H. Lee, T. Sato and L. C. Smith, Phys. Rev. C **75**, 015205 (2007).
[26] J. S. Ball and T.-W. Chiu, Phys. Rev. D **22**, 2542 (1980).

Journal of Physics: Conference Series **426** (2013) 012015 doi:10.1088/1742-6596/426/1/012015

Production and Dalitz decays of baryon resonances in proton-proton interaction at $\sqrt{s} = 3.16\ GeV$ with HADES

Adrian Dybczak for the HADES collaboration
Jagiellonian University, Cracow, Poland
E-mail: adybczak@gmail.com

Abstract. One of the physics goals of HADES is to investigate spectral modifications of light vector mesons in strongly interacting matter via their dielectron (e^+e^-) decay channel. Theoretical models predict such modifications due to strong meson-baryon resonance coupling which can be also probed in elementary collisions. In 2007 electron-positron pair production has been measured in p+p reactions at a beam kinetic energy of 3.5 GeV. We present analysis results of the exclusive channels $pp\pi^0$, $pn\pi^+$, ppe^+e^-, which indicate contributions of higher resonances. In order to estimate production cross sections of the baryonic resonances, the results have been compared to Monte Carlo calculations based on a resonance model assuming an incoherent sum of various four-star resonances. To obtain the corresponding e^+e^- yields, two models of the width parametrization are applied in the simulation.

1. Introduction

The High Acceptance Di-Electron Spectrometer (HADES) [1] operates at the GSI Helmholtzzentrum für Schwerionenforschung in Darmstadt, Germany. One of the physics goals of HADES is to investigate spectral modifications of light vector mesons in strongly interacting matter via the dilepton (e^+e^-) decay channel. The HADES collaboration has set up an experimental program to measure dilepton spectra in p+p and p+n collisions [2]. In 2007, electron-positron pair production has been measured in elementary collisions of two protons with 3.5 GeV projectile kinetic energy. One of the basic observables in this measurement is the inclusive e^+e^- mass distribution, shown in Fig.1 after Combinatorial Background (CB) subtraction. The expected e^+e^- production channels are Dalitz decays of π^0, η, ω mesons and of the $\Delta(1232)$ resonance. In the high-mass region, two-body decays of ρ/ω play a role. Indeed, the experimental data can be described by PLUTO [3] simulations (see Fig. 1) assuming the aforementioned components, except the mass region below the vector meson pole $M_{inv}^{e^+e^-} \in (0.5 - 0.7) MeV/c^2$. However, the Dalitz decay channel of baryonic resonances is included only for $\Delta(1232)$ here. In order to study the contributions of higher lying baryonic resonances the ppe^+e^-, $pp\pi^0$, $pn\pi^+$ and $pp\omega(\eta) \rightarrow \pi^+\pi^-\pi^0$ exclusive channels have been analysed. The ppe^+e^- channel has been selected by a condition on the pe^+e^- missing mass $\sim 0.938\ MeV/c^2$. The respective e^+e^- invariant mass spectrum is shown in Fig. 2. The mass region below the vector meson pole is not described by a simulation including only ω and η channels, with production cross sections constrainted by the hadronic channels mentioned above (see [7]). In this region, the contribution of baryonic resonances is expected to be enhanced due

Journal of Physics: Conference Series **426** (2013) 012015 doi:10.1088/1742-6596/426/1/012015

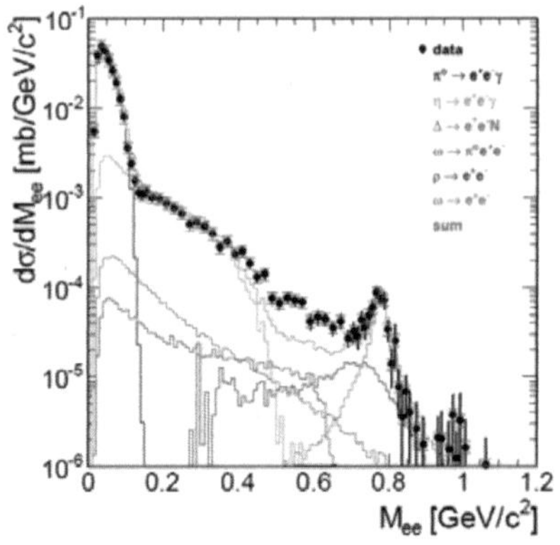

Figure 1. Efficiency corrected inclusive invariant mass distribution of dielectrons inside the geometrical acceptance of HADES. e^+e^- - signal pairs (black dots), after combinatorial background subtraction for p(3.5GeV)+p collisions with corresponding simulation cocktail components. The data are normalized to the measured p+p elastic events.

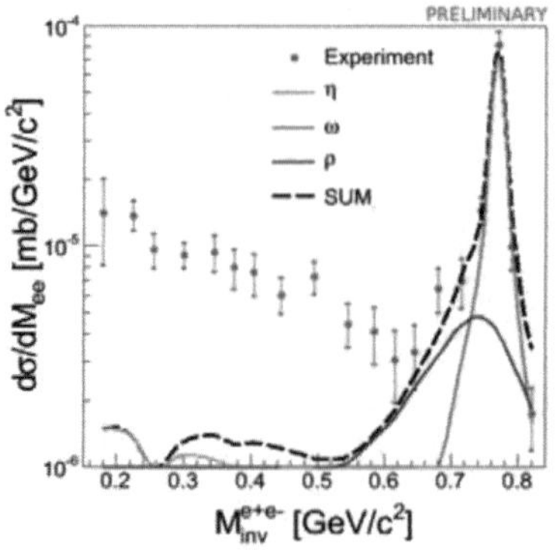

Figure 2. Efficiency corrected invariant mass distribution of dielectrons for exclusive ppe^+e^- channel inside the geometrical acceptance of HADES after combinatorial background subtraction. The data are normalized to p+p elastic events and compared to simulation cocktail which includes the sources given in the legend.

to the coupling to the ρ meson. In order to estimate production cross sections of the resonances which decay in the $pn\pi^+$ and $pp\pi^0$ final states, the results have been compared to Monte Carlo calculations based on the resonance model [4] assuming an incoherent sum of various resonances. The differential Dalitz decay width is calculated for each resonance as a function of the dielectron invariant mass, using different electromagnetic transition form factor models.

2. Analysis of $pp \to pn\pi^+$ and $pp \to pp\pi^0$ final states

In both final states only events with two positive tracks were selected. Protons and positive pions were distinguished using measured momentum and particle velocity of reconstructed tracks. The channels were selected via a condition on missing mass of neutral particles (n and π^0, respectively). The signal was obtained as the number of counts above the fitted continuous background. In the case of $pp\pi^0$, the elastic scattering and multi-pion contribution was first subtracted to enhance the purity of the signal. The resonance cross sections were estimated by simultaneous fits of invariant masses ($M_{inv}^{p\pi^+}$, $M_{inv}^{n\pi^+}$, $M_{inv}^{p\pi^0}$) and of angular distributions ($cos\theta_{CM}^{p\pi^+}$, $cos\theta_{CM}^{n\pi^+}$, $cos\theta_{CM}^{p\pi^0}$) in several bins of the $N-\pi$ invariant mass. An iterative procedure was applied: in the first step, an isotropic angular distribution has been assumed for resonance production. Since experimental distributions show a clear forward-backward peaking, in the next step a modified distribution of the form $d\sigma/dt \sim A/t^{\alpha}$ has been used. The quantity t is Mandelstam variable and A, α are fit parameters depending on resonance mass, respectively. Furthermore, the isospin relations connecting the contributions of a given resonance in $pn\pi^+$ and $pp\pi^0$ channels have been preserved. In Figs. 3a, 3b, 4a and 4b the results obtained after acceptance/efficiency corrections assuming the above model are presented. The $pn\pi^+$ channel allows to distinguish doubly and singly charged resonances.

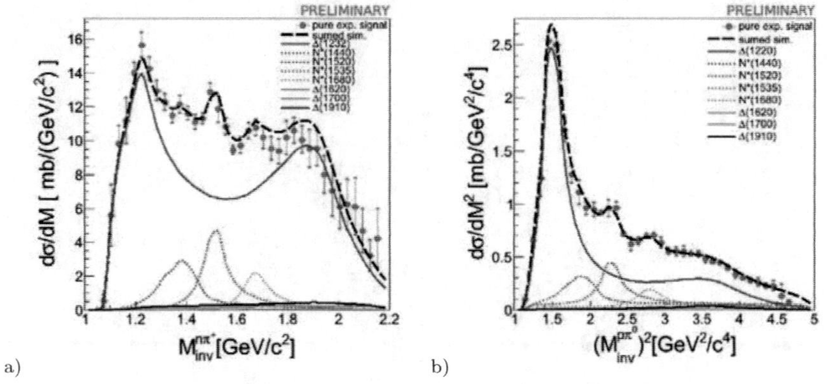

Figure 3. Acceptance and efficiency corrected cross sections as a function of $n\pi^+$ (a) and $p\pi^0$ (b) invariant mass. Simulation components are depicted by curves.

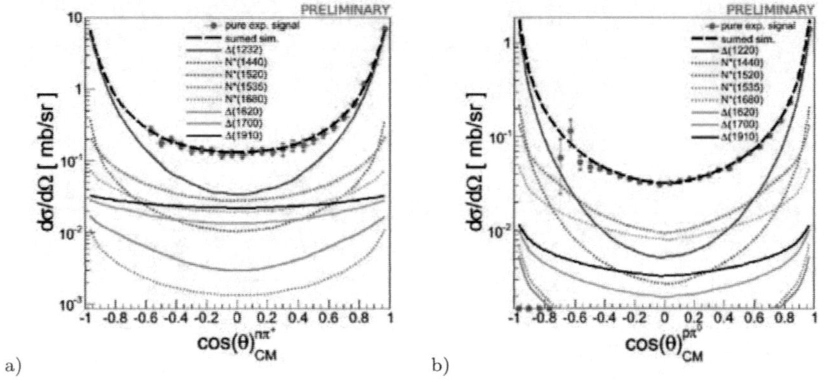

Figure 4. Acceptance and efficiency corrected spectra of $p\pi^+$ (a) and $p\pi^0$ (b) angular distribution in the centre of mass. Simulation components are depicted by curves.

3. Analysis of $pp \rightarrow ppe^+e^-$ final state

Five different models have been compared to the data inside the HADES acceptance including momentum resolution in reconstruction process. Three models ("QED", "Wan/Iachello", "eVMD") are compared to the data with the following assumptions:

- Cross sections for resonances production have been obtained from the analysis of $pp\pi^0/pn\pi^+$ final states presented above.
- Two parametrizations of e^+e mass dependent width of decay in Dalitz mode has been used in "QED"("Wan/Iachello") and "eVMD" models.
- Cross sections for η, $\omega, /\rho$ production have been obtained from an independent analysis of $pp \rightarrow pp\pi^+\pi^-\pi^0$ final state (see [7]).
- The grey band shows the uncertainty of mesonic and baryonic resonances cross sections estimates.

In Figs. 5a and 5b, the result of e^+e^- production obtained from "QED" model [5] has been compared to the data. This model does not have an electromagnetic form factor for different baryonic resonances which decay in the Dalitz mode $R \rightarrow pe^+e^-$.

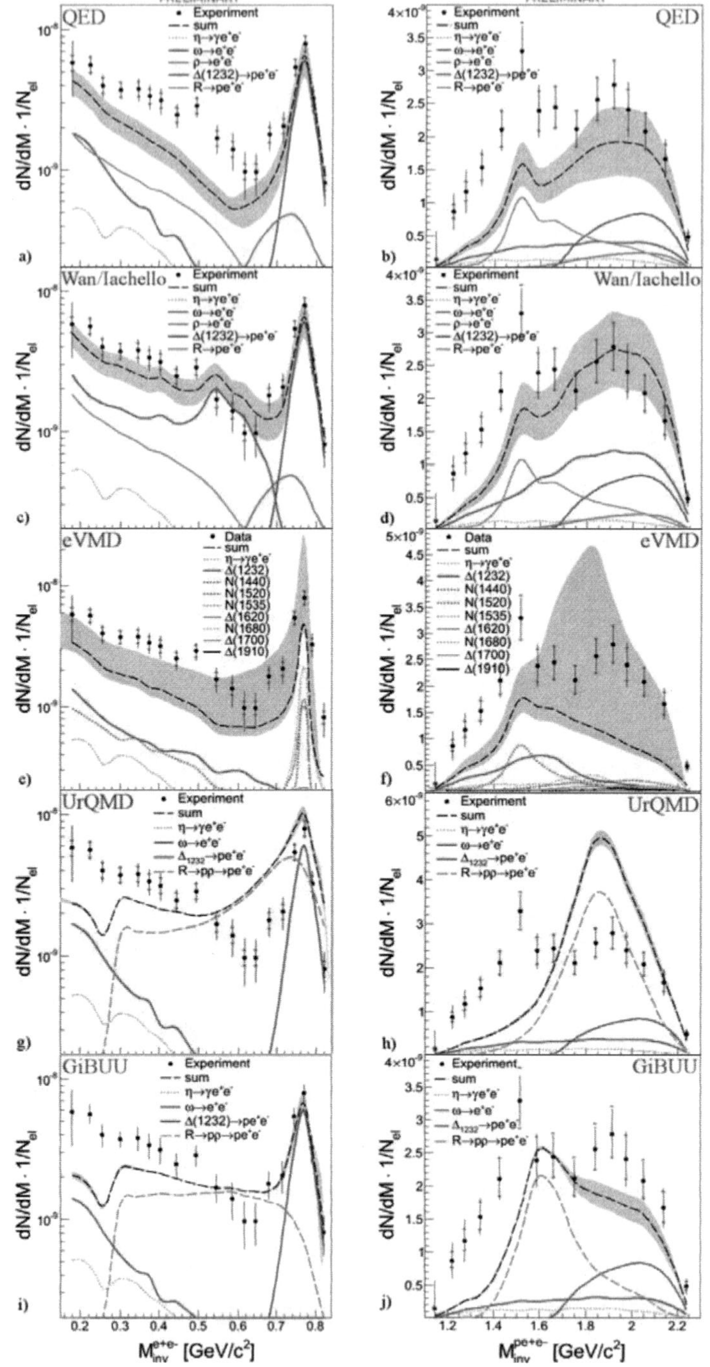

Figure 5. Invariant mass spectrum of e^+e^- and pe^+e^- inside the HADES acceptance. Simulations are based on models "QED"(a,b), "eVMD"(c,d), "Wan/Iachello"(e,f), "UrQMD"(g,h), "GiBUU"(i,j).

"Wan/Iachello" model is a modification of "QED" which assumes a non-constant electromagnetic form factor for the $\Delta(1232)$ resonance Dalitz decay. It causes an enhancement in the range $M_{e^+e^-} \in (0.3 - 0.6)\ GeV/c^2$ (see Fig. 5c) which leads to a better description of data. In addition, $M_{inv}^{pe^+e^-}$ spectrum is not overestimated by this calculation (see Fig. 5d).

Figures 5e and 5f show a comparison of data with predictions of the "eVMD" model [6] where the transition form factors take into account the interferences between the coupling of the baryonic resonance to different vector meson excited states (ρ', ρ'', ω', ...). Even considering the large uncertainty, this model cannot describe the data, in particular data are underestimated below and overestimated above the region $M_{inv}^{pe^+e^-} = 1.6\ GeV/c^2$.

Transport models calculations like UrQMD [8] and GiBUU [9] models are compared to the data with the following assumptions:

- Only $\Delta(1232)$ decays in a Dalitz mode; the rest of baryonic resonances decay in the channel $R \to N\rho$.
- Cross sections for resonance production are taken from model calculations.
- Cross sections for η, ω production have been obtained from an independent analysis of $pp \to pp\pi^+\pi^-\pi^0$ final state (see [7]).
- The grey band shows uncertainty of η and ω cross sections estimations.

"UrQMD" model [8] favors higher lying resonances which populates high values of $N\rho$ masses leading to an overestimate of the data (see Figs. 5g and 5h). In addition, the $M_{inv}^{pe^+e^-}$ spectrum cannot be described at $M_{inv}^{pe^+e^-} < 1.7\ GeV/c^2$.

The composition of baryonic resonance production in "GiBUU" model (see [9]) based on aforementioned resonance model [4] where the dominant contribution is due to the coupling of the ρ meson to the N(1520) resonance, gives a better description of both the $M_{inv}^{pe^+e^-}$ and $M_{inv}^{e^+e^-}$ invariant mass spectra (see Fig. 5i). However, like in the UrQMD model, the yield is underestimated for $M_{inv}^{e^+e^-} < 0.25\ GeV/c^2$, due to the cut at $2M_\pi$ in the ρ mass distribution (see Figs. 5h and 5j).

4. Conclusions and outlook

The applied resonance model of baryon resonance production describes the production of pions in the $pp\pi^0$ and $pn\pi^+$ channels. The obtained cross sections for the various resonances fulfill all isospin relations. The estimated cross section for the $\Delta(1232)$ production is in a good agreement with observations of other experiments. The obtained cross sections for the total π^0 and π^+ yields are also in a agreement with previous measurements.

The best descriptions of the experimentally measured invariant mass spectrum inside HADES acceptance without overshooting other observables, are obtained by "Wan/Iachello" and "GiBUU" models. The Wan/Iachello model, where a form factor is applied only in the case of the $N - \Delta(1232)$ transition is just an attempt to test the sensitivity of the data to the electromagnetic transition form factor models. Agreement with "GiBUU" model points to the importance of the coupling of the ρ meson to the baryonic resonances.

References

[1] G. Agakichiev et al., *Eur. Phys. J. A*, **41**, 243 (2009).
[2] "Dielectron production in pp, pd, AA collisions" - HADES proposal 2004.
[3] I. Froehlich et al., arXiv:0708.2382v2 [nucl-ex].
[4] S. Teis et al. , *Z. Phys. A* **356**, 421 (1997).
[5] M. Zetenyi and G. Wolf, *Heavy Ion Phys.* **17**, 27 (2003).
[6] M.I. Krivoruchenko and B.V. Martemyanov, *Annals Phys.* **296**, 299 (2002).
[7] K. Teilab, *nt.J.Mod.Phys.* **A26**, 694 (2011).
[8] S. A. Bass et al., arXiv:nucl-th/9803035v2.
[9] J. Weil, H. van Hees and U. Mosel, arXiv:1203.3557v2.

FAIRNESS 2012: FAIR NExt Generation of ScientistS 2012 IOP Publishing
Journal of Physics: Conference Series **426** (2013) 012016 doi:10.1088/1742-6596/426/1/012016

Measurement of time-like baryon electromagnetic form factors in processes with initial state radiation

Cristina Morales Morales and Frank E Maas
Helmholtz-Institut Mainz, Johannes Gutenberg-Universität, 55099 Mainz, Germany
E-mail: morales@kph.uni-mainz.de

Abstract. Initial state radiation processes can be effectively used to measure e^+e^- annihilation at high luminosity storage rings, such as the $B-$factory PEP-II in Stanford and the tau-charm factory BEPC-II in Beijing. The BaBar Collaboration has measured with unprecedented accuracy the channels $e^+e^- \to p\overline{p}$, $\Lambda\overline{\Lambda}$, $\Sigma^0\overline{\Sigma}^0$, $\Lambda\overline{\Sigma}^0$, produced with initial state radiation at $\sqrt{s} = 10.6$ GeV. BES-III aims to collect a luminosity of 10 fb^{-1} at $\sqrt{s} = 3.77$ GeV in the next years. The measurements of the baryon electromagnetic form factors published by BaBar are summarized here together with the expectations of BES-III for the same channels.

1. Introduction

Electromagnetic form factors (FFs) account for the non point-like structure of hadrons. The vertex operator $\Gamma^\mu(q)$ describing the hadronic current in the Feynman diagrams of Fig. 1 can be written in terms of the so called Dirac and Pauli FFs, F_1 and F_2:

$$\Gamma^\mu(q^2) = \gamma^\mu F_1(q^2) + \frac{i\sigma^{\mu\nu}q_\nu}{2m_N}F_2(q^2), \tag{1}$$

with m_N the mass of the nucleon N or spin-1/2 baryon. The FFs are analytic functions of the momentum transfer q^2. They are real for the space-like (SL) region ($q^2 < 0$) and complex in the (time-like) TL region ($q^2 > 0$) for $q^2 > 4m_\pi^2$. The use of the so-called Sachs FFs has become conventional:

$$G_E(q^2) = F_1(q^2) + \frac{q^2}{4m^2}F_2(q^2), \qquad G_M(q^2) = F_1(q^2) + F_2(q^2), \tag{2}$$

with $G_E(0) = G_M(0)/\mu_N = 1$ and μ_N the nucleon magnetic moment. Form factors for $q^2 < 0$ are determined by elastic scattering of electrons from hadrons available as targets. Form factors for $q^2 > 0$ are measured in annihilation processes $e^+e^- \leftrightarrow N\overline{N}$.

2. Initial State Radiation

The process of e^+e^--annihilation can be accompanied by the emission of one or several photons from the initial state (ISR) giving rise to annihilations into a hadron pair $e^+e^- \to h^+h^-\gamma_{\rm ISR}$. The following equation relates the differential cross section of $e^+e^- \to N\overline{N}\gamma_{ISR}$ and the cross section of $e^+e^- \to N\overline{N}$:

$$\frac{d\sigma_{e^+e^- \to N\overline{N}\gamma_{ISR}}}{dq^2} = \frac{1}{s}\cdot W(s, x, \theta_\gamma^*)\cdot\sigma(q^2), \tag{3}$$

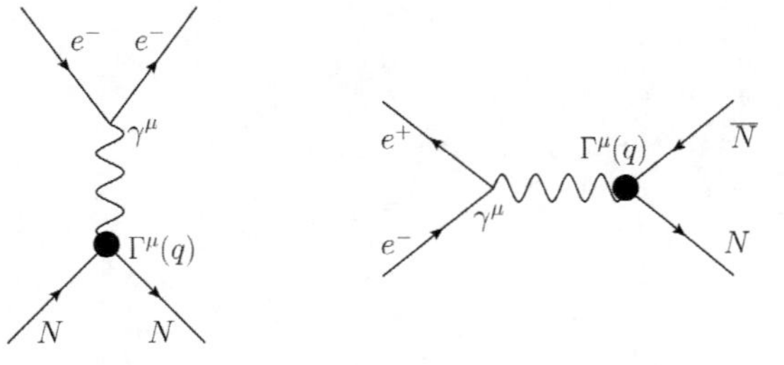

Figure 1. Lowest-order Feynman diagrams for $e^-N \to e^-N$ (left) and for $e^-e^+ \to N\overline{N}$ (right).

where $q^2 = m^2$ is the momentum transfer of the virtual photon, $x = 2E^*_\gamma/\sqrt{s} = 1 - m^2/s$, $\sqrt{s}$ is the center of mass energy of the collider, and E^*_γ and θ^*_γ are the energy and polar angle of the ISR photon in the e^+e^- center of mass (c.m.). The radiator funtion $W(s, x, \theta^*_\gamma)$ [1], describes the probability of the ISR photon emission

$$W(s, x, \theta^*_\gamma) = \frac{\alpha}{\pi x} \cdot \left(\frac{2 - 2x + x^2}{\sin^2\theta^*_\gamma} - \frac{x^2}{2} \right), \tag{4}$$

with α the electromagnetic fine structure constant. This function favors the emission of the ISR photon at small c.m. polar angle and with low energy. The differential cross section of the annihilation process $e^+e^- \to N\overline{N}$ in c.m. reads [2]

$$\frac{d\sigma(q^2, \theta^*_N)}{d\Omega} = \frac{\alpha^2 \beta C}{4q^2} \left[(1 + \cos^2\theta^*_N)|G_M(q^2)|^2 + \frac{1}{\tau}\sin^2\theta^*_N |G_E(q^2)|^2 \right], \tag{5}$$

where θ^*_N is the polar angle of the nucleon in c.m., $\tau = 4m^2/q^2$, $\beta = \sqrt{1 - 1/\tau}$, $C = y/(1 - \exp(-y))$, $y = \pi\alpha/\beta$ and the Coulomb factor C accounts for the electromagnetic $N\overline{N}$ interactions [3]. Angular integration of the previous equation gives the total cross section:

$$\sigma(q^2) = \frac{4\pi\alpha^2\beta C}{3q^2} \left[|G_M(q^2)|^2 + \frac{1}{2\tau}|G_E(q^2)|^2 \right]. \tag{6}$$

From the measurement of $q^2 = m^2$ for the process $e^+e^- \to N\overline{N}\ \gamma_{ISR}$, Eq. 5 and Eq. 6 for the process $e^+e^- \to N\overline{N}$ can be extracted for the q^2 range starting from the $N\overline{N}$ threshold $4m^2_N$ and up to $\sqrt{s}$. The analysis of the angular differential cross section of Eq. 5 allows the independent extraction of G_E and G_M for different momentum transfers in a single experiment. The rate for ISR production of a given channel is suppressed by at least a factor α/π with respect to the cross section without ISR photon (Eq. 3). However, this loss of cross section is compensated in high luminosity experiments like BaBar [4] and BES-III [5], located in e^+e^- storage rings.

3. Tagged and untagged ISR measurements in BaBar and BES-III

There are two approaches for studying ISR events. In the first approach, the untagged one, the ISR photon is not detected, but all the final hadrons must be detected and fully reconstructed. In the second approach, the tagged one, the ISR photon and the hadrons must be detected and fully reconstructed. Since the differential cross section for ISR events increases significantly for photons with small θ^*_γ (Eq. 4), the untagged approach offers much larger statistics than the tagged one. The fact that the ISR photon in BES-III is much less energetic than in BaBar for the hadronic invariant masses of interest, opens up the possibility for untagged measurements

Journal of Physics: Conference Series **426** (2013) 012016 doi:10.1088/1742-6596/426/1/012016

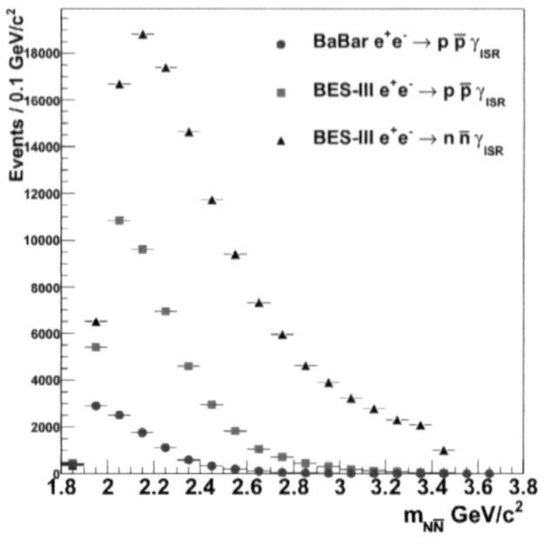

Figure 2. Expected yield of $p\overline{p}$ (red) and $n\overline{n}$ (black) ISR events in BES-III and of $p\overline{p}$ ISR events in BaBar (blue).

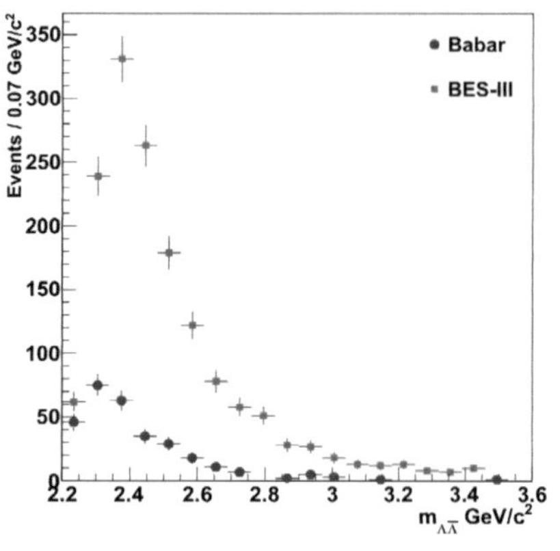

Figure 3. Expected yield of $\Lambda\overline{\Lambda}$ ISR events in BES-III (red) and in BaBar (blue).

in BES-III. This is not possible for BaBar, where the hadrons of the final state are highly boosted in the opposite direction to the ISR photon and can only reach the detector if the ISR photon also does. Fig. 2 shows the expected BES-III statistics for the $p\overline{p}$ and $n\overline{n}$ ISR channels (sum of tagged and untagged events). Fig. 3 shows the corresponding statistics for the $\Lambda\overline{\Lambda}$ ISR channel. An integrated luminosity of 10 fb^{-1} and $\sqrt{s} = 3.77$ GeV has been assumed here. For comparison, the event yield for the BaBar $p\overline{p}$ and $\Lambda\overline{\Lambda}$ ISR channels (only tagged events) for an integrated luminosity of 232 fb^{-1} and $\sqrt{s} = 10.6$ GeV is shown as well. The Phokhara event generator [6, 7] was used for these simulations and the geometrical acceptance in the c.m. polar angle of the detectors used was $22° < \theta^* < 159°$ for the hadrons and the ISR photon in BES-III, and $47° < \theta^* < 157°$ for the hadrons and $40° < \theta^* < 157°$ for the ISR photon in BaBar.

4. BaBar measurement of $e^+e^- \to p\overline{p}$ via ISR

BaBar analyzed a data sample corresponding to 232 fb^{-1} from which 4025 $e^+e^- \to p\overline{p}\,\gamma_{\rm ISR}$ candidates were selected [8]. Events were tagged by detecting a photon radiated at polar angle $22° < \theta < 137°$ in the laboratory frame. After all selection criteria, a background contamination of about 5% for $m_{p\overline{p}} < 2.5{\rm GeV}/c^2$, increasing with $m_{p\overline{p}}$ and consistent with 100% above 4.5 GeV$/c^2$ was estimated. The main background source were $e^+e^- \to p\overline{p}\pi^0$ events with a soft photon lost or two overlapping γ. From the measurement of $\sigma(e^+e^- \to p\overline{p})$, the effective FF defined as:

$$|F(q^2)|^2 = \frac{2\tau|GM(q^2)|^2 + |G_E(q^2)|^2}{2\tau + 1} \qquad (7)$$

was extracted. This effective FF can be directly compared to previous measurements of $|G_M(q^2)|$, all obtained under the working hypothesis that $|G_E(q^2)| = |G_M(q^2)|$ (Fig. 4). A general consistency among all these data was observed. In particular the precise data by BaBar and by the PS170 experiment at LEAR [9], show a similar rise of the effective FF when the energy approaches the $p\overline{p}$ threshold. There are several possible explanations for this feature [10, 11, 12, 13]. A similar behavior has been observed in $m_{p\overline{p}}$ in other processes with different dynamics [14, 15, 16, 17, 18]. The dashed line in Fig. 4 is the result of a QCD inspired fit [19]. The fit curve is about a factor 2 higher than the corresponding curve in the SL region, while QCD and analycity predict that asymptotically $|G^p_{E,M}(q^2)| = |G^p_{E,M}(-q^2)|$.

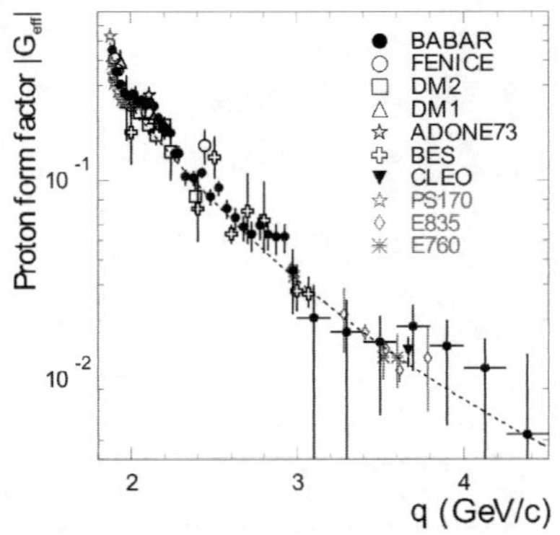

Figure 4. Effective proton FF measured by BaBar and previous experiments.

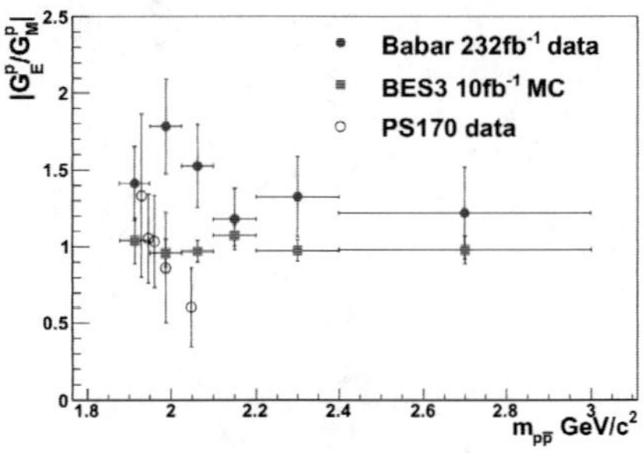

Figure 5. Ratio $|G_E^p/G_M^p|$ measured by BaBar (blue) and PS170 (black), and expected by BES-III ($G_E^p/G_M^p = 1$, red).

The ratio of electric to magnetic form factor was extracted from the analysis of the distribution of the proton helicity angle in the $p\bar{p}$ rest frame. The mass region from $p\bar{p}$ threshold up to 3 GeV/c^2 was divided in six intervals. For each one of these bins, BaBar measured values of $|G_E/G_M|$ greater than unity, in disagreement with the previous results from experiment PS170 (Fig 5). The PS170 data had a non complete angular acceptance and strong angular dependence of the detection efficiency. In the same figure, the expectations for the expected BES-III statistics under the assumption that $G_E^p/G_M^p = 1$ are also shown.

5. BaBar measurements of strange-baryon form factors via ISR

BaBar measured the production via ISR of the following strange baryon pairs: $e^+e^- \rightarrow \Lambda\bar{\Lambda}, \Sigma^0\bar{\Sigma}^0, \Lambda\bar{\Sigma}^0(\Sigma^0\bar{\Lambda})$ by tagging the ISR photon [20]. Fig. 6 shows a summary of the strange-baryon effective FFs measured by BaBar. It is seen that the Λ, Σ^0, and $\Sigma^0\Lambda$ FFs are of the same order. The same figure reports, by comparison, the effective proton FF measured by BaBar. A fit to the Λ FF with the power-law function $F(Q^2) \sim Q^{-n}$ returned $n \simeq 9$, showing that the asymptotic regime $n = 4$ predicted by perturbative-QCD is not reached in the energy range below 3 GeV. The ratio $|G_E^\Lambda/G_M^\Lambda|$ was measured from the angular distribution of the produced Λ in the $\Lambda\bar{\Lambda}$ channel for two different mass intervals. The results are shown in Fig. 7 together with BES-III expectations for this channel. The values are consistent with unity within large errors. BaBar also tried to extract the relative phase Φ, between G_E^Λ and G_M^Λ by measuring the polarization of the outgoing Λ, perpendicular to the scattering plane of the e^+e^- process. Under the assumption $|G_E^\Lambda| = |G_M^\Lambda|$, the measured lambda polarization was translated in an interval for the relative phase between the two FFs: $-0.76 < \sin\Phi < 0.98$.

6. Prospects for time-like baryon FFs measurements in BES-III

The BEPC-II accelerator is designed to have the highest instantaneous luminosity on the $\Psi(3770)$ resonance. Peak luminosities of $6 \cdot 10^{32}$ cm$^{-2}s^{-1}$ have been achieved ($\sim$ 60% of design value). Background conditions appear to be optimal for data taken on the $\Psi(3770)$ resonance, on which 2.9 fb^{-1} of integrated luminosity have already been taken and for which 10 fb^{-1} are expected in the coming years. Feasibility studies show, that the expected statistics at BES-III will be competitive with the existing BaBar measurements (see Figs. 5 and 6). A determination of the FFs of the neutron seems feasible as well at BES-III. The precision remains to be investigated.

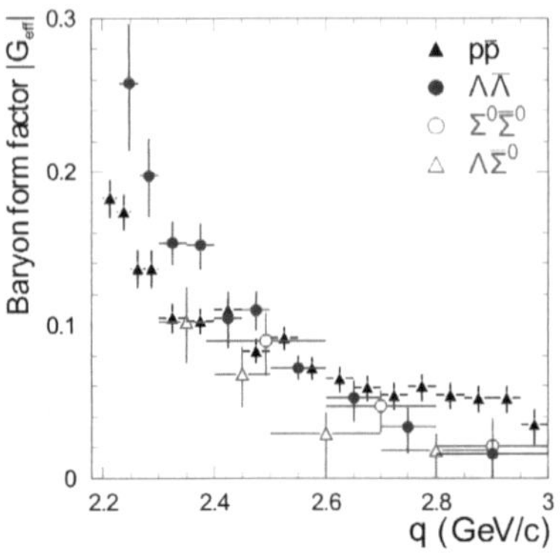

Figure 6. Baryon FFs measured by BaBar.

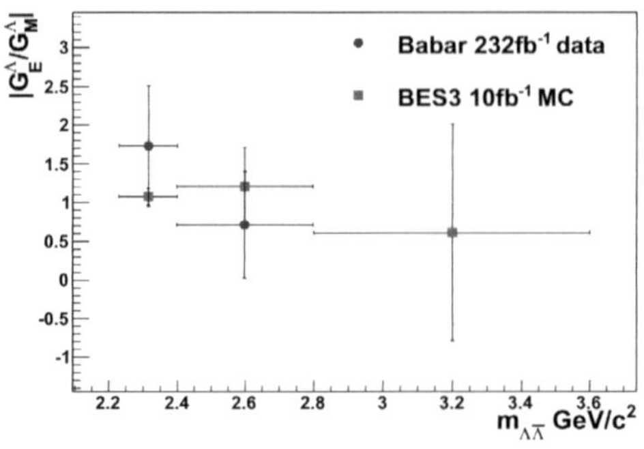

Figure 7. Ratio $|G_E^\Lambda/G_M^\Lambda|$ measured by BaBar (blue) and expected by BES-III ($G_E^p/G_M^p = 1$, red).

7. Conclusions

Compared to the SL sector, where precision measurements of FFs have been achieved on the percent level, the data base of TL FFs is rather scarce. A major improvement has been achieved by introducing the initial state radiation technique at BaBar obtaining an effective proton form factor with unprecedented precision and covering a wide q^2 range. The BaBar result concerning the ratio of the proton electric to the magnetic FFs is still limited by statistical uncertainties, and it is in conflict with a previous measurement at PS170. It is expected, that BES-III improves the achieved accuracy of BaBar by a factor 3 to 4 in the proton and hyperons FFs by using also the ISR technique at the $\Psi(3770)$ resonance. Moreover, at the BEPC-II accelerator a new precision measurement of the time-like baryon FFs over a wide energy range between 2.0 to 4.5 GeV is foreseen [21]. From new facilities at Novosibirsk and FAIR/Darmstadt we can expect significantly improved results [22].

References

[1] Bonneau G and Martin F 1971 *Nucl. Phys.* B **27** 381
[2] Zichichi A, Berman S M, Cabibbo N and Gatto R 1962 *Nuovo Cimento* **24** 170
[3] Tzara C 1970 *Nucl. Phys.* B **18** 216-252
[4] Aubert B *et al* 2002 *Nucl. Instr. and Meth.* A **479** 1
[5] Asner D M *et al* 2009 *Int. J. of Mod. Phys.* A **24** Issue: 1 supp
[6] Czyż H, Kühn J H, Nowak E and Rodrigo G 2004 *Eur. Phys. J.* C **35** 527-536
[7] Czyż H, Grzelinska A and Kühn J H 2007 *Phys. Rev.* D **75** 074026
[8] Aubert B *et al* 2006 *Phys. Rev.* D **73** 012005
[9] Bardin G, Burgun G, Calabrese R, Capon G, Carlin R (1994) *et al.Nucl. Phys.* B **411** 3-32
[10] Dmitriev V and Milstein A 2008 *Phys. Lett.* B **658** 13-16
[11] Aubert B *et al* 2006 *Phys. Rev.* D **73** 052003; Frabetti P *et al Phys. Lett.* B **514**, 240-246
[12] Frabetti P *et al Phys. Lett.* B **514**, 240-246
[13] Baldini R 2012 *Nucl. Phys.* B (Proc. Suppl.) **225227** 211215
[14] Aubert B *et al* 2005 *Phys. Rev.* D **72** 051101
[15] Aubert B *et al* 2006 *Phys. Rev.* D **74** 051101
[16] Abe K *et al* 2002 *Phys. Rev. Lett.* **88** 181803
[17] Abe K *et al* 2002 *Phys. Rev. Lett.* **89** 151802
[18] Bai J Z *et al* 2003 *Phys. Rev. Lett.* **91** 022001
[19] Chernyak V, Zhitnitsky A, Serbo V 1977, *JETP Lett.* **26** 594-597
[20] Aubert B *et al* 2007 *Phys. Rev.* D **76** 092006
[21] Denig A, Salmé G 2012 *arxiv*:1210.4689v1 [hep-ex]
[22] Sudol M *et al* 2010 *Eur. Phys. J.* A **44** 373

Journal of Physics: Conference Series **426** (2013) 012017 doi:10.1088/1742-6596/426/1/012017

Charmed hadron spectroscopy on the lattice for $N_f = 2+1$ flavours

Gunnar Bali, Sara Collins and Paula Pérez-Rubio[1]

Institut für Theoretische Physik, Universtät Regensburg, D-93040 Regensburg, Germany

E-mail: gunnar.bali@physik.uni-r.de, sara.collins@physik.uni-r.de, paula.perez-rubio@physik.uni-r.de

Abstract. We study the spectra of charmonia, charmed mesons, singly and doubly charmed baryons using Lattice QCD, with 2+1 flavours of fermions. In the case of mesons, we include higher spin states, while for baryons, both positive and negative parity channels were investigated. By means of the variational method, we were able to extract a clean signal from the correlation functions and information about the excited states.

1. Introduction

In recent years the spectroscopy of hadrons containing charm quarks has received special attention. Concerning charmonium-like states, since 2002, new resonances have been found that are unlikely to be conventional $\bar{c}c$ states (see ref, [1] for a summary of quarkonium states). The triggering point was the discovery of the $X(3872)$ resonance by the Belle collaboration [2], later confirmed by BaBar, CDF and D0, whose mass lies close to the $D\bar{D}^*$ threshold. It has been widely studied and no consensus on its inner structure has been reached so far. After this observation, a plethora of puzzling chamonium-like states emerged from the experiments, the most interesting being the ones lying close to the open charm thresholds. These new states could correspond to loosely bound hadronic molecules, hybrid states or tetraquark states.

We find a similar situation in open charm meson spectroscopy, although less states have been found. In 2002, the BaBar Collaboration discovered a positive parity scalar meson, $D_{sJ}(2317)$ [3] which was confirmed soon after by CLEO, [4]. In the same experiment, CLEO observed a 1^+ state, $D_{sJ}(2460)$. Both resonances are very narrow and lie right below the DK and D^*K thresholds, respectively. Recently, more new states were observed, for example the $D^*_{s1}(2700)$ and $D^*_{sJ}(2860)$.

Concerning charmed baryons, there exist 17 experimentally well established singly charmed baryons [5], most of which have been found in the last decade, in the e^+e^- colliders and at Fermilab. The charm quark is sufficiently massive for the states to be described as a combination of a heavy quark and a light di-quark and their properties can be understood in terms of Heavy Quark Effective Theory (HQET). Moreover, doubly-charmed baryons are also interesting, since they combine two scales of QCD: the size of the two heavy quark (QQ) system and Λ^{-1} (where Λ is a typical binding energy). Two possible pictures are shown in Figure 1. In the charmonium-like picture (left), the di-quark is formed by a heavy and a light quark. The resulting object will interact with the remaining heavy Q, as if it was a charmonium system. The radius of the QQ

[1] Speaker.

system is much larger than Λ^{-1}. In the HQET picture (right), the QQ diquark system binds itself into the $\bar{3}$ representation of SU(3). In that case the radius of the heavy-heavy diquark is smaller than Λ^{-1}.

Figure 1. Structure of a doubly charmed baryon: Charmonium-like picture (left), HQET picture (right).

In the next few years, more results are expected to appear in currently running experiments, e.g. Belle, BES-III and LHC and the future PANDA experiment at the FAIR facility. In order to study these systems from a theoretical point of view, we need a non perturbative approach to QCD. Lattice QCD achieves this through a Monte Carlo evaluation of the path integral after a discretisation of space time on a finite lattice with lattice spacing a. To keep systematic errors under control, we need a^{-1} to be much larger than the relevant physical scales of the problem, the size of the box L to be larger than the typical size of the hadrons and an extrapolation to the physical quark masses to be carried out (simulating physical msses is computationally expensive).

2. Methods and computational details

We have employed $N_f = 2{+}1$ gauge configurations generated using the tree level $O(a^2)$ Symanzik improved Wilson action for the gluonic degrees of freedom. The fermionic action uses non-perturbatively improved Wilson fermions with stout links in the derivative terms (SLiNC [6]). The quark masses were first tuned to the $SU(3)_{\text{flavour}}$ -symmetric point where the flavour singlet mass average $m_q = (m_u + m_d + m_s)/3$ takes its physical value. Then, the different quark masses are varied while keeping the singlet quark mass fixed [7, 8]. At present, there is only one β value available, $\beta = 5.5$, corresponding to $a \sim 0.0795$ fm.

κ_l	κ_s	a fm	# meas (meson)	# meas (baryon)	M_π (MeV)
0.12090	0.12090	0.0795(3)	941	$--$	442
0.12104	0.12062	0.0795(3)	450	450	348

Table 1. Details of the configurations used so far in this study.

Let $\hat{O}_1$ and $\hat{O}_2$ be two interpolating operators overlapping with the state we are interested in. A correlation function can then be defined,

$$\begin{aligned} C(\hat{\mathcal{O}}_i, \hat{\mathcal{O}}_j, t) &= \langle \hat{\mathcal{O}}_j(0)\hat{\mathcal{O}}_i^\dagger(t)\rangle = \lim_{T\to\infty} \frac{1}{Z(T)} \mathrm{Tr}\left[e^{-(T-t)\hat{H}}\hat{\mathcal{O}}_j e^{-t\hat{H}}\hat{\mathcal{O}}_i^\dagger\right] = \\ &= \sum_n \langle 0|\hat{\mathcal{O}}_j|n\rangle\langle n|\hat{\mathcal{O}}_i^\dagger|0\rangle e^{-E_n t}, \quad Z(T) = \mathrm{e}^{-\mathrm{T}\hat{\mathrm{H}}}, \end{aligned} \tag{1}$$

with T being the temporal extent of the lattice. Every correlation function contains a tower of states. Since we are primarly interested in the lower lying states, the contributions for higher excitations can represent a problem if they do not die out sufficiently fast. The variational method [9] allows us to disentangle different states. The idea is to choose a basis of operators $\hat{O}_i$

and construct a cross correlation matrix $[C(t)]_{ij} = C(\hat{O}_i, \hat{O}_j, t)$. One then solves the generalised eigenvalue problem (GEVP),

$$C^{-1/2}(t_0)C(t)C^{-1/2}(t_0)v^\alpha(t,t_0) = \lambda^\alpha(t,t_0)v^\alpha(t,t_0). \tag{2}$$

It can be shown that the eigenvalues behave like,

$$\lambda^\alpha(t,t_0) \propto e^{-(t-t_0)E_\alpha}\left[1 + O\left(e^{-\Delta E_\alpha(t-t_0)}\right)\right] \quad \Delta E_\alpha = E_{\alpha'} - E_\alpha \text{ with } \alpha' > \alpha. \tag{3}$$

For both mesons and baryons we use a basis of three operators. Each operator has the same Fock structure but different spatial extent. The latter is achieved by applying gauge invariant smearing on the quark fields.

3. Mesons with hidden and open charm

To compute the meson spectra, we used a subset of the operators given in Ref [10], which have the general form,

$$O(x) = \bar{q}_1(x)\Gamma D q_2(x), \tag{4}$$

where the indices denote flavour, Γ is a combination of gamma matrices and D represents a covariant derivative operator, containing zero, one or two derivatives depending on the spatial angular momentum of the particle under consideration (we have explored states with $L \leq 2$). The results obtained for the spectra of the D_s and charmonium systems are shown in Figure 2, for the flavour symmetric ensemble ($\kappa_l = \kappa_s$). We could resolve the first excited state as well as the ground state for all operators. In most charmonium states and for two D_s operators, the second excited state was also extracted. We can see that the experimental spectra are qualitatively reproduced. Before making a more detailed comparison, results at different light quark masses and volumes are required. This work is in progress. The good signals obtained for radially excited and non-zero angular momentum states gives us confidence in our methods. For the D_s system, we will study the mixing of the 0^+ and 1^+ states with the DK and DK^* molecules, respectively. Analogously, we will study molecule mixing for the charmonium-like 1^{++} state, lying close to the $(D^0)^*\bar{D}^0$ threshold .

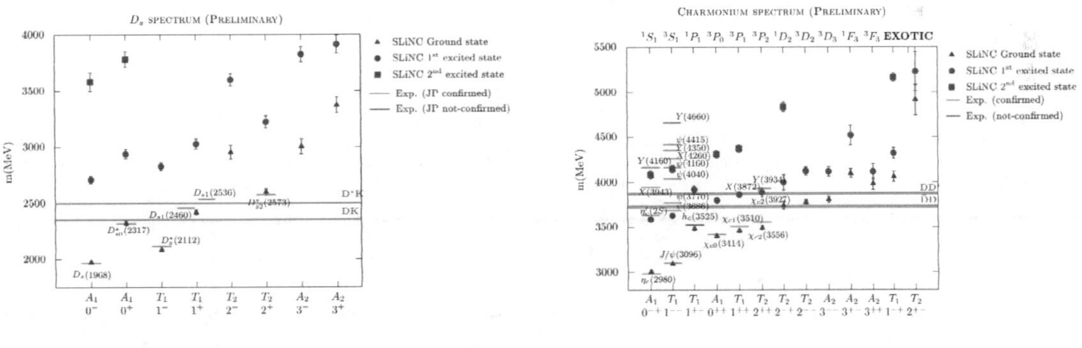

Figure 2. D_s (left) and charmonium (right) spectra from $N_f = 2+1$ configurations. The points are results from the flavour symmetric ensemble. Error bars are statistical only.

4. Charmed baryons

4.1. Interpolating operators

Even though flavour symmetry is not an exact symmetry of nature, $SU(2)_{\text{flavour}}$ is reasonably well respected. If we include the strange quark, since $m_s - m_u < \Lambda_{\text{QCD}}$, we still observe

Journal of Physics: Conference Series **426** (2013) 012017 doi:10.1088/1742-6596/426/1/012017

an approximate flavour symmetry ($SU(3)_{\text{flavour}}$) in the baryon spectrum. Pushing things further, when including the charm quark, flavour symmetry is not a good symmetry of the system. In principle, it is not clear that the interpolating operators falling into the irreducible representations of $SU(4)_{\text{flavour}}$, cf. Figure 3 have significant overlaps with the actual baryon states. However, empirically, we have found that this is indeed the case.

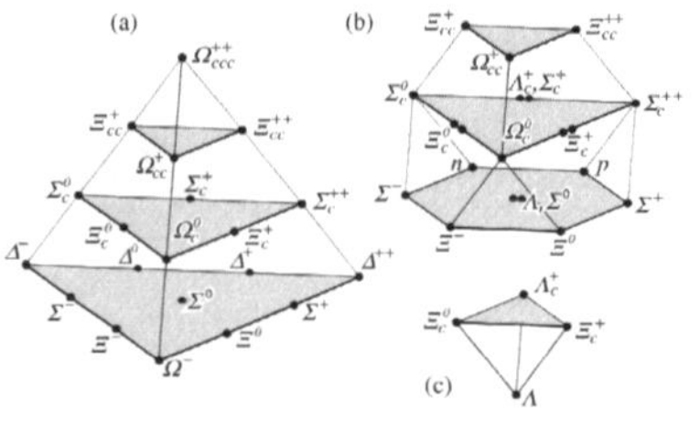

Figure 3. SU(4) multiplets of baryons containing u, d, s, c quarks. a) Totally symmetric 20-plet with the SU(3) decuplet in the lowest level. b) 20-plet with the SU(3) octet in the lowest level. c) Totally antisymmetric quadruplet ($\bar{4}$). The last one can only occur for spatial angular momentum with odd values. Figure from [5].

Alternatively, following HQET, we can use operators that describe a heavy baryon as a light di-quark in the presence of a heavy quark. They are listed in Table 2. Comparing results from the two sets of operators can help us understand the inner structure of the baryons.

Singly charmed					$J^P = \frac{1}{2}^+$	$J^P = \frac{3}{2}^+$
(S)	**(I)**	$\mathbf{s_d^{\pi_d}}$	**(qq)Q**	$\mathcal{O}$	**Name**	**Name**
(0)	(0)	$(0)^+$	$(ud)c$	$\mathcal{O}_5 = \epsilon_{abc}(u^{aT}C\gamma_5 d^b)c^c$	Λ_c	
(0)	(1)	$(1)^+$	$(uu)c$	$\mathcal{O}_\mu = \epsilon_{abc}(u^{aT}C\gamma_\mu u^b)c^c$	Σ_c	Σ_c^*
(−1)	$(\frac{1}{2})$	$(0)^+$	$(us)c$	$\mathcal{O}_5 = \epsilon_{abc}(u^{aT}C\gamma_5 s^b)c^c$	Ξ_c	
(−1)	$(\frac{1}{2})$	$(1)^+$	$(us)c$	$\mathcal{O}'_\mu = \epsilon_{abc}(u^{aT}C\gamma_\mu s^b)c^c$	Ξ'_c	Ξ_c^*
(−2)	(0)	$(1)^+$	$(ss)c$	$\mathcal{O}_\mu = \epsilon_{abc}(s^{aT}C\gamma_\mu s^b)c^c$	Ω_c	Ω_c^*

Doubly charmed					$J^P = \frac{1}{2}^+$	$J^P = \frac{3}{2}^+$
(S)	**(I)**	$\mathbf{s_d^{\pi_d}}$	**(QQ)q**	$\mathcal{O}$	**Name**	**Name**
(0)	(0)	$(1)^+$	$(cc)u$	$\mathcal{O}_\mu = \epsilon_{abc}(c^{aT}C\gamma_\mu c^b)u^c$	Ξ_{cc}	Ξ_{cc}^*
(−1)	$(\frac{1}{2})$	$(1)^+$	$(cc)s$	$\mathcal{O}_\mu = \epsilon_{abc}(c^{aT}C\gamma_\mu c^b)s^c$	Ω_{cc}	Ω_{cc}^*

Table 2. Summary of quantum numbers of heavy baryons and interpolating operators following the HQET approach. Some of these operators were suggested in [11] for lattice calculations.

4.2. Results

The singly and doubly charmed baryon spectra for the $\kappa_l = 0.12104, \kappa_s = 0.12062$ ensemble are shown in Figure 4. On the left hand side, we present our preliminary singly (above) and doubly (below) charmed baryon spectra for the two bases of operators chosen, including positive and negative parity channels. As was the case for meson spectroscopy, the simulations are still at an early stage. In our results, we can see that the mass differences between baryons containing u, d quarks and the ones with s quarks are smaller than the experimental values. This is to be expected as we are far from the physical point in terms of the masses of the quarks: the singlet quark m_q is tuned to the physical value which means that the u, d and s quark masses are respectively heavier and lighter than their physical values. On the right hand side, we can see a summary of lattice results for the singly and doubly charmed spectra, with different systematics [12, 13, 14, 15, 16, 17]. We can see that, overall, lattice results agree with experiment.

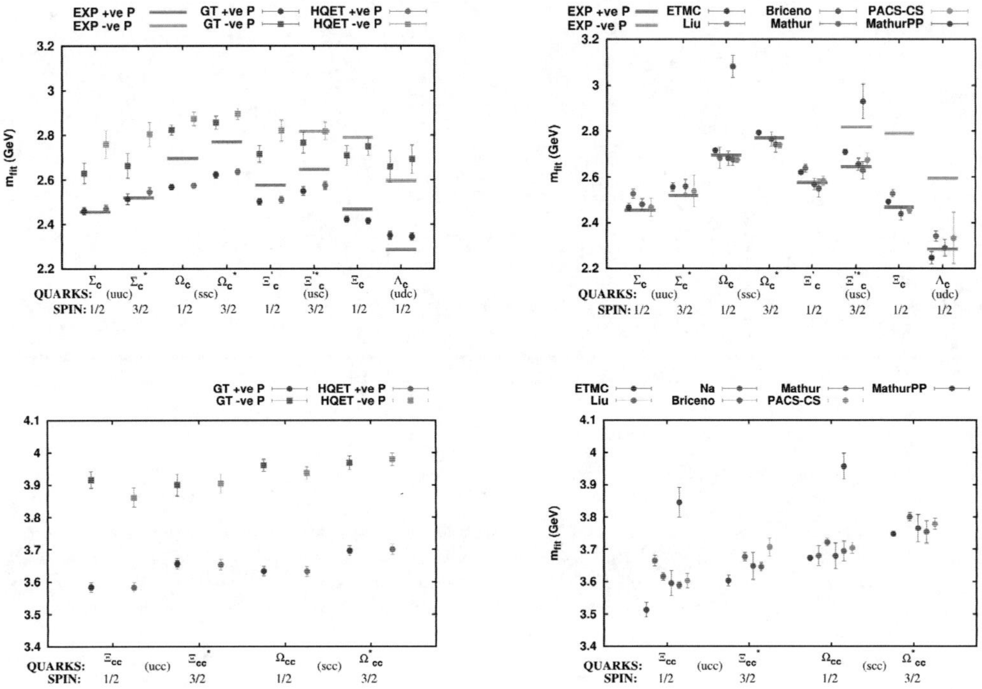

Figure 4. Singly charmed (top) and doubly charmed (bottom) low lying spectrum. On the left hand size, results from the SLiNC configurations are shown ($M_\pi = 348$MeV). Errors are statistical only. On the right hand size, a summary of lattice results is presented.

5. Acknowledgements

The numerical calculations were performed on the SGI Altix ICE machines at HLRN (Berlin-Hannover, Germany). We have made use of the Chroma software package [18] for some of the analysis. This work was supported by the EU ITN STRONGnet and the DFG SFB/TR 55.

6. References

[1] Brambilla N, Eidelman S, Heltsley B, Vogt R, Bodwin G *et al.* 2011 *Eur.Phys.J.* **C71** 1534
[2] Choi S *et al.* (Belle Collaboration) 2003 *Phys.Rev.Lett.* **91** 262001
[3] Aubert B *et al.* (BABAR Collaboration) 2003 *Phys.Rev.Lett.* **90** 242001
[4] Besson D *et al.* (CLEO Collaboration) 2003 *Phys.Rev.* **D68** 032002
[5] Beringer J *et al.* (Particle Data Group) 2012 *Phys.Rev.* **D86** 010001
[6] Cundy N, Gockeler M, Horsley R, Kaltenbrunner T, Kennedy A *et al.* 2009 *Phys.Rev.* **D79** 094507
[7] Bietenholz W, Bornyakov V, Cundy N, Gockeler M, Horsley R *et al.* 2010 *Phys.Lett.* **B690** 436–441
[8] Bietenholz W, Bornyakov V, Gockeler M, Horsley R, Lockhart W *et al.* 2011
[9] Luscher M and Wolff U 1990 *Nucl.Phys.* **B339** 222–252
[10] Dudek J J, Edwards R G, Mathur N and Richards D G 2008 *Phys.Rev.* **D77** 034501
[11] Bowler K *et al.* (UKQCD Collaboration) 1996 *Phys.Rev.* **D54** 3619–3633
[12] Na H and Gottlieb S A 2007 *PoS* **LAT2007** 124
[13] Liu L, Lin H W, Orginos K and Walker-Loud A 2010 *Phys.Rev.* **D81** 094505
[14] Alexandrou C, Carbonell J, Christaras D, Drach V, Gravina M *et al.* 2012
[15] Briceno R A, Lin H W and Bolton D R 2012
[16] Basak S, Datta S, Padmanath M, Majumdar P and Mathur N 2012 (*Preprint* 1211.6277)
[17] Namekawa Y (PACS-CS collaboration) 2012 (*Preprint* 1212.0073)
[18] Edwards R G and Joo B (SciDAC Collaboration, LHPC Collaboration, UKQCD Collaboration) 2005 *Nucl.Phys.Proc.Suppl.* **140** 832

Journal of Physics: Conference Series **426** (2013) 012018 doi:10.1088/1742-6596/426/1/012018

Phenomenology of the pseudoscalar glueball with a mass of 2.6 GeV

Walaa I. Eshraim and Stanislaus Janowski

Institute for Theoretical Physics, Goethe-University, Max-von-Laue-Str. 1, 60438 Frankfurt am Main, Germany

E-mail: weshraim@th.physik.uni-frankfurt.de; janowski@th.physik.uni-frankfurt.de

Abstract. We calculate the branching ratios of the lightest pseudoscalar glueball, $J^{PC} = 0^{-+}$, with a mass of 2.6 GeV predicted by lattice QCD by using a chiral Lagrangian. This study is relevant in view of the planned PANDA experiment at the upcoming FAIR facility, which will measure the energy range in which the pseudoscalar glueball is predicted to exist.

1. Introduction

Quantum Chromodynamics (QCD) is the theory of strongly interacting matter. In QCD one expects the existence of glueballs, which are bound states made of two or more gluons, the gauge bosons of QCD, and are invariant under gauge transformations. The glueballs are important objects in order to understand the nonperturbative behaviour of QCD as well as the hadronic spectroscopy [1].

Calculations of lattice QCD in the pure Yang-Mills sector predict a complete glueball spectrum [2], where the lightest glueball state is a scalar-isoscalar, $J^{PC} = 0^{++}$, and its mass is about 1.6 GeV. This energy region has been studied in a variety of effective approaches, e.g. Refs. [3, 4]. The second lightest glueball state has been predicted to be a tensor state, $J^{PC} = 2^{++}$, with a mass of 2.2 GeV, e.g. Ref. [5] for a related phenomenological discussion. The third lightest glueball state predicted by lattice QCD is a pseudoscalar state, $J^{PC} = 0^{-+}$, and has a mass of about 2.6 GeV. This pseudoscalar glueball, denoted as $\tilde{G} \equiv gg$, is studied here: we calculate its decays into three pseudoscalar mesons, $\tilde{G} \to PPP$, and a pseudoscalar and a scalar meson, $\tilde{G} \to PS$. The interaction between the pseudoscalar glueball and the ordinary scalar and pseudoscalar mesons is described by the effective chiral Lagrangian introduced in Refs. [6, 7]. The branching ratios represent a parameter free prediction of our approach.

The PANDA experiment at the FAIR facility in Darmstadt is currently under construction and will use an 1.5 GeV antiproton beam hitting a proton target at rest [8]. Therefore a center of mass energy higher than $\sim$2.5 GeV will be reached and the 2.6 GeV pseudoscalar glueball can be directly produced as an intermediate state [7].

2. The chiral Lagrangian

The effective chiral Lagrangian which couples the pseudoscalar glueball $\tilde{G} \equiv gg$ with quantum numbers $J^{PC} = 0^{-+}$ to the ordinary scalar and pseudoscalar mesons reads [6, 7, 9]:

$$\mathcal{L}_{\tilde{G}}^{int} = ic_{\tilde{G}\Phi}\tilde{G}\left(\det\Phi - \det\Phi^{\dagger}\right), \quad (1)$$

where $c_{\tilde{G}\Phi}$ is a coupling constant and Φ is a multiplet containing the ordinary scalar and pseudoscalar mesons. In this study we consider three flavours, $N_f = 3$, thus $c_{\tilde{G}\Phi}$ is dimensionless and the multiplet Φ reads [10]:

$$\Phi = \frac{1}{\sqrt{2}} \begin{pmatrix} \frac{(\sigma_N + a_0^0) + i(\eta_N + \pi^0)}{\sqrt{2}} & a_0^+ + i\pi^+ & K_S^+ + iK^+ \\ a_0^- + i\pi^- & \frac{(\sigma_N - a_0^0) + i(\eta_N - \pi^0)}{\sqrt{2}} & K_S^0 + iK^0 \\ K_S^- + iK^- & \bar{K}_S^0 + i\bar{K}^0 & \sigma_S + i\eta_S \end{pmatrix} . \tag{2}$$

We first study the symmetries of the effective Lagrangian (1). The pseudoscalar glueball $\tilde{G}$ is made of gluons and is therefore chirally invariant. The multiplet Φ transforms under the chiral symmetry as $\Phi \to U_L \Phi U_R^\dagger$, where $U_{L(R)} = e^{-i\theta^a_{L(R)} t^a}$ is an element of $U(3)_{R(L)}$. Performing these transformations on the determinant of Φ it is easy to prove that this object is invariant under $SU(3)_L \times SU(3)_R$, but not under the axial $U_A(1)$ transformation:

$$\det\Phi \to \det U_A \Phi U_A = e^{-i\theta_A^0 \sqrt{2N_f}} \det\Phi \neq \det\Phi . \tag{3}$$

This is in agreement with the chiral anomaly. Consequently the effective Lagrangian (1) possesses only the $SU(3)_R \times SU(3)_L$ symmetry. Further essential symmetries of the strong interacting matter are the parity $\mathcal{P}$ and charge conjugation $\mathcal{C}$. The pseudoscalar glueball and the multiplet Φ transform under parity as

$$\tilde{G}(t,\vec{x}) \to -\tilde{G}(t,-\vec{x}) , \ \Phi(t,\vec{x}) \to \Phi^\dagger(t,-\vec{x}) , \tag{4}$$

and under charge conjugation as

$$\tilde{G} \to \tilde{G} , \ \Phi \to \Phi^T . \tag{5}$$

Performing these discrete transformations $\mathcal{P}$ and $\mathcal{C}$ on the effective Lagrangian (1) leave it unchanged. In conclusion, one can say that the symmetries of the effective Lagrangian (1) are in agreement with the symmetries of the QCD Lagrangian.

Let us now consider the assignment of the ordinary mesonic d.o.f. in Eq. (1) or (2). In the scalar sector the field $\vec{a}_0$ corresponds to the physical isotriplet state $a_0(1450)$ and the scalar kaon field K_S to the physical isodoublet state $K_0^\star(1430)$ [11]. The fields $\sigma_N \equiv (\bar{u}u + \bar{d}d)/\sqrt{2}$ and $\sigma_S \equiv \bar{s}s$ are the bare nonstrange and strange isoscalars and they correspond to the physical isoscalars $f_0(1370)$ and $f_0(1710)$, respectively [10]. Scalars below 1 GeV are predominantly tetraquarks or mesonic molecular states, see Refs. [12, 13] and references therein. As shown in Ref. [10] the mixing of the bare fields σ_N and σ_S is small and is neglected at this stage. In the pseudoscalar sector we assign the fields $\vec{\pi}$ and K to the physical pion isotriplet and the kaon isodoublet [11]. The bare quark-antiquark fields $\eta_N \equiv (\bar{u}u + \bar{d}d)/\sqrt{2}$ and $\eta_S \equiv \bar{s}s$ are the nonstrange and strange mixing contributions of the physical states η and η' [10]. In the effective Lagrangian (1) there exist a mixing between the bare pseudoscalar glueball $\tilde{G}$ and the both bare fields η_N and η_S, but, due to the large mass difference between the pseudoscalar glueball and the pseudoscalar quark-antiquark fields, it turns out that its mixing is very small and is therefore negligible.

In accordance with the spontaneous breaking of the chiral symmetry we shift the scalar-isoscalar fields by their vacuum expectation values $\sigma_N \to \sigma_N + \phi_N$ and $\sigma_S \to \sigma_S + \phi_S$, where ϕ_N and ϕ_S are the corresponding chiral condensates. In order to be consistent with the full effective chiral Lagrangian of the extended Linear Sigma Model [10, 14, 15] we have to consider the shifting of the axialvector fields and thus to redefine the wave function of the pseudoscalar fields

$$\vec{\pi} \to Z_\pi \vec{\pi} , \ K \to Z_K K , \ \eta_{N,S} \to Z_{\eta_{N,S}} \eta_{N,S} , \tag{6}$$

where Z_i are the renormalization constants of the corresponding wave functions [10].

3. Results

The unknown coupling constant $c_{\tilde{G}\Phi}$ of the effective chiral Lagrangian (1) is the only free parameter of our approach. In order to make parameter free predictions regarding the future PANDA experiment at the FAIR facility we calculated the branching ratios of the two- and three-body decays of the pseudoscalar glueball $\tilde{G}$. When evaluating the decay widths from Eq. (1), we used, in agreement with lattice QCD, the mass of the pseudoscalar glueball $M_{\tilde{G}} = 2.6$ GeV [2, 6, 7]. Our results for the decay channels $\tilde{G} \to PPP$ and $\tilde{G} \to PS$ are presented in Table 1 and Table 2 respectively, where $\Gamma^{tot}_{\tilde{G}} = \Gamma_{\tilde{G}\to PPP} + \Gamma_{\tilde{G}\to PS}$ is the total decay width, [6, 7].

Quantity	Value
$\Gamma_{\tilde{G}\to KK\eta}/\Gamma^{tot}_{\tilde{G}}$	0.049
$\Gamma_{\tilde{G}\to KK\eta'}/\Gamma^{tot}_{\tilde{G}}$	0.019
$\Gamma_{\tilde{G}\to\eta\eta\eta}/\Gamma^{tot}_{\tilde{G}}$	0.016
$\Gamma_{\tilde{G}\to\eta\eta\eta'}/\Gamma^{tot}_{\tilde{G}}$	0.0017
$\Gamma_{\tilde{G}\to\eta\eta'\eta'}/\Gamma^{tot}_{\tilde{G}}$	0.00013
$\Gamma_{\tilde{G}\to KK\pi}/\Gamma^{tot}_{\tilde{G}}$	0.47
$\Gamma_{\tilde{G}\to\eta\pi\pi}/\Gamma^{tot}_{\tilde{G}}$	0.16
$\Gamma_{\tilde{G}\to\eta'\pi\pi}/\Gamma^{tot}_{\tilde{G}}$	0.095

Table 1. Branching ratios for the decay of the pseudoscalar glueball $\tilde{G}$ into three pseudoscalar mesons.

Quantity	Value
$\Gamma_{\tilde{G}\to KK_S}/\Gamma^{tot}_{\tilde{G}}$	0.060
$\Gamma_{\tilde{G}\to a_0\pi}/\Gamma^{tot}_{\tilde{G}}$	0.083
$\Gamma_{\tilde{G}\to\eta\sigma_N}/\Gamma^{tot}_{\tilde{G}}$	0.0000026
$\Gamma_{\tilde{G}\to\eta\sigma_S}/\Gamma^{tot}_{\tilde{G}}$	0.012
$\Gamma_{\tilde{G}\to\eta'\sigma_N}/\Gamma^{tot}_{\tilde{G}}$	0.039

Table 2. Branching ratios for the decay of the pseudoscalar glueball $\tilde{G}$ into a scalar and a pseudoscalar meson.

The largest contribution to the total decay width is given by the following decay channels: $KK\pi$, which contributes to almost 50%, as well as $\eta\pi\pi$ and $\eta'\pi\pi$, where each one contributes about 10%. Although the decays of the pseudoscalar glueball into the scalars-isoscalars contribute only 6% to the total decay width, they should be treated carefully because they mix with each other and its assignment to the physical states is up to now unclear. In order to make more precisely statements in the scalar-isoscalar sector detailed calculations are necessary and are presently ongoing. A further interesting outcome of our approach is that the decay of the pseudoscalar glueball into three pions, $\tilde{G} \to \pi\pi\pi$, is not allowed.

4. Conclusions

We have presented a chirally invariant three-flavour effective Lagrangian with scalar and pseudoscalar quark-antiquark states and a pseudoscalar glueball. We have calculated two- and three-body decay processes of the pseudoscalar glueball with a mass of 2.6 GeV, as evaluated by lattice QCD. We predict that the decay channel $\tilde{G} \to KK\pi$ is predominant and $\tilde{G} \to \eta\pi\pi$ as well as $\tilde{G} \to \eta'\pi\pi$ are the next dominant ones [6, 7]. Moreover, the decay channel $\tilde{G} \to \pi\pi\pi$ is not allowed.

The results presented in this work can be tested in the upcoming PANDA experiment at the FAIR facility [8].

Acknowledgments

The authors thank Francesco Giacosa and Dirk H. Rischke for useful discussions. W.E. acknowledges support from DAAD and HGS-HIRe, S.J. acknowledges support from H-QM and HGS-HIRe.

References

[1] C. Amsler and N. A. Tornqvist, Phys. Rept. **389**, 61 (2004). E. Klempt and A. Zaitsev, Phys. Rept. **454**, 1 (2007) [arXiv:0708.4016 [hep-ph]]. F. Giacosa, Phys. Rev. D **80** , 074028 (2009) [arXiv:0903.4481 [hep-ph]].

[2] C. Morningstar and M. J. Peardon, AIP Conf. Proc. **688**, 220 (2004) [arXiv:nucl-th/0309068]. M. Loan, X. Q. Luo and Z. H. Luo, Int. J. Mod. Phys. A **21**, 2905 (2006) [arXiv:hep-lat/0503038]. Y. Chen *et al.*, Phys. Rev. D **73**, 014516 (2006) [arXiv:hep-lat/0510074]. E. Gregory, A. Irving, B. Lucini, C. McNeile, A. Rago, C. Richards and E. Rinaldi, arXiv:1208.1858 [hep-lat].

[3] C. Amsler and F. E. Close, Phys. Rev. D **53**, 295 (1996) [arXiv:hep-ph/9507326]. W. J. Lee and D. Weingarten, Phys. Rev. D **61**, 014015 (2000) [arXiv:hep-lat/9910008]. F. Giacosa, T. Gutsche, V. E. Lyubovitskij and A. Faessler, Phys. Rev. D **72**, 094006 (2005) [arXiv:hep-ph/0509247]. H. Y. Cheng, C. K. Chua and K. F. Liu, Phys. Rev. D **74**, 094005 (2006) [arXiv:hep-ph/0607206].

[4] S. Janowski, D. Parganlija, F. Giacosa and D. H. Rischke, Phys. Rev. D **84**, 054007 (2011) [arXiv:1103.3238 [hep-ph]]. S. Janowski, Acta Phys. Polon. Supp. **5**, 1071 (2012) [arXiv:1208.5397 [hep-ph]].

[5] F. Giacosa, T. Gutsche, V. E. Lyubovitskij, A. Faessler, Phys. Rev. D **72**, 114021 (2005) [hep-ph/0511171]. L. Burakovsky and J. T. Goldman, Phys. Rev. D **57**, 2879 (1998) [hep-ph/9703271].

[6] W. I. Eshraim, S. Janowski, F. Giacosa and D. H. Rischke, arXiv:1208.6474 [hep-ph].

[7] W. I. Eshraim, S. Janowski, A. Peters, K. Neuschwander and F. Giacosa, Acta Phys. Polon. Supp. **5**, 1101 (2012) [arXiv:1209.3976 [hep-ph]].

[8] M. F. M. Lutz *et al.* [PANDA Collaboration], arXiv:0903.3905 [hep-ex].

[9] C. Rosenzweig, A. Salomone and J. Schechter, Phys. Rev. D **24**, 2545 (1981). K. Kawarabayashi and N. Ohta, Nucl. Phys. B **175**, 477 (1980).

[10] D. Parganlija, P. Kovacs, G. Wolf, F. Giacosa and D. H. Rischke, arXiv:1208.0585 [hep-ph].

[11] K. Nakamura *et al.* (Particle Data Group), J. Phys. G **37**, 075021 (2010).

[12] R. L. Jaffe, Phys. Rev. D **15**, 267 (1977). R. L. Jaffe, Phys. Rev. D **15**, 281 (1977). L. Maiani, F. Piccinini, A. D. Polosa and V. Riquer, Phys. Rev. Lett. **93**, 212002 (2004) [arXiv:hep-ph/0407017]. F. Giacosa, Phys. Rev. D **74**, 014028 (2006) [arXiv:hep-ph/0605191]. A. H. Fariborz, R. Jora and J. Schechter, Phys. Rev. D **72**, 034001 (2005) [arXiv:hep-ph/0506170]. M. Napsuciale and S. Rodriguez, Phys. Rev. D **70**, 094043 (2004). F. Giacosa, G. Pagliara, Nucl. Phys. A **833**, 138-155 (2010) [arXiv:0905.3706 [hep-ph]]. F. Giacosa, Phys. Rev. D **75**, 054007 (2007) [arXiv:hep-ph/0611388].

[13] E. van Beveren, T. A. Rijken, K. Metzger, C. Dullemond, G. Rupp and J. E. Ribeiro, Z. Phys. C **30**, 615 (1986) [arXiv:0710.4067 [hep-ph]]. N. A. Tornqvist, Z. Phys. C **68**, 647 (1995) [arXiv:hep-ph/9504372]. M. Boglione and M. R. Pennington, Phys. Rev. D **65**, 114010 (2002) [arXiv:hep-ph/0203149]. E. van Beveren, D. V. Bugg, F. Kleefeld and G. Rupp, Phys. Lett. B **641**, 265 (2006) [arXiv:hep-ph/0606022]. J. R. Pelaez, Phys. Rev. Lett. **92**, 102001 (2004) [arXiv:hep-ph/0309292]. J. A. Oller and E. Oset, Nucl. Phys. A **620**, 438 (1997) [Erratum-ibid. A **652**, 407 (1999)] [arXiv:hep-ph/9702314].

[14] S. Gallas, F. Giacosa and D. H. Rischke, Phys. Rev. D **82**, 014004 (2010) [arXiv:0907.5084 [hep-ph]].

[15] D. Parganlija, F. Giacosa and D. H. Rischke, Phys. Rev. D **82**, 054024 (2010) [arXiv:1003.4934 [hep-ph]].

FAIRNESS 2012: FAIR NExt Generation of ScientistS 2012 IOP Publishing
Journal of Physics: Conference Series **426** (2013) 012019 doi:10.1088/1742-6596/426/1/012019

Scalar Mesons and FAIR

Denis Parganlija
Institute for Theoretical Physics, Vienna University of Technology, Wiedner Hauptstr. 8-10, A-1040 Vienna, Austria
E-mail: denisp@hep.itp.tuwien.ac.at

Abstract. I review issues related to the scalar-meson phenomenology considered of interest for the future PANDA experiments at FAIR.

1. Introduction

Scalar mesons are degrees of freedom of QCD (Quantum Chromodynamics) that possess quantum numbers $J^{PC} = 0^{++}$, where J is the total spin, P represents parity and C is the charge conjugation. Mesons that possess isospin $I = 0$ are referred to as isoscalars. Scalars with $I = 0$ represent QCD degrees of freedom with some peculiar features: they are defined as carrying quantum numbers of vacuum but are identifiable as resonances; as mesons, they should be of $\bar{q}q$ (quarkonium) structure but are too many to simply represent $\bar{q}q$ objects; and they are strongly mixed with each other so that disentangling them is in the majority of cases a highly non-trival task. However, they are also of extreme importance for QCD since their vacuum expectation values represent the cause of the spontaneous breaking of the chiral symmetry in QCD, and this "Higgs mechanism" is in turn expected to generate Goldstone bosons of QCD: pions, kaons and others. Thus understanding isoscalar mesons is a fundamental question of QCD, and one that has by no means been closed after decades of research and debate.

1.1. Scalar Mesons and Experiment

Listings of the Particle Data Group (PDG) cite the existence of five $IJ^{PC} = 00^{++}$ states in the energy region up to 1.8 GeV: $f_0(500)$ or σ, $f_0(980)$, $f_0(1370)$, $f_0(1500)$ and $f_0(1710)$ [1]. Let us review basic experimental data regarding these resonances.

- Historically, debates regarding scalar mesons have been mostly focused on the $f_0(500)$ resonance (also known as σ). The latter is characterised by a decay width that is of the same magnitude as the resonance mass – according to PDG, it has a mass of (400-500) MeV and a decay width of (400-700) MeV. The existence of this state was suggested in linear sigma models long ago: the state was introduced theoretically as the putative chiral partner of the pion [2]. Since the pion is a well-established quarkonium state, then the chiral-partner state of the pion had to be of the same structure. Additionally, the pion is a pseudo-Goldstone bosone – its mass is the lowest in the meson spectrum. Therefore, naive expectations were that (*i*) the chiral partner of the pion would be the lightest scalar state with $I = 0$, i.e., the σ meson, and consequently (*ii*) that the σ meson was a $\bar{q}q$ state. However, currently it is by no means clear that σ is truly a $\bar{q}q$ state, implying that its interpretation as the chiral partner of the pion is in doubt. We will come back to this issue later in this manuscript.

- First hints about the existence of the $f_0(980)$ resonance were discovered in analyses of $\pi\pi$ scattering data found to exhibit a rapid drop in the cross-section in the energy region between 950 MeV and 980 MeV (i.e., close to the KK threshold) in Saclay data on antiproton-proton [3] and Berkley data on pion-proton collisions [4]. The most precise modern-data analyses result in pole mass and pole decay widths values in the close vicinity of the mentioned energy region – see, e.g., Ref. [5] where the pole mass was determined as $m_{f_0(980)} = (996 \pm 7)$ MeV and the pole decay width as $\Gamma_{f_0(980)} = 50^{+20}_{-12}$ MeV.
- The $f_0(1370)$ resonance is a broad enhancemenet in the $\pi\pi$, 4π and, to a lesser extent, KK and $\eta\eta$ channels in the energy region of approximately 1.3 GeV. It is characterised by emergence of two peaks, respectively in the 2π and 4π decay channels [6], the disentaglement of which requires a simultaneous analysis of $\pi\pi \to \pi\pi$ and $\pi\pi \to 4\pi$ scattering data. The analysis of Ref. [6] demonstrates that the 2π decay channel is dominant in the energy region up to 1.35 GeV whereas the energy region thereafter is dominated by the opening of the $\rho\rho$ threshold leading to a strong 4π decay pattern. Note that $f_0(1370)$ represents a single resonance despite possessing two peaks since a combined dispersive analysis of both 2π and 4π channels yields only one pole. Historically, early hints of the $f_0(1370)$ existence were observed in $\bar{p}p$ collisions at CERN in 1969 with a possible $\rho\rho$ enhancement claimed at 1.4 GeV [7]. Current edition of the Particle Data Group listings accumulates all available data on $f_0(1370)$ estimating an interval for, rather than stating exact values of, its mass and decay width: $m_{f_0(1370)} =$ (1200-1500) MeV and $\Gamma_{f_0(1370)} =$ (200-500) MeV [1].
- The discovery of the $f_0(1500)$ resonance originated in search for the scalar glueball state. This resonance is found mostly in pion final states from nucleon-nucleon (or antinucleon-nucleon) and pion-nucleon scattering processes. If such processes produce four pions, then $f_0(1500)$ is reconstructed from $\rho\rho$ final states in the $2(\pi^+\pi^-)$ channel and from $\sigma\sigma$ final states in the $2(\pi^+\pi^-)$ or $2(\pi^0\pi^0)$ channels. The resonance is therefore at least partly reconstructed in channels containing a double Pomeron exchange rendering the state a glueball candidate [20]. The PDG cites a world-average mass $m_{f_0(1500)} = (1505 \pm 6)$ MeV and decay width $\Gamma_{f_0(1710)} = (109 \pm 7)$ MeV [1].
- The $f_0(1710)$ resonance is characterised by a predominant KK decay channel, marking a clear point of distinction between this resonance and the above mentioned ones. The earliest evidence for the $f_0(1710)$ resonance was obtained from the decay $J/\psi \to \gamma\eta\eta$ at the SLAC Crystal Ball detector from e^+e^- annihilation [8]; a resonance with a mass of (1640 ± 50) MeV, a decay width of 220^{+100}_{-70} MeV and the charge conjugation $C = +1$ was found (the latter was because the resonance had been produced in a radiative J/ψ decay but, interestingly, no final determination of the total spin and parity was possible from these first data). Subsequent analyses, such as those in Ref. [9], determined the resonance to be of $J^P = 0^+$ nature. The PDG cites a world-average mass $m_{f_0(1710)} = (1720 \pm 6)$ MeV and a decay width $\Gamma_{f_0(1710)} = (135 \pm 8)$ MeV [1].

1.2. Scalar Mesons and Theory

Assuming a $\bar{q}q$ structure for mesons (where q denotes a constituent quark) it is possible to construct two $IJ^{PC} = 00^{++}$ states for three quark flavours (up u, down d and strange s) once the approximate isospin symmetry in the nonstrange-quark sector is considered ($u = d$): $f_0^N \equiv (\bar{u}u+\bar{d}d)/\sqrt{2}$ and $f_0^S \equiv \bar{s}s$. The latter are per construction pure states. They can obviously mix by virtue of carrying the same quantum numbers; mixed states are expected to be those appearing in the physical spectrum. Clearly, two pure states will produce the same number of mixed states – thus only two of the five previously discussed physical states can conceivably be described in this approach. Hence the natural questions regard (*i*) location of isoscalar $\bar{q}q$ states in the physical spectrum and (*ii*) structure of the remaining isoscalar states. Answers to these questions are complicated by the mentioned mixing of isoscalar states. An isoscalar (or indeed any other) meson need not necessarily be of $\bar{q}q$ structure – it can also possess admixtures (or even be predominantly composed) of tetraquark ($\bar{q}\bar{q}qq$), glueball (bound states of QCD gauge bosons – gluons) or molecular (e.g., $\pi\pi$ or KK) contributions. Note that the hadronic spectrum is additionally influenced by confinement via Regge trajectories; expectation value of the Polyakov loop (confinement order parameter, "a loop of confining string") and the chiral condensate mix. Highly excited scalar mesons are all mixtures of spin waves, quarkonium or gluonium states, string excitations and isospin waves.

Neglecting confinement effects, for the cases of the above-stated isoscalar mesons I consider the following to be of importance:

- As indicated above, the $f_0(500)$ resonance is usually considered as the natural option for the chiral partner of the pion, implying that $f_0(500)$ possesses the $\bar{q}q$ structure as suggested, e.g., by Ref. [10]. However, a $\bar{q}q$ scalar state possesses the intrinsic angular momentum $L = 1$ as well as the relative spin of the quarks $S = 1$. For this reason one could also easily expect the scalar $\bar{q}q$ state to be in the region above 1 GeV. This would imply that (*i*) isoscalars above 1 GeV possess the $\bar{q}q$ structure and (*ii*) $f_0(500)$ represents a tetraquark state or a $\pi\pi$ bound state. For details on both statements see, e.g., Refs. [11, 12, 13].
- As already discussed, the $f_0(980)$ resonance is close to the kaon-kaon threshold rendering a structure analysis for this state rather difficult. The resonance may be interpreted as a quarkonium [14], as a $\bar{q}^2q^2$ state [15], as a KK bound state [16], as a glueball [17] or even as an $\eta\eta$ bound state [18].
- The $f_0(1370)$ resonance appears to be a suitable candidate for a non-strange $\bar{q}q$ state [12, 13] or a dynamically generated state [19].
- The $f_0(1500)$ resonance seems to represent a (predominant) glueball state [20, 21] although claims have also been made that, contrarily, $f_0(1710)$ is of predominantly glueball nature [22]. Alternatively, $f_0(1710)$ can also be interpreted as a predominantly $\bar{s}s$ state [13].

Hence the current state of knowledge allows us to only claim with certitude that all the above mentioned states mix [23]; some of them even overlap due to large decay widths [1]. As yet, no definitive statement is possible about their structure. The situation is thus rather complicated – and it is enhanced even further by the possible existence of a sixth isoscalar state, located very proximal to the set of the already mentioned isoscalars: the $f_0(1790)$ resonance.

2. The $f_0(1790)$ Resonance and FAIR

The existence of the putative new $f_0(1790)$ resonance has been claimed by the BES II Collaboration in 2004 [24]. The mass and decay width of the resonance were determined by the Collaboration as $m_{f_0(1790)} = 1790^{+40}_{-30}$ and $\Gamma_{f_0(1790)} = 270^{+60}_{-30}$ MeV. The stated large decay width implies a strong overlap of this resonance with $f_0(1710)$; however, there is a clear point of distinction between the two resonances: $f_0(1710)$ is reconstructed predominantly in the kaon decay channels whereas $f_0(1790)$ is reconstructed predominantly in the pion decay channels. There are four basic production mechanisms for $f_0(1710)$ and $f_0(1790)$ via J/ψ decays [24]:

- (*i*) $J/\psi \to \varphi K^+K^-$,
- (*ii*) $J/\psi \to \varphi\pi^+\pi^-$,
- (*iii*) $J/\psi \to \omega K^+K^-$,
- (*iv*) $J/\psi \to \omega\pi^+\pi^-$.

Reactions (*i*) and (*iii*) allow for reconstruction of $f_0(1710)$ [1] whereas reactions (*ii*) and (*iv*) allow for reconstruction of $f_0(1790)$. Importantly, assuming $f_0(1710)$ and $f_0(1790)$ to be the same resonance leads to a contradiction: such a resonance would have to possess a pion-to-kaon-decay ratio of 1.82 ± 0.33 according to reactions (*i*) and (*ii*) and, simultaneously, the pion-to-kaon-decay ratio < 0.11 according to reactions (*iii*) and (*iv*) [24]. There can be no one resonance with these simultaneous features – thus BES II data suggest $f_0(1710)$ and $f_0(1790)$ to represent two distinct resonances.

Let us finally discuss the relevance of the deliberations in this paper for FAIR. The proposed PANDA experiments at FAIR will study interactions between antiprotons and fixed-target protons and nuclei in the momentum range of 1.5-15 GeV. A stated goal of the PANDA Collaboration is the exploration of light-hadron spectroscopy, including search for glueballs and multiquark states [25]. In this regard I would like to suggest the following courses of action in the particular case of possible mixing scenarios in the scalar sector:

- Any viable search for members of a glueball spectrum must include the ground state as otherwise the spectrum would be incomplete. The PANDA Collaboration is in a unique position to greatly advance our search for glueballs since (*i*) the Collaboration can build upon extensive experimental work performed in the last decades and (*ii*) the technical specifications of the PANDA detector allow for a search for glueballs with unprecedented accuracy in comparison with similar undertakings in the past.
  However, the glueball state will inevitably overlap with other states that possess the same quantum numbers (and may be of $\bar{q}q$, $\bar{q}\bar{q}qq$ or molecular structure) and the disentanglement of various signals from each of the mentioned resonances will represent a formidable task.

- Current theoretical results [21, 22] suggest the scalar glueball to be positioned at (1.5-1.7) GeV, i.e., in the vicinity of the established resonances $f_0(1370)$, $f_0(1500)$ and $f_0(1710)$ as well as the newly proposed $f_0(1790)$ state. Consequently, a viable search for the scalar glueball will have to consider overlap of the glueball not only with $f_0(1370)$, $f_0(1500)$ and $f_0(1710)$ *but also* with $f_0(1790)$. In other words: a viable search for the glueball ground state at PANDA must consider the possibility of the existence of $f_0(1790)$. If $f_0(1790)$ is not found at PANDA, then BES results may be put in doubt; but if $f_0(1790)$ is ascertained then its signal will have to be disentangled from the glueball signal in order for the glueball search to be successful, making the search even more challenging and thus even more interesting.

3. Conclusions

Scalar mesons represent an interesting challenge for the future PANDA experiments at FAIR: the planned search for the glueball state in the $IJ^{PC} = 00^{++}$ channel will have to consider mixing/overlap between non-strange $\bar{q}q$, strange $\bar{q}q$, non-strange and strange $\bar{q}\bar{q}qq$ as well as possible molecular-type states. Additionally, obtaining a clear signal for the glueball will inevitably require the confirmation (or negation) of the existence of a putative new isoscalar state referred to as $f_0(1790)$ by the BES II Collaboration [24]. If $f_0(1790)$ exists, it will most certainly overlap with the glueball. Thus a viable search for a glueball state simultaneously implies a search for $f_0(1790)$: ignoring $f_0(1790)$ will strongly distort any signal for the glueball. For this reason, the search for a glueball at PANDA is more than a search for merely one state – it is a search where several very close-by states have to be considered simultaneously.

References

[1] J. Beringer *et al.* (Particle Data Group), Phys. Rev. D**86**, 010001 (2012).
[2] J. S. Schwinger, Annals Phys. **2**, 407 (1957); M. Gell-Mann and M. Levy, Nuovo Cim. **16**, 705 (1960); S. Weinberg, Phys. Rev. Lett. **18**, 188 (1967).
[3] R. Bizzarri, M. Foster, P. Gavillet, G. Labrosse, L. Montanet, R. Salmeron, P. Villemoes, C. Ghesquiere *et al.*, Nucl. Phys. B **14**, 169 (1969).

[4] M. Alston-Garnjost, A. Barbaro-Galtieri, S. M. Flatte, J. H. Friedman, G. R. Lynch, S. D. Protopopescu, M. S. Rabin, F. T. Solmitz, Phys. Lett. B **36**, 152 (1971); S. M. Flatte, M. Alston-Garnjost, A. Barbaro-Galtieri, J. H. Friedman, G. R. Lynch, S. D. Protopopescu, M. S. Rabin, F. T. Solmitz, Phys. Lett. **B38**, 232 (1972); S. D. Protopopescu, M. Alston-Garnjost, A. Barbaro-Galtieri, S. M. Flatte, J. H. Friedman, T. A. Lasinski, G. R. Lynch, M. S. Rabin *et al.*, Phys. Rev. D **7**, 1279 (1973).

[5] R. Garcia-Martin, R. Kaminski, J. R. Pelaez and J. Ruiz de Elvira, Phys. Rev. Lett. **107**, 072001 (2011) [arXiv:1107.1635 [hep-ph]].

[6] D. V. Bugg, Eur. Phys. J. C **52**, 55 (2007) [arXiv:0706.1341 [hep-ex]].

[7] R. A. Donald *et al.*, Nucl. Phys. B **11**, 551 (1969).

[8] C. Edwards, R. Partridge, C. Peck, F. Porter, D. Antreasyan, Y. F. Gu, W. S. Kollmann, M. Richardson *et al.*, Phys. Rev. Lett. **48**, 458 (1982).

[9] D. Alde *et al.* [Serpukhov-Brussels-Los Alamos-Annecy(LAPP) Collaboration], Phys. Lett. B **182**, 105 (1986); D. Alde *et al.* [IFVE-Brussels-Annecy-Los Alamos Collaboration], Phys. Lett. **B284**, 457 (1992); T. A. Armstrong *et al.* [E760 Collaboration], Phys. Lett. B **307**, 394 (1993); D. V. Bugg, I. Scott, B. S. Zou, V. V. Anisovich, A. V. Sarantsev, T. H. Burnett and S. Sutlief, Phys. Lett. B **353**, 378 (1995); D. Barberis *et al.* [WA102 Collaboration], Phys. Lett. B **462**, 462 (1999) [arXiv:hep-ex/9907055]; S. Chekanov *et al.* [ZEUS Collaboration], Phys. Rev. Lett. **101**, 112003 (2008) [arXiv:0806.0807 [hep-ex]].

[10] Y. Nambu and G. Jona-Lasinio, Phys. Rev. **122**, 345 (1961); R. Delbourgo and M. D. Scadron, Phys. Rev. Lett. **48**, 379 (1982); R. Delbourgo and M. D. Scadron, Mod. Phys. Lett. A **10**, 251 (1995) [hep-ph/9910242]; R. Delbourgo and M. D. Scadron, Int. J. Mod. Phys. A **13**, 657 (1998) [hep-ph/9807504].

[11] P. Ko and S. Rudaz, Phys. Rev. D **50**, 6877 (1994); J. R. Pelaez, Mod. Phys. Lett. A **19**, 2879 (2004) [hep-ph/0411107]; D. Parganlija, F. Giacosa and D. H. Rischke, AIP Conf. Proc. **1030**, 160 (2008) [arXiv:0804.3949 [hep-ph]]; D. Parganlija, F. Giacosa and D. H. Rischke, PoS C**ONFINEMENT8**, 070 (2008) [arXiv:0812.2183 [hep-ph]]; D. Parganlija, F. Giacosa and D. H. Rischke, arXiv:0911.3996 [nucl-th]; J. Ruiz de Elvira, J. R. Pelaez, M. R. Pennington and D. J. Wilson, Phys. Rev. D **84**, 096006 (2011) [arXiv:1009.6204 [hep-ph]]; D. Parganlija, F. Giacosa, P. Kovacs and G. Wolf, AIP Conf. Proc. **1343**, 328 (2011) [arXiv:1011.6104 [hep-ph]]; P. Kovacs, G. Wolf, F. Giacosa and D. Parganlija, arXiv:1102.4732 [hep-ph]; D. Parganlija, Phys. Scripta T **150**, 014029 (2012) [arXiv:1109.4331 [hep-ph]]; W. Heupel, G. Eichmann and C. S. Fischer, arXiv:1206.5129 [hep-ph]; D. Parganlija, P. Kovacs, G. Wolf, F. Giacosa and D. H. Rischke, arXiv:1208.2054 [hep-ph].

[12] D. Parganlija, F. Giacosa, D. H. Rischke, Phys. Rev. D **82**, 054024 (2010) [arXiv:1003.4934 [hep-ph]]; D. Parganlija, F. Giacosa and D. H. Rischke, Acta Phys. Polon. Supp. **3**, 963 (2010) [arXiv:1004.4817 [hep-ph]]; D. Parganlija, F. Giacosa, D. H. Rischke, P. Kovacs and G. Wolf, Int. J. Mod. Phys. A **26**, 607 (2011) [arXiv:1009.2250 [hep-ph]]; D. Parganlija, Acta Phys. Polon. Supp. **4**, 727 (2011) [arXiv:1105.3647 [hep-ph]].

[13] D. Parganlija, P. Kovacs, G. Wolf, F. Giacosa and D. H. Rischke, arXiv:1208.0585 [hep-ph].

[14] D. Morgan, Phys. Lett. B **51**, 71 (1974); A. Bramon and E. Masso, Phys. Lett. B **93**, 65 (1980) [Erratum-ibid. B **107**, 455 (1981)]; M. D. Scadron, Phys. Rev. D **26**, 239 (1982); E. van Beveren, T. A. Rijken, K. Metzger, C. Dullemond, G. Rupp and J. E. Ribeiro, Z. Phys. C **30**, 615 (1986) [arXiv:0710.4067 [hep-ph]].

[15] R. L. Jaffe, Phys. Rev. D **15**, 267 (1977).

[16] T. Barnes, Phys. Lett. B **165**, 434 (1985); J. D. Weinstein and N. Isgur, Phys. Rev. D **41**, 2236 (1990); M. P. Locher, V. E. Markushin and H. Q. Zheng, Eur. Phys. J. C **4**, 317 (1998) [arXiv:hep-ph/9705230].

[17] R. L. Jaffe, K. Johnson, Phys. Lett. **B60**, 201 (1976); D. Robson, Nucl. Phys. B **130**, 328 (1977).

[18] Yu. S. Surovtsev, P. Bydzovsky, R. Kaminski and M. Nagy, Phys. Rev. D **81**, 016001 (2010) [arXiv:0811.0906 [nucl-th]].

[19] R. Molina, D. Nicmorus and E. Oset, Phys. Rev. D **78**, 114018 (2008) [arXiv:0809.2233 [hep-ph]]; L. S. Geng and E. Oset, Phys. Rev. D **79**, 074009 (2009) [arXiv:0812.1199 [hep-ph]].

[20] F. E. Close, Rept. Prog. Phys. **51**, 833 (1988); S. Janowski, D. Parganlija, F. Giacosa and D. H. Rischke, Phys. Rev. D **84**, 054007 (2011) [arXiv:1103.3238 [hep-ph]].

[21] C. Morningstar and M. J. Peardon, AIP Conf. Proc. **688**, 220 (2004) [arXiv:nucl-th/0309068]; Y. Chen *et al.*, Phys. Rev. D **73**, 014516 (2006) [arXiv:hep-lat/0510074]; S. Janowski, D. Parganlija, F. Giacosa and D. H. Rischke, Phys. Rev. D **84**, 054007 (2011) [arXiv:1103.3238 [hep-ph]].

[22] E. Klempt and A. Zaitsev, Phys. Rept. **454**, 1 (2007) [arXiv:0708.4016 [hep-ph]]; M. Albaladejo and J. A. Oller, Phys. Rev. Lett. **101**, 252002 (2008) [arXiv:0801.4929 [hep-ph]].

[23] D. Black, A. H. Fariborz, F. Sannino and J. Schechter, Phys. Rev. D **59**, 074026 (1999) [hep-ph/9808415]; A. H. Fariborz, R. Jora and J. Schechter, Phys. Rev. D **72**, 034001 (2005) [hep-ph/0506170]; F. Giacosa, Phys. Rev. D **75**, 054007 (2007) [arXiv:hep-ph/0611388].

[24] M. Ablikim *et al.* [BES Collaboration], Phys. Lett. B **607**, 243 (2005) [arXiv:hep-ex/0411001].

[25] M. F. M. Lutz *et al.* [PANDA Collaboration], arXiv:0903.3905 [hep-ex].

The technological concept of the Compressed Baryonic Matter (CBM) experiment

M. Deveaux on behalf of the CBM-collaboration
Goethe Universität Frankfurt am Main, Max-von-Laue-Str. 1, 60438 Frankfurt/M, Germany
E-mail: deveaux@physik.uni-frankfurt.de

Abstract. The Compressed Baryonic Matter (CBM) experiment is to explore the properties of strongly interacting matter in the regime of highest net baryon densities. It aims to find experimental evidence for numerous predicted effects like a first order phase transition between hadronic and partonic matter, the existence of a critical endpoint of this phase transition and the expected onset of chiral symmetry restoration. The 8-45 AGeV heavy ion beam needed to create the hot and dense matter in the fixed target experiment will be provided by the SIS100 and the SIS300 synchrotron of the future FAIR facility in Darmstadt, Germany.
The paper provides an introduction into the measurement challenges and the technological concept of CBM-experiment from an instrumentalist's point of view.

1. The phase diagram of nuclear matter

A sketch of the phase diagram of strongly interacting matter is displayed in Figure 1. This phase diagram shows the state of matter as a function of the temperature and the baryo-chemical potential (μ_b). The temperature (T) is defined via the energy of the particles per

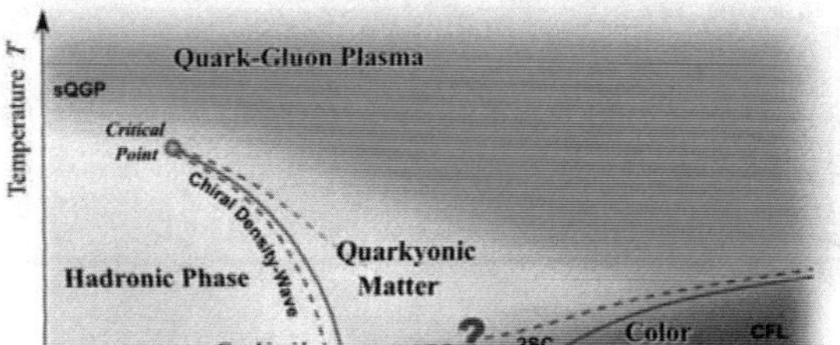

Figure 1. Simplified view of the phase diagram of hadronic matter. From [1].

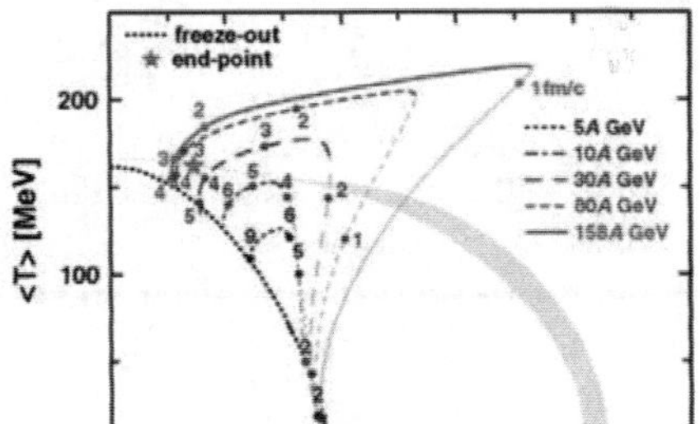

Figure 2. Evolution of a nuclear collision. The thin lines refer to the evolution prior to thermalization, the thick lines to an evolution in "thermal equilibrium". From [3].

degree of freedom and can be determined by measuring their kinetic energy spectrum. The μ_b is the energy needed to add one baryon to the volume (V) of the system. Although both are connected, it should not be confused with the (net) baryon density $\rho_B = B/V$. The baryon number $B = 1/3\ (n_q - n_{\bar{q}})$ of a system with n_q quarks and $n_{\bar{q}}$ anti-quarks is conserved during heavy ion collisions.

The phase diagram is divided into two major regimes[1], the confined hadronic matter and the postulated deconfined quark gluon plasma (QGP). In simple terms, QGP is created if individual hadrons start to overlap and the previously confined quarks may travel within the common potential. The necessary particle density may be created by compression of existing hadrons (low T, high μ_b) or by an excessive thermal generation of particles (high T, low μ_b). The phase transition between the hadronic matter and the QGP is considered to be of first order in the region of high baryonic densities. A critical endpoint separates this regime from the cross-over studied by ALICE and STAR in the regimes of highest temperatures and lowest baryonic densities. The existence and position of the first order phase transition and the endpoint are debated controversially as nowadays theoretical predictions underlie substantial uncertainties [2] and as no experimental observation was reported so far.

Experimentally, the extremely hot and dense nuclear matter needed for exploring the phase diagram is generated in heavy ion collisions. In a simplified picture, a heavy ion collision starts with head-on collisions of the nuclei, which may create particles of highest mass. Hereafter, the system reaches a kind of thermal equilibrium while its temperature and pressure increase further and QGP may be formed. After reaching the maximum density and T, the fireball expands and cools down. A phase transition from QGP to hadronic matter, the so-called hadronization, occurs. Once T turns too low for thermal particle formation, the chemical equilibrium breaks and the relative number of particles remains constant except for usual particle decays. If the density shrinks too far to force frequent elastic collisions, the thermal equilibrium breaks and the kinetic properties of the particles stay unmodified. Both processes are referred to as chemical and thermal freeze out. They fix the T and μ_b observed when interpreting the properties of the decay products of the collision by means of statistical thermodynamics. The T derived from the chemical composition is named T_{chem}, the T taken from the kinetic properties is called T_{kin}.

Note that some of the hadrons/quarks formed in head-on collisions may be too heavy to join the chemical equilibrium. Moreover, some heavy particles might leave this equilibrium slightly earlier than light ones.

The μ_b and ρ_b increase when colliding heavier ions. Model calculations like the one displayed in Figure 2 suggest that an increased the beam energy turns into a higher maximum T and μ_b during the collision. However, the system tends to pass the phase border already before reaching thermal equilibrium. As a phase is defined only in equilibrium, a transition in the strict term occures only during expansion. The T and μ_b needed for observing a first order phase transition during expansion is found at intermediate beam energies, which motivates the choice of the CBM beam energy. Note that the detailed numbers of the evolution of the collision are model dependent, a compilation of different results is found in [4].

The "thermal" expansion of the fireball has a strong impact on the chemical composition and kinetic properties of its strongly interacting particles. The decay products of the fireball show therefore in first order the T and μ_b reached in the moment of freeze-out. To study the earlier evolution of the collision, one has to identify so-called diagnostic probes, which conserve information through the expansion of the fireball.

Numerous probes were proposed for CBM [5]. A non-conclusive list includes leptons and photons, which are not subject to the strong interaction and leave the fireball mostly

[1] Discussing the additional potential phases shown in Figure 1, e.g. the quarkyonic matter, is beyond the scope of this work. A related review is found in [1].

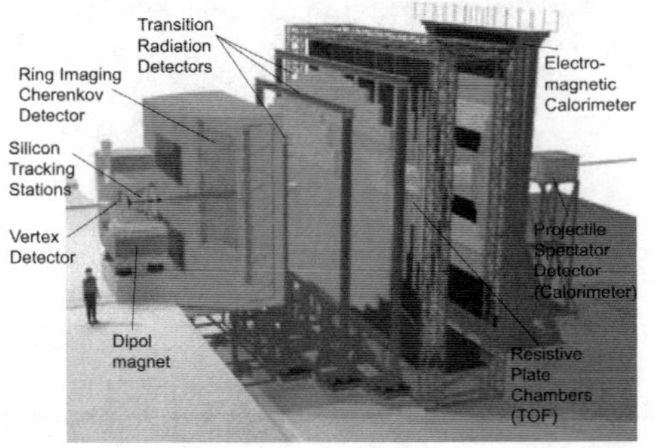

Figure 3. Conceptual design of CBM serving as combined e^+/e^-, $\gamma-$ and hadron spectrometer.

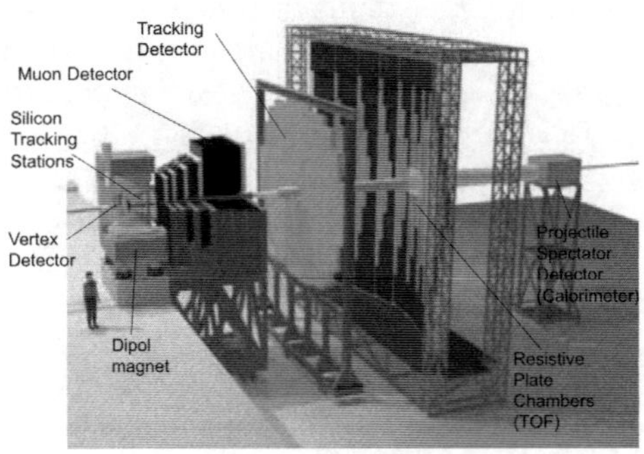

Figure 4. Conceptual design of CBM serving as μ^+/μ^- spectrometer. See text.

undisturbed. Moreover, particles containing the heavy charm quarks are suited as those quarks are created only in the early collisions and hardly destroyed during expansion. The collective properties of the fireball, e.g. its pressure distribution during the collision, may be probed by assessing the collective motion (flow) of all particles relative to the so-called collision plane. This plane is spanned by the beam direction and the impact parameter of the colliding hadrons.

A common feature of all probes is the difficulty to interpret the encoded information. Light vector mesons decaying into lepton pairs within the fireball may for example provide information about possible modifications of the meson mass in the medium. However, to name only one trivial complication, one has hard times to distinguish, at which time of the collision the meson was formed and decaying. Instead, one integrates over the full evolution of the fireball.

The data is interpreted by comparing the observations with results from theoretical models, which are to reproduce the evolution of the collision. This method is limited as the heavy ion collision itself is not accessible to observation while the decay products observed in the detectors provide only reduced information. This gives room for numerous theoretical concepts, which differ substantially concerning their physics content but predict, within the uncertainties of available experimental data, similar decay products. In the CBM energy range, this room is particularly wide as most existing data was taken with elder experiments, which were technologically restricted to the measurement of only few of the potential probes and observables.

2. The concept of CBM

Accounting for the above mentioned complication and profiting from the rapid progresses of sensor and computer technology during the last two decades, CBM aims to measure as many different probes as accessible to nowadays most powerful detector technologies. The results will constrain the degrees of freedom in nowadays theoretical models. This will hopefully allow for rejecting a substantial number of models and assumptions, which would sharpen our understanding of the physics of hot and dense nuclear matter quite substantially.

CBM will measure e.g. strange and multi-strange hadrons, open charm[2] and prompt gamma rays. The particularly interesting light vector mesons and charmonium will be observed via both, the e^+/e^- and μ^+/μ^- decay channels. This is as the e^+/e^- channel appears better suited for light vector mesons while the μ^+/μ^- channel seems more appropriate to observe charmonium.

[2] Open charm particles combine one charm quark with light quarks, charmonium contains a $c/\bar{c}$-pair.

Moreover, the interpretation of di-lepton spectra requires a particularly good quantitative understanding of the background. Accessing the same physics in the same acceptance but with two independent measurements will ease to spot potential systematic errors in the background subtraction and such increase the reliability of the data and its interpretation.

As many of the probes are rare (e.g. few 10^{-6} $D^0 \rightarrow K^- + \pi^+$ decays per central Au-Au collision), CBM is designed as a dedicated high rate experiment. It will operate at the SIS100 and the SIS300 synchrotron of the future FAIR facility, which will both provide slowly extracted spills of 10-100s with a beam intensity of ($\gtrsim 10^9$/s). The design goal of CBM is to obtain a rate of $\sim 10^7$ heavy ion collisions per second by means of a 1% target. The wish to handle this collision rate and the need to reach the instrumental sensitivity required for identifying rare probes determine the layout of the experiment call for a design at the edge of nowadays technologies. Intense and fruitful R&D had to be carried out to transform the ambitious plans to a working device. Today, numerous technical details are still under (cost-)optimization. However, the global detector concept has converged.

As shown in Figure 3 and 4, CBM is designed as a facility integrating a hadron-, an electron-, a γ- and a muon-spectrometer[3]. All systems are extensions of a central detector, the Silicon Tracking System (STS)[7, 8]. This silicon strip detector is located in an 1 Tm magnetic field which bends the trajectories of charged particles by means of the Lorentz force. The STS measures the impact points of the particles in 8 consecutive sensing layers, reconstructs the trajectory and determines the particle momentum p from the bending radius. Its only 300 μm thin, double sided sensors have a very fine (58 μm) strip pitch and a stereo angle of 8°. The sensors are connected via ultra-thin capton cables with readout chips located outside the detector acceptance. This solution allows for a measurement accuracy of $\Delta p/p \approx 1\%$.

The ability to identify hadrons is of particular interest when reconstructing strange and open charm particles from their decay products. To do so, the velocity (v) of the hadrons is determined by measuring their time of flight while passing a known distance. Knowing v and p, the rest mass of the particles can be determined. The Time Of Flight (TOF) system of CBM [9] will be composed from a diamond detector serving as start counter. The stop counter will rely on Resistive Plate Chambers (RPC) with enhanced rate capability. Both detectors together reach a time resolution of 80 ps, which should allow to identify protons. Moreover, it should separate pions and kaons up to $p \approx 6$ GeV/c. Above this value, both particles are highly relativistic and their difference in v becomes marginal.

Strange and open charm particles decay before reaching a detector and must be reconstructed from their daughter particles. To distinguish those from other decay products of the fireball, one separates their decay point (secondary vertex, SV) from the decay point of the fireball (primary vertex, PV). This is done with a Micro Vertex Detector (MVD) [10], which helps to extrapolate the measured particle trajectories back to the vicinity of the target and to search for intersection points. Those points are interpreted as decay vertex. If besides a PV, a SV is found, one concludes on the decay of a short lived particle. To separate the PV and the SV of the short lifed open charm particles, one needs a vertex resolution of ~ 50 μm along the beam axis. To reach it, the MVD has to show a very good spatial resolution ($\lesssim 5$ μm). Moreover, it must be very light. Otherwise it would scatter the impinging particles and loose its accuracy for track extrapolation.

The MVD will consist of 3-4 layers, each equipped with the 50 μm thick CMOS Monolitic Active Pixel Sensors [11]. It will operate in the vacuum of the target chamber of CBM and its heat is evacuated with a cooling support made from CVD diamond. The material budget of the crucial first station including cables and support will be $\sim 0.3\%$ X_0 only[4]. As it is exposed to

[3] This work focuses on the full CBM-setup as foreseen for SIS300. A reduced start version will be used for early experiments at SIS100. This setup and the related physics cases are discussed in Ref. [6].

[4] A radiation length (X_0) of material forces ultra relativistic e^- penetraing it to emit but $1/e$ of their energy as

extreme track densities, the MVD is limited to $\sim 10^5$ coll./s for the startup of CBM, which is sufficient for open charm measurements.

The e^+/e^- spectrometer is to reconstruct charmonium and light vector mesons by computing the invariant mass of e+/e- pairs. To control the background, one has to distinguish the electrons very precisely from pions. To reach the ambitioned pion suppression factor of 10^{-4}, a complementary Ring Imaging Cherenkov Detector (RICH) [12] and Transition Radiation Detectors (TRD) [13] are used.

The RICH identifies highly relativistic particles via the production of Cherenkov light. This light is emitted in a cone, if the speed of a charged particle exceeds the speed of light in a medium, which was chosen as CO_2 for CBM. The photons are focused with a concave glass mirror and the ~ 20 photons of the light rings obtained are observed with a camera made from Multi Anode Photomultiplier Tubes (MAPMT). The RICH of CBM allows to reject pions with $p \lesssim 8$ GeV. For $p > 4,65$ GeV, the velocity dependence of the ring radius is exploited.

Highly relativistic charged particles emit photons like X-rays when passing the interface between materials with high and low refraction index. The TRDs of CBM recognize those fast particles (namely electrons with $p > 1.5$ GeV) by detecting this so-called transition radiation. The TRDs of CBM are formed from radiators like for example stacks of plastic foils, which create a high number of interfaces in the path of flight of the particles. The X-rays created are detected with a gas detector, which is also sensitive to the particle itself. One separates slow and fast particles by measuring the energy deposit in its gas volume, which varies in the presence or absence of the photon. Different energy deposit turns into different pulse highs and shapes in the output signal of the detector. As the creation and absorption of the X-ray is somewhat statistic, about 10 consecutive TRD-layers are needed to obtain significant results. To match the requirements in speed (250 ns charge collection time), no or only few cm drift distance in the gas detector is foreseen. The technical details of the TRD are not yet fixed as several alternative concepts were tested successfully.

The muon detector (MUCH) [14] of CBM is to reconstruct light vector mesons and charmonium via their μ^+/μ^- decays. It will profit from the ability of the muons to penetrate thick materials and identify them by stopping all other particles with 6 iron absorbers with a total thickness of 225 cm. They are interrupted by up to 6 sensor planes made from three layers of triple GEM [15] detectors with pad readout. Those are to measure the tracks of identified muons and to match them back to the related track found in the STS. The MUCH is put into the place of the RICH (1.20m from the target) by means of a rail system. This hampers CBM from measuring μ^+/μ^--decays simultaneously with other decays but suppresses efficiently the the background from secondary muons generated in $\pi^\pm \rightarrow \mu^\pm + \nu_{mu}$ decays.

The electro-magnetic calorimeter (E-Cal) of CBM [16] is to measure direct photons and reconstruct the π^0 and η - decays needed for normalizing the e^+/e^- spectra. It will be a sampling calorimeter composed from 1 mm thick lead absorbers and 1 mm scintillator sensors. The absorbers force γ and $e^\pm$ to produce showers of secondary particles. The scintillators located between the absorbers record the energy of the secondary particles and the sum of all secondaries scales with the energy of the initial particle. The spatial segmentation of 3×3 cm^2, 6×6 cm^2 and 12×12 cm^2 will provide a good spatial resolution. The ambitioned energy resolution is $\sigma(E)/E = 4 - 6\%/\sqrt{E/\mathrm{GeV}}$.

The second calorimeter of CBM is the segmented projectile spectator detector (PSD) [16]. The PSD is to measure nuclear fragments produced in the collision and such to determine the related event plane. Moreover, the energy deposit will allow determining the number of nuclei, which did not participate in the collision, which is crucial for the study of event-to-event fluctuations. The PSD comprises 44 individual modules each consisting of 60 lead/scintillator

bremsstrahlung. The related multiple scattering angle is determined by the material thickness devided by X_0.

layers with a surface of 20×20 cm^2. Its energy resolution will be $\sigma(E)/E = 56.1\%/\sqrt{E}$.

CBM aims to combine high collision rates with complex trigger signatures like displaced decay vertices in the case of open charm. In traditional trigger concepts, high rates call for fast trigger decisions while searching for the vertices needs substantial reconstruction efforts and thus sizable computing time. To solve this issue, a self triggered readout is used. The latter consists of FEE-chips, which autonomously recognize particle hits, add time information and send them to a central DAQ in the data push mode. The data is received and sorted by a concentrator and event building network. Hereafter, information related to a defined period in time is send to a CPU, which buffers it in its RAM during processing. Such, complicated radiation tolerant data buffers in the sensors are replaced by custom of the shelf PC-RAM.

However, this concept requires a unusually high network bandwidth of ~1TB/s and the estimated computing power of several 10^4 computing cores for real time event reconstruction[17]. To address this issue , a first prototype computer named LOEWE-CSC was built and equipped with 2 PFlops computing power based on 20900 cores and 778 GPGPUs. This allowed reaching rank 22 in the worldwide list of supercomputers. Reconstruction algorithms profiting from modern multi-core and GPGPU environments were written and tested off-line with simulated data. The simulation and data analysis framework of CBM (CBMRoot) is to be extended to control the real time algorithms at least off-line. This will allow to compare measurement results with simulated data interpreted by the identical reconstruction algorithms and further improve the reliability of the CBM results.

3. Summary and Conclusion

The CBM-experiment will explore the phase diagram of nuclear matter at beam energies of 8-40 AGeV. This region should allow for observing the predicted first order phase transition between hadronic matter and QGP. Moreover, the critical endpoint of this phase transition might be observed. CBM integrates an electron-, hadron-, γ- and μ-spectrometer in order to constrain theoretical models with a maximum amount of experimental probes. The technological concept of the experiment has been developed and demonstrated with various demonstrators.

4. Acknowledgements

This work was supported by BMBF (06FY9099I and 06FY7113I), GSI and HIC for FAIR.

References

[1] K. Fukushima and T. Hatsuda, Rep. Prog. Phys. 74(2011) 014001(28p)
[2] P. de Forcrand and O. Philipsen, PoS LAT2006 (2006) 130
[3] Yu. B. Ivanov, et al., PHYSICAL REVIEW C 73, 044904 (2006)
[4] I. C. Arsene, et al., PHYSICAL REVIEW C 75, 034902 (2007)
[5] B. Friman, et al. (Eds.), “The CBM Physics Book”, Lect. Notes in Phys 814, Springer (2011) , doi 10.1007/978-3-642-13293-3
[6] P. Senger and V. Friese (Eds.), “Nuclear Matter Physics at SIS-100”, CBM Report 2012-01, Darmstadt 2012
[7] I. Sorokin, et al., NIM-A(2012),http://dx.doi.org/10.1016/j.nima.2012.10.118
[8] J. M. Heuser, et al., NIM-A 568 (2006) 258262
[9] I. Deppner, et al., NIM-A 661(2012)S121S124
[10] M. Deveaux, et al., POS (Vertex 2008) 028
[11] M. Deveaux, et al., 2011 JINST 6 C02004 doi:10.1088/1748-0221/6/02/C02004
[12] C. Höhne, et al., NIM-A 639 (2011) 294297
[13] A. Andronic, et al., NIM-A 563 (2006) 349354
[14] A. K. Dubey, et al., NIM-A (2012),http://dx.doi.org/10.1016/j.nima.2012.10.043
[15] F. Sauli, NIM-A 386 (1997) 531-534
[16] Yu. M. Zaitsev, et al., Atomic Energy, Vol. 112, No. 2, June, 2012
[17] J. de Cuveland, et al., Journal of Physics: Conference Series 331 (2011) 022006

Journal of Physics: Conference Series **426** (2013) 012021 doi:10.1088/1742-6596/426/1/012021

The phase diagram of $N_f = 2$ and $N_f = 2 + 1$ QCD from quark and gluon propagators

Jan Luecker and Christian S. Fischer

Institut für Theoretische Physik, Justus-Liebig-Universität Gießen, Heinrich-Buff-Ring 16, D-35392 Gießen, Germany

Abstract. In this talk we present results for the propagators and the phase diagram of QCD, obtained from a truncated set of Dyson-Schwinger equations at finite temperature T and quark chemical potential μ. We include back-coupling effects of the dressed quark and gluon propagators, which also allows us to study the influence of strange quarks. For the phase diagram we find a critical end-point at $\mu/T \approx 1.9$ and nearby chiral and deconfinement phase transitions.

1. Introduction

The most reliable source of information on quantum chromodynamics (QCD) at finite temperature has been lattice QCD which finds a first order phase transition in the quenched case and a crossover when dynamical quarks are included [1, 2]. In the latter case the order parameter for confinement, the Polyakov loop, shows a crossover transition nearby the chiral phase transition. However, at finite density lattice studies are hindered by the fermion sign problem. This led to a large industry of studies in effective models such as the Polyakov-loop extended Nambu–Jona-Lasinio [3, 4] and quark-meson models [5, 6, 7]. From these studies the general expectation that the crossover turns into a first order phase transition at a critical end-point (CEP) has emerged, with the position and even the existence of the CEP under debate. When employing effective models one has to trade the QCD degrees of freedom for effective ones. In contrast to that we use Dyson-Schwinger equations (DSEs) to study QCD and its degrees of freedom, quarks and gluons, directly. We take advantage of this possibility by introducing a truncation where the back-reaction of quarks onto gluons is explicitly taken into account. Here we summarize our results published in Ref. [14].

2. Order parameters from Dyson-Schwinger equations

The main objects of interest in this work are the propagators for the fully dressed quark and gluon, which are given, respectively, as

$$S(p) = [i(\omega_n + i\mu)\gamma_4 C(p) + i\vec{p}\vec{\gamma}A(p) + B(p)]^{-1}, \quad (1)$$

$$D_{\mu\nu}(p) = P^L_{\mu\nu}(p)\frac{Z^L(p)}{p^2} + P^T_{\mu\nu}(p)\frac{Z^T(p)}{p^2}, \quad (2)$$

where $p = (\vec{p}, \omega_p)$. The quark is dressed by the scalar functions A, B and C which depend on $\vec{p}^2$ and ω_p only. The in-medium gluon has two dressing functions Z_L and Z_T for the components

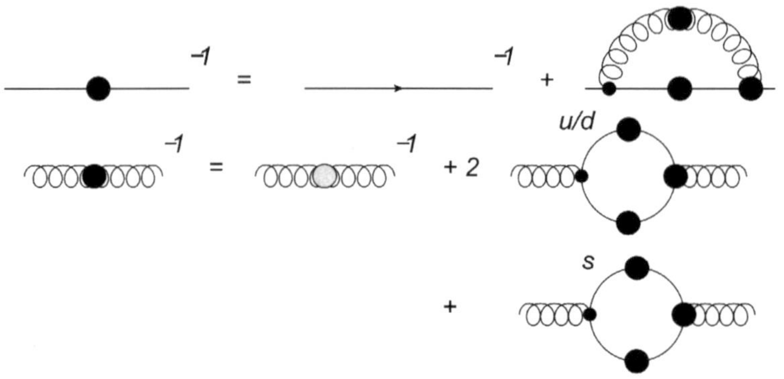

Figure 1. The coupled Dyson-Schwinger equations for the quark and gluon propagators.

longitudinal and transversal to the medium. $P^L_{\mu\nu}$ and $P^T_{\mu\nu}$ are the corresponding projectors. In the quark propagator we neglected a fourth dressing functions which comes with $\vec{p}\vec{\gamma}\gamma_4$ and can be shown by explicit calculations to have virtually no influence on the results.

2.1. Quark condensate

In the quark propagator, Eq. (1), the presence of chiral symmetry breaking is directly reflected in a non-vanishing scalar dressing funtion B. A derived order parameter for chiral symmetry breaking is the condensate

$$\langle\bar{\psi}\psi\rangle = Z_2 T \sum_n \int \frac{d^3p}{(2\pi)^3} \mathrm{Tr}_D \left[S(p)\right], \tag{3}$$

which is calculated from the propagator once the quark DSE is solved.

2.2. Dressed Polyakov loop

It is a much more non-trivial task to find an order parameter for confinement. The desirable object would be the Polyakov loop which can be connected to the free energy of a static quark, and is thus a description for confinement. Since we can not directly access the Polyakov loop from solutions of DSEs we will use the so-called dual condensates. They are defined as

$$\Sigma_n = \int_0^{2\pi} \frac{d\varphi}{2\pi} e^{-i\varphi n} \langle\bar{\psi}\psi\rangle_\varphi, \tag{4}$$

where $\langle\bar{\psi}\psi\rangle_\varphi$ is the quark condensate evaluated at generalized $U(1)$-valued boundary conditions $\psi(\vec{x}, 1/T) = e^{i\varphi}\psi(\vec{x}, 0)$ with $\varphi \in [0, 2\pi[$. In [8, 9, 10] it has been shown that $\Sigma_{\pm 1}$ contains the ordinary Polyakov loop as well as contributions from spatial detours. It has therefore been called the dressed Polyakov loop. Most notably it is sensible to center symmetry, just like the conventional Polyakov loop. It is therefore an order parameter for confinement that is accessible by continuum functional methods, since only the dressed quark propagator is necessary as an input [11].

With Eqs. (3,4) we have identified order parameters that rely only on the quark propagator. We will now define the truncation scheme that we will use to get our hands on not only the quark but also the gluon propagator.

3. Truncation scheme

The upper diagram in Fig. (1) shows the quark DSE. This equation defines the fully dressed quark propagator. On the right hand side we have the self-energy which depends on the fully dressed gluon and quark-gluon vertex. We need a prescription for these objects in order to get a closed set of equations, this will be given in the following. The challenge of in-medium Dyson-Schwinger equations is to include the relevant medium effects in the applied truncation scheme. We will here use the same construction for the quark-gluon vertex as in [11, 12, 13, 14]:

$$\Gamma_\mu = \gamma_\mu \cdot \Gamma(p^2, k^2, q^2) \cdot \left(\delta_{\mu,4}\frac{C(p)+C(k)}{2} + \delta_{\mu,i}\frac{A(p)+A(k)}{2}\right), \quad (5)$$

$$\Gamma(p^2, k^2, q^2) = \frac{d_1}{d_2+q^2} + \frac{q^2}{\Lambda^2+q^2}\left(\frac{\beta_0\alpha(\mu)\ln[q^2/\Lambda^2+1]}{4\pi}\right)^{2\delta}, \quad (6)$$

here q is the gluon momentum while p and k are the quark momenta. This *ansatz* contains the first part of the Ball-Chiu construction and is thus dependent on the medium. On the other hand, the dressing function Γ is constant in T and μ. This works well up to T_c but leads to an over-estimation of chiral symmetry breaking in the quark-gluon plasma as will be shown below.

The parameters are fixed to reproduce the pion decay constant in the vacuum and turn out to be $d_1 = 7.5$ GeV^2, $d_2 = 0.5$ GeV^2 and $\Lambda = 1.4$ GeV. In the UV we have $\delta = -9\frac{Nc}{44N_c-8N_f}$ and $\beta_0 = \frac{11N_c-2N_f}{3}$.

For the gluon propagator our strategy is to start with quenched QCD. Here, lattice calculations are up to now the most reliable source for the Landau gauge and fully temperature dependent propagator [15]. We have used this input to calculate the first (second) order phase transition in $SU(3)$ ($SU(2)$) gauge theory in [16]. To be able to correctly describe the crossover in unquenched QCD and to go to finite chemical potential, we add the quark loop from the gluon DSE to the quenched propagator. In Fig. (1) the resulting set of equations that we have solved is shown diagrammatically, where the first term on the right hand side of the gluon DSE denotes the quenched propagator. A strange-quark loop is included for the $N_f = 2{+}1$ case. This procedure for unquenching the gluon propagator neglects all quark loops inside the Yang-Mills self-energies. In the vacuum this leads to an error on the 5 percent level.

The quark loop at finite temperature can be split into longitudinal (L) and transversal (T) parts:

$$\Pi_{\mu\nu}(p) = P^T_{\mu\nu}\Pi^T(p) + P^L_{\mu\nu}\Pi^L(p), \quad (7)$$

where one notes that $\lim_{p\to 0} p^2\Pi^T(p) = 0$ but

$$\lim_{p\to 0} p^2\Pi^L(p) = \frac{m^2_{Th.}}{2}, \quad (8)$$

i.e. the longitudinal part produces a thermal (or Debye) screening mass. In the color superconducting phase the transversal part would generate a Meissner mass in a similar way.

4. Results

We present results from a calculation of the quark condensates and the thermal mass for $N_f = 2+1$ QCD in the left side of Fig. (2).

What we find from the light quark condensate is a crossover for small chemical potential, which becomes stronger with rising μ and eventually turns into a first order phase transition. The strange quark condensate reflects this change due to the coupling to the light quark via the

Journal of Physics: Conference Series **426** (2013) 012021 doi:10.1088/1742-6596/426/1/012021

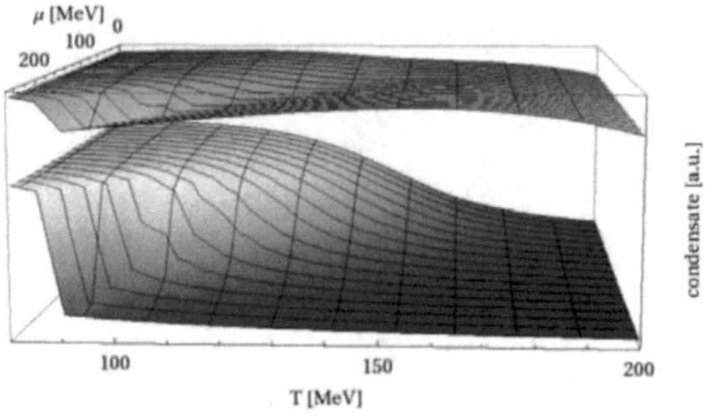

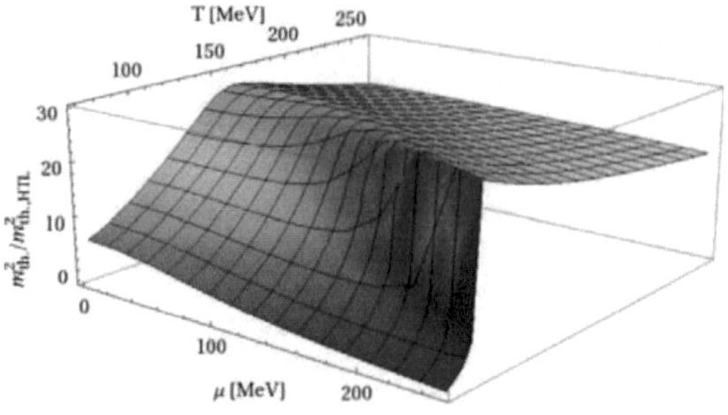

Figure 2. The left figure shows the light and strange quark condensates as the lower and upper surfaces, respectively. The right figure shows the thermal mass contribution from the quark loop, normalized by $T^2 + \frac{\pi}{3}\mu^2$, the asymptotic value. Both figures are adopted from Ref. [14].

gluon propagator. However, it continues to decrease at larger temperatures due to the larger strange quark mass.

The contribution to the thermal gluon mass due to the quark loop, Eq. (8), is shown to the right in Fig. (2), normalized by its asymptotic HTL/HDL behaviour. It is small in the hadronic phase, where it is suppressed by the inverse quark mass. Around the phase transition it rises strongly and approaches the asymptotic value above T_c. For large μ, at the first order phase transition, we observe a jump, which is inherited from the jump in the quark propagator. The quark loop reduces the gluon strength, and therefore also the quark self-energy. A smaller self energy in turn leads to a larger quark loop, and therefore the back-coupling of quark and gluon has an accelerating effect on the phase transition. This acceleration will be important to reproduce lattice data, which we do in the following.

From the light and strange quark condensates we can obtain the quantity

$$\Delta_{l,s} = \langle\bar{\psi}\psi\rangle_l - \frac{m_l}{m_s}\langle\bar{\psi}\psi\rangle_s, \tag{9}$$

which is finite since the divergent term $m\Lambda^2$ in $\langle\bar{\psi}\psi\rangle$ cancels. It can therefore be used in comparisons with different approaches like lattice QCD. This comparison is shown in the left part of Fig. (3). It shows a rather good agreement up to T_c and an overestimation of the remaining chiral symmetry breaking in our approach above T_c. This we attribute to the constant vertex strength, which should be temperature dependent in a refined model. Nonetheless we find a critical temperature of $T_c \approx 156$ MeV in good agreement with the lattice [1], and a similar steepness of the crossover.

Finally we show the main result of Ref. [14], the phase diagram of $N_f = 2$ and $N_f = 2+1$ QCD in the right part of Fig. (3). We find a critical end-point at $(\mu, T) \approx (190\text{ MeV}, 100\text{ MeV})$ for $N_f = 2+1$, which is certainly outside the area that can be accessed by extrapolations from lattice data. The (pseudo-)critical temperature for deconfinement lies slightly above that of chiral restoration in the crossover regime, and coincides close to and at the CEP. Note that to the right of the CEP, the realized phase is found by the pressure difference between the chirally symmetric and broken phase. Since we do not have access to the pressure we show only the spinodals of the coexistence region and we do not define a deconfinement temperature there. A naïve calculation of the dual condensate in the first order region leads to the same critical temperature as for the chiral transition. Fig. (3) also shows in lighter colors the phase diagram for the two flavor case, where $T_c \approx 200$ MeV. The impact of the strange quark is evidently a

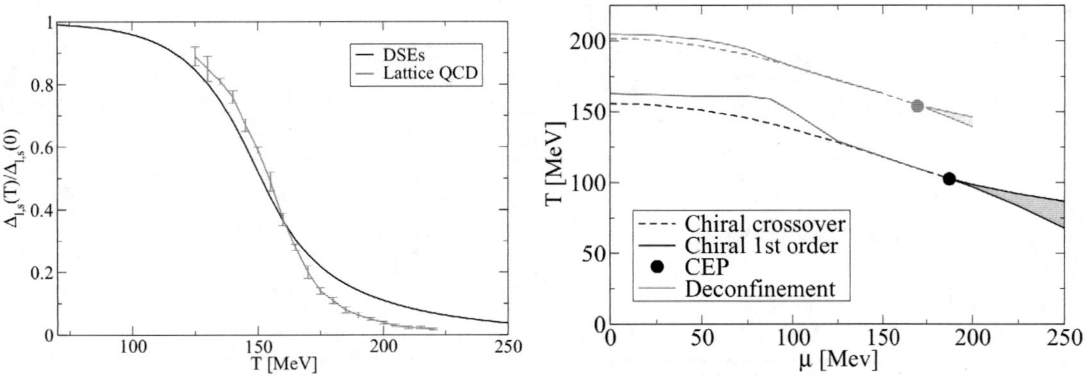

Figure 3. In the left figure we compare the regulated and normalized quark condensate to corresponding results from lattice QCD [1]. In the right figure we show the phase diagram for $N_f = 2+1$ QCD. In lighter colors we also show the results for $N_f = 2$ QCD (upper lines). Both figures are adopted from Ref. [14].

reduction of T_c by about 50 MeV while the CEP moves only slightly to larger μ.

5. Summary

We discussed a truncation scheme to the Dyson-Schwinger equations of QCD that explicitly couples the quark and gluon propagators. This makes the gluon sensitive to the chiral phase transition via the thermal mass contribution from the quark loop. In this approximation we showed that the condensate is in reasonable agreement with lattice simulations. For the phase diagram we find a CEP at $\mu/T \approx 1.9$, and nearby chiral and deconfinement phase transitions.

Acknowledgments

We thank the organisers of the FAIRNESS workshop 2012. This work has been supported by the Helmholtz Young Investigator Grant VH-NG-332 and the Helmholtz International Center for FAIR within the LOEWE program of the State of Hesse.

References

[1] S. Borsanyi *et al.* [Wuppertal-Budapest Collaboration], JHEP **1009** (2010) 073 [arXiv:1005.3508 [hep-lat]].
[2] A. Bazavov, T. Bhattacharya, M. Cheng, C. DeTar, H. T. Ding, S. Gottlieb, R. Gupta and P. Hegde *et al.*, arXiv:1111.1710 [hep-lat].
[3] K. Fukushima, Phys. Lett. **B591** (2004) 277-284. [hep-ph/0310121].
[4] C. Ratti, M. A. Thaler, W. Weise, Phys. Rev. **D73** (2006) 014019. [hep-ph/0506234].
[5] B. -J. Schaefer, J. M. Pawlowski, J. Wambach, Phys. Rev. **D76** (2007) 074023. [arXiv:0704.3234 [hep-ph]].
[6] V. Skokov, B. Stokic, B. Friman and K. Redlich, Phys. Rev. C **82** (2010) 015206 [arXiv:1004.2665 [hep-ph]]; V. Skokov, B. Friman and K. Redlich, Phys. Rev. C **83** (2011) 054904 [arXiv:1008.4570 [hep-ph]].
[7] T. K. Herbst, J. M. Pawlowski, B. -J. Schaefer, Phys. Lett. **B696** (2011) 58-67. [arXiv:1008.0081 [hep-ph]].
[8] C. Gattringer, Phys. Rev. Lett. **97** (2006) 032003.
[9] F. Synatschke, A. Wipf and C. Wozar, Phys. Rev. D **75** (2007) 114003;
[10] E. Bilgici, F. Bruckmann, C. Gattringer and C. Hagen, Phys. Rev. D **77** (2008) 094007.
[11] C. S. Fischer, Phys. Rev. Lett. **103** (2009) 052003 [arXiv:0904.2700 [hep-ph]].
[12] C. S. Fischer and J. A. Mueller, Phys. Rev. D **80** (2009) 074029 [arXiv:0908.0007 [hep-ph]].
[13] C. S. Fischer, J. Luecker and J. A. Mueller, Phys. Lett. B **702** (2011) 438 [arXiv:1104.1564 [hep-ph]].
[14] C. S. Fischer and J. Luecker, Phys. Lett. B in press [arXiv:1206.5191 [hep-ph]].
[15] C. S. Fischer, A. Maas and J. A. Muller, Eur. Phys. J. C **68** (2010) 165 [arXiv:1003.1960 [hep-ph]].
[16] J. Luecker and C. S. Fischer, Prog. Part. Nucl. Phys. **67** (2012) 200 [arXiv:1111.0180 [hep-ph]].

Status and first Results of the CBM TRD Prototype Development

Andreas Arend for the CBM TRD Group

Goethe-Universität, Institut für Kernphysik, Max-von-Laue-Str. 1, 60438 Frankfurt am Main

E-mail: `arend@ikf.uni-frankfurt.de`

Abstract. The measurement of decay products from rare particles in the CBM experiment defines the requirements for the used detector systems. This report describes the approaches for the Transition Radiation Detector to fulfil these requirements and presents the current status of the prototype development. First results from beam tests with these prototypes are discussed.

1. Introduction

The Compressed Baryonic Matter (CBM) experiment at the future Facility for Antiproton and Ion Research (FAIR) is aimed to measure rare observables in an environment of unprecedented high particle flux. The measurement of the D-mesons and the J/ψ as well as its excited states defines the experimental requirements of this experiment. Almost all listed particles in the left panel of Figure 1 decay into electrons and positrons or pions. The precise and fast discrimination between these decay products is an essential key feature of CBM. The purpose of the Transition Radiation Detector (TRD) is to provide this capability. The employed Multi-Wire Proportional Chambers (MWPC) additionally provide sufficient position resolution for the experiment-wide particle tracking and can be designed to be fast enough for the extreme high event rates at FAIR. The future CBM experiment is planed to be modular so that two different setups can be realized: an electron identification setup with very efficient electron/pion separation, and tracking capabilities and a muon detection setup where the TRD only contributes to tracking between the muon chambers (MUCH) and the time-of-flight detector (TOF). The SIS100 StartUp-Version of CBM also includes one TRD station for track matching to TOF. Both setups are depicted in the right part of Figure 1. In the current design of the TRD for SIS300, the active area covered by the detector is $585\,\mathrm{m}^2$ where 800.000 channels are read out. The TRD is divided in 3 stations with 10 layers in total. The first two stations consist of 4 layers each. The TRD is designed for a track density of 600 charged particles in an angular range of 2.5° to 25°at an event rate of 10 MHz. The misidentification of a pion as an electron has to be smaller than 1% at 90% electron efficiency for electrons with a momentum larger than $p = 1.5\,\mathrm{GeV}/c$. The typical position resolution of the used MWPC design, in the order of $250\,\mu\mathrm{m}$, is sufficiently precise for track reconstruction and matching to other detector systems.

2. Detector concepts with and without dedicated drift region

The TRD consists of two main detector parts: the radiator which generates the transition radiation (TR) photon depending on the γ-factor of the charged particle passing through, and

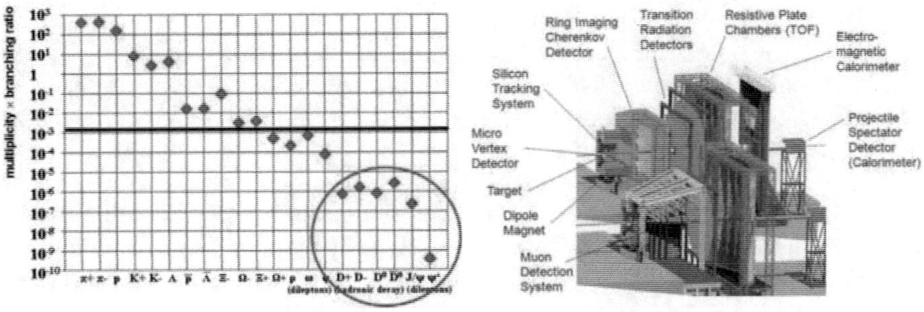

Figure 1. Left: Expected production rate of observables in Au+Au-collisions (E_{beam} = 25 GeV). Right: Schematic view of the planned CBM experiment at SIS300: muon detection setup in front, electron identification setup in the back. [1]

the MWPC which absorbs the TR photon, in addition to the ionization signal of the charged particle. The symmetric detector design without a dedicated drift region only provides an amplification region and therefore only requires one wire plane which results in an easy-to-construct detector. The signal generation is, compared to a MWPC with dedicated drift, faster. This conceptual design is followed by the CBM TRD group at Institut für Kernphysik Frankfurt. The detector design with drift region consists of a MWPC with two separate regions in the detector: a dedicated drift and an amplification region. It requires two wire planes, the intrinsic signal generation is slower due to the additional drift region compared to the design without drift zone. The advantage of this concept is that the MWPC and the readout pad geometry are decoupled from the overall thickness of the detector which allows a more flexible design. This type of TRD has been build for the ALICE TRD[3] and it is followed by the CBM TRD group at Institut für Kernphysik Münster.

3. Prototype Construction

Based on results from previous test beam campaigns [4] and simulations [5] prototypes without drift region are built with dimensions of the active gas volume of 150 × 150 × [4+4; 5+5; 6+6] mm^3. The aluminium frame was covered with 20 μm aluminized mylar foil as entrance window. The anode wires are 20 μm gold plated tungsten wires with 2.5 mm spacing. The pad geometry is 5 × 50 mm^2. With these MWPCs different radiator types have been tested. The sandwich-type radiator as used in the ALICE TRD made of 8 mm *ROHACELL* foam sheets coated with 25 μm aluminized mylar foil on carbon fibre sheets, and 40 mm polypropylene fibres was used as reference. A foam radiator made of polyethylene foam (which is used as commercial packing material) with a bubble size of 0.7 mm was chosen to provide around 350 transitions. Also regular foil radiators made of 20 μm polypropylene foil glued to an 0.5 mm aluminium frame as spacing between foils has been built in combinable stacks of 50 transitions each, so that up to 350 foil layers can be used. The detectors are read out by *Self-triggered Pulse Amplification and Digitization asIC* (SPADIC) in version 0.3 combined with the *SUSIBO*[6] readout board. The 8 bit current-sensitive ADC reads out 8 channels with 45 time bins each, at a sampling frequency of 25 MHz.

4. Results from in beam Test Measurements

The constructed prototypes with amplification region only and the prototypes with an additional dedicated drift region were tested in October 2011 at the CERN PS[7] providing a mixed electron and pion beam with particle momenta of 2 - 10 GeV/c. The raw signal read out by the SPADIC is corrected for baseline variations using empty events. Correlated noise is calculated using a covariance matrix method and subtracted from the raw signals. The integral of the noise-corrected signal in the hit region (time bins 10 to 44 of the SPADIC signals) is clustered in the pad direction. The sum of the three largest pad signals in the cluster is defined as the total deposited charge q_{tot}. Figure 2 shows the total deposited charge for the Frankfurt 4+4 mm-prototype (only amplification region) and the Münster prototype with 3 mm distance from readout pads to anode wires, 3 mm distance of the anode wires to the cathode wires and 6 mm distance of the cathode to the entrance window (MS336). Both prototypes have been equipped with the same ALICE-type sandwich radiator. It has to be emphasized that independent analyses using different signal cuts are applied. The normalized spectra of the total

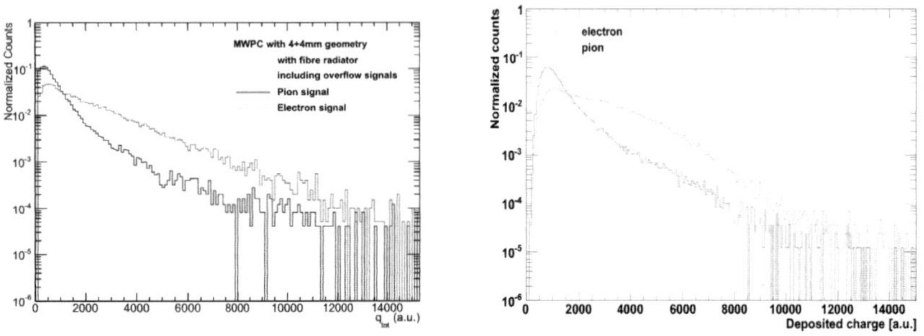

Figure 2. Total deposited charge in the FFM 4+4 mm (left) and MS336 (right) prototypes. Electrons are shown as red line, pions in black.

deposited charge serve as the input to a likelihood method to determine the electron and pion discrimination capability of the respective radiator and MWPC prototype combination. The normalized counts are interpreted as the probability P_e and P_π of electrons/pions depositing the measured charge. Using this measurement for one single detector layer one can extrapolate the electron likelihood L for multiple layers from

$$L = \frac{P_e}{P_e + P_\pi} \quad with \quad P_e = \prod_{i=1}^{N} P\left(q_{tot,i}|e\right) \ , \quad P_\pi = \prod_{i=1}^{N} P\left(q_{tot,i}|\pi\right) \tag{1}$$

where N is the number of extrapolated layers. For an electron efficiency of 90%, the corresponding integrated percentage of pions misidentified as electrons is shown in Figure 3 for different radiators as function of the number of extrapolated layers. One can achieve a pion misidentification of less than 1% for more than 6 detector layers with regular radiators (foil radiators) or more than 8 detector layers with foam-based radiators. The pad response function has been measured for all detector prototypes. Figure 4 shows the result for the FFM 4+4 mm and FFM 5+5 mm prototypes. It can be clearly seen that the pad geometry was not optimized for the given wire geometry. The central pad carries only a fraction of roughly 60%, the remaining charge is distributed on the neighbouring pads. When applying a simple straight line tracking algorithm, which only interpolates between three identical MWPC layers and calculates the distance between the reconstructed (center of gravity) an the measured position without any alignment and without any external reference, a position resolution of 291±10 μm can be achieved for the FFM 5+5 mm prototype. Using an iterative approach to correct also for spatial uncertainties a position resolution of 200±50 μm for the MS336 prototype was reached.

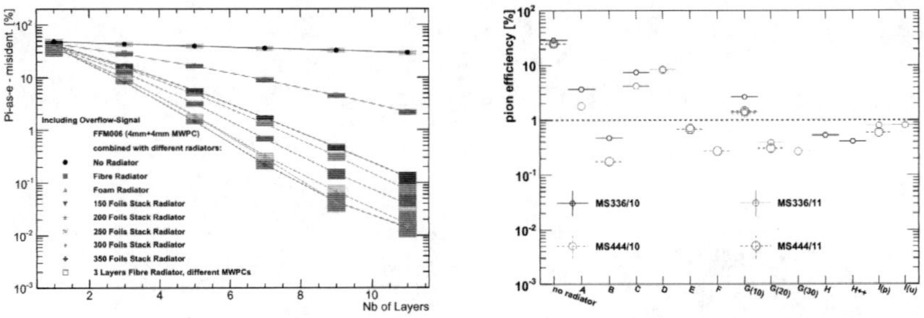

Figure 3. Fraction of misidentified pions for the FFM 4+4mm, as a function of the number of layers (left), and in MS336 (right) for 10 extrapolated layers and different radiator types (A is ALICE Type Radiator, B is a regular foil radiator with 250 layers, C to F are different regular foil radiators, G are fibre mats of variable thickness, I is mixed foam and foil-based radiator and H is a foam radiator)[8].

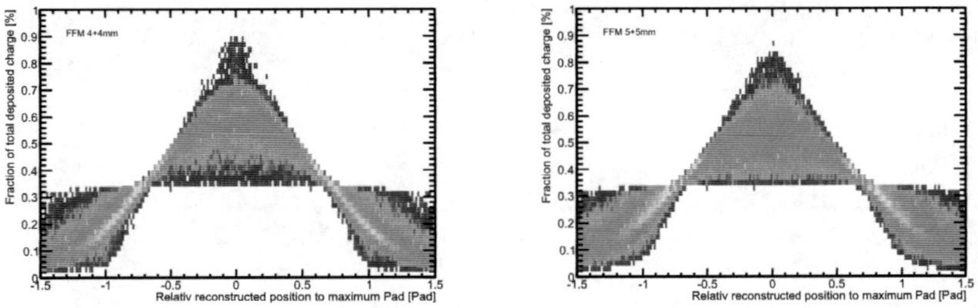

Figure 4. Pad response function for the FFM 4+4 mm (left) and FFM 5+5 mm (right) prototypes. The reconstructed position in the MWPC relative to the maximal pad is shown on the x-axis, the y-axis shows the fraction of the total charge in the corresponding pad.

5. Future Developments

Based on the results of the test beam time in 2011, the development of full-size prototypes (600 × 600 mm^2) for the inner part of the TRD in the first two stations has started. Detector layout with and without dedicated drift region are considered. A common pad plane was designed which provides different sizes of readout pads to accommodate all considered prototype geometries. The prototype with amplification region only is designed in a modular way. The detector body consists of an aluminium frame and a honeycomb support structure holding the pad plane and the read out electronics. Distance ledges and the anode wires are attached to this main part. The detector is closed by a cover which holds a thin aluminized mylar foil as entrance window, as well as an optional support structure. A technical drawing of the prototype is shown in the left part of figure 5. The procedure for stretching the entrance foil is based on a thermal expansion[10]. The foil is fixed to a plexiglass frame, which is heated, the resulting expansion causes a mechanical stretching of the foil. Simulations of the deformation of a 600 × 600 mm^2 large entrance window and the mechanical stress of MWPC body were performed using the ABAQUS software framework [11]. A mylar foil stretched at a plexiglass temperature of 60 °C gets bulged by 0.076 mm at an overpressure of 0.01 mbar. These simulations were verified by measurements on a full-size prototype window. The deformation of the entrance window modifies the gas gain of the MWPC. These variations were simulated with the GARFIELD software package [12]. The right pad of figure 5 shows the relative gain change as a function of the distance of the entrance window to the anode wires for three detector geometries with amplification region only. According to these simulations, a gas gain variation of less than 10% can be achieved if the gas system limits differential pressure variations to less than $0.01mbar$.

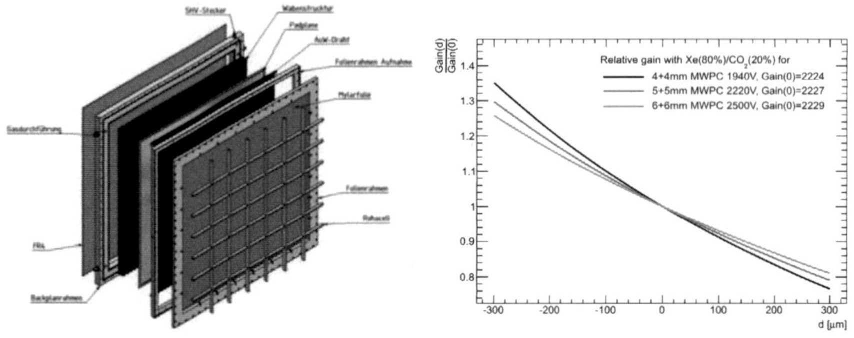

Figure 5. Left: Technical drawing of a full-size prototype. Right: Relative gain change depending on the displacement of the entrance window for three detector types with amplification region only.

6. Summary and Outlook

Results from beam test show that small prototypes with and without additional drift region fullfill the requirements of the CBM TRD for electron / pion separation and position resolution. Based on the results, no obvious preferences for final detector concept could be given. Large scale prototypes will be studied in upcoming beam tests. The final detector setup, including the number of stations and layers, the choice of the radiator, and the requirements for infrastructure and supply have to be evaluated. A Technical Design Report is scheduled for submission in 2014.

References

[1] The Compressed Baryonic Matter Experiment at FAIR. Johann M. Heuser (for the CBM Collaboration), GSI Helmholtz Center for Heavy Ion Research GmbH, Darmstadt, Germany, Quark Matter 2012 Proceedings, to be published.
[2] The CBM experiment at SIS-100. P. Senger and the CBM Collaboration, GSI Helmholtz Center for Heavy Ion Research GmbH, Darmstadt, Germany, CBM Progress Report 2011, Darmstadt 2012.
[3] ALICE TRD Technical Design Report, ALICE Colaboration, CERN/LHCC 2001-021, 2001
[4] First beam test of the Frankfurt prototypes for the CBM-TRD, W. Yu et al., CBM Progress Report 2010, Darmstadt 2011, p. 44.
[5] Study on electron-pion discrimination with the CBM Transition Radiation Detector, P. Reichelt, et al., CBM Progress Report 2010, Darmstadt 2011, p. 39.
[6] CBM-TRD readout with the SPADIC amplifier / digitizer chip, T. Armbruster, et al., CBM Progress Report 2010, Darmstadt 2011, p. 45.
[7] Common CBM beam test of the RICH, TRD and TOF subsystems at the CERN PS, C. Bergmann, et al., CBM Progress Report 2011, Darmstadt 2012, p. 4.
[8] Development, Simulation and Test of Transition Radiation Detector Prototypes for the Compressed Baryonic Matter Experiment at the Facility for Antiproton and Ion Research, PhD Thesis, C. Bergmann, to be published.
[9] Performance of CBM TRD Prototypes from Münster, C. Bergmann, DPG Spring Meeting 2012 (Mainz), 2012.
[10] Michael Staib, et al, RD51-Note- 2011-004
[11] http://www.3ds.com/products/simulia/portfolio/abaqus/overview/
[12] http://garfield.web.cern.ch/garfield/

Journal of Physics: Conference Series **426** (2013) 012023 doi:10.1088/1742-6596/426/1/012023

J/ψ Production in Pb–Pb collisions at $\sqrt{s_{\rm NN}}$=2.76 TeV

Ionut-Cristian Arsene for the ALICE Collaboration

Research Division and ExtreMe Matter Institute EMMI, GSI Helmholtzzentrum für Schwerionenforschung, Darmstadt, Germany

E-mail: i.c.arsene@gsi.de

Abstract. We report on the J/ψ production in Pb–Pb collisions at $\sqrt{s_{\rm NN}}$=2.76 TeV recorded with ALICE during the 2010 and 2011 LHC runs. The J/ψ candidates are reconstructed using the e^+e^- and $\mu^+\mu^-$ decay channels at mid- and forward rapidity, respectively. We show the J/ψ nuclear modification factor $R_{\rm AA}$ as a function of centrality, transverse momentum and rapidity. Our results are compared to the lower energy measurements from PHENIX and to model calculations.

1. Introduction

The Quark Gluon Plasma (QGP) is a state of deconfined nuclear matter formed in relativistic heavy ion collisions. One of the promising proposed probes to study such a state is the suppression of heavy quarkonia (bound states of $c\bar{c}$ and $b\bar{b}$ pairs) via Debye screening [1]. Due to their large mass, the pair creation of heavy quarks occurs in the initial partonic stage of the collision, thus enabling them to experience the entire history of the collision. Without any cold nuclear matter (CNM) effect, the total production cross-section of these pairs in nuclear collisions should correspond to the one in nucleon-nucleon collisions scaled with the number of binary nucleon-nucleon collisions ($N_{\rm coll}$). It is expected that the production probabilities of quarkonium states are affected by the hot and dense QGP which makes them a sensitive tool for its properties. At the lower collision energies from SPS and RHIC, a strong J/ψ suppression is observed. However, the interpretation of these results is still under debate due to insufficient understanding of the CNM effects which break the $N_{\rm coll}$ scaling. It is expected that the measurements in Pb–Pb collisions at LHC energies will shed more light on this subject. Shadowing, the modification of the free nucleon gluon distribution functions, is the dominant contribution to the CNM effects at LHC energies and will be measured in p-Pb collisions.

In these proceedings we report on measurements of the inclusive J/ψ production as a function of centrality, transverse momentum ($p_{\rm T}$) and rapidity (y) in Pb–Pb collisions at $\sqrt{s_{\rm NN}}$=2.76 TeV recorded using the ALICE setup during the 2010 and 2011 runs. The results are presented using the nuclear modification factor ($R_{\rm AA}$) which is defined as

$$R_{\rm AA} = \frac{1}{N_{\rm coll}} \times \frac{\sigma_{J/\psi}^{Pb-Pb}}{\sigma_{J/\psi}^{pp}} \quad (1)$$

where $\sigma_{J/\psi}^{Pb-Pb}$ and $\sigma_{J/\psi}^{pp}$ are the inclusive J/ψ cross-sections in Pb–Pb and pp collisions, respectively. $N_{\rm coll}$ is obtained using Glauber model calculations.

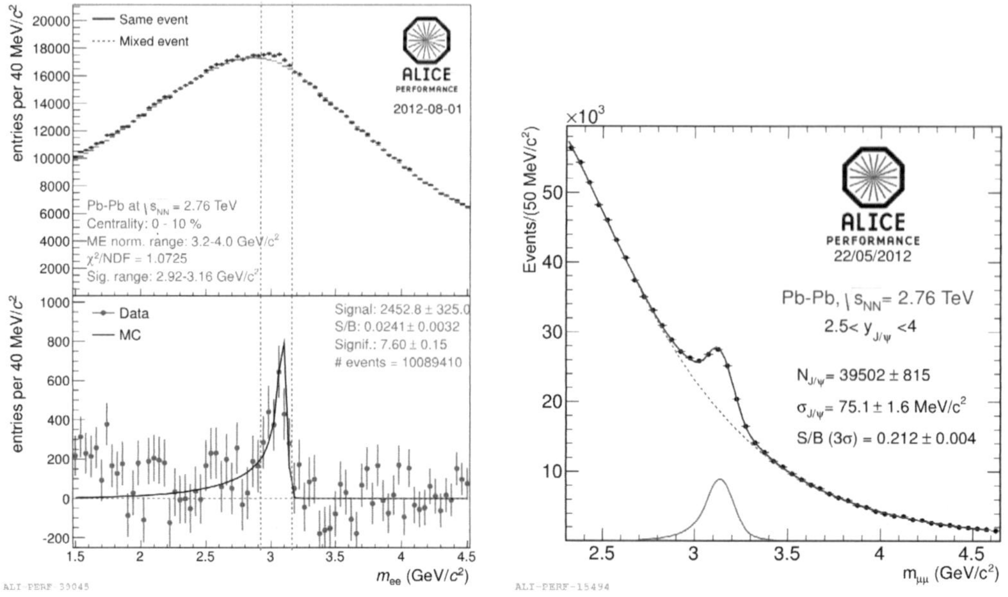

Figure 1. Invariant mass distribution for e^+e^-(left) in central (0-10%) and $\mu^+\mu^-$(right) pairs in minimum bias Pb–Pb collisions.

2. Experimental setup and data analysis

A detailed description of the ALICE experimental setup is provided in [2]. It consists of several types of tracking and particle identification detectors covering the mid-rapidity region ($|\eta| < 0.9$) and a forward muon spectrometer covering the range $2.5 < \eta < 4.0$. Additional smaller detectors are used for triggering and event characterization. At mid-rapidity, the J/ψ are reconstructed in the di-electron decay channel. The electrons are tracked using the Inner Tracking System (ITS) and the Time Projection Chamber (TPC). Electron identification is done using the specific energy loss in the TPC gas. At forward rapidity, the J/ψ are measured in the di-muon decay channel, with the muons being tracked, using five tracking stations, which are placed behind the hadron absorber. A fast triggering system placed behind an additional iron wall ensures further background rejection and increases the inspected luminosity. The J/ψ coverage at both mid-rapidity and forward rapidity extends down to zero transverse momentum. The present measurements do not allow for a separation between direct and feed-down J/ψ from the decays of higher mass charmonium states or from B mesons, so in the following we always refer to inclusive J/ψ. Detailed descriptions of this analysis are given in [3, 4, 5, 6, 7].

For the di-electron channel, the event sample used for this work consists of 30 million events, the majority of them being centrality-triggered collisions. For the di-muon channel we used 17.7 million di-muon triggered events. The analyzed events correspond to an integrated luminosity of $15\mu b^{-1}$ in the di-electron channel and $70\mu b^{-1}$ in the di-muon channel.

The J/ψ signal is extracted from the invariant mass distribution constructed from opposite-sign (OS) pairs of electron or muon candidates (see Fig.1). The dominant background contribution in the OS invariant mass distribution is coming from uncorrelated pairs. For the di-electron channel, the background is constructed using the event mixing technique. Within the statistical uncertainties a good agreement is found between the background-subtracted distribution and the signal shape as expected from our Monte-Carlo (MC) simulations (bottom-

Journal of Physics: Conference Series **426** (2013) 012023 doi:10.1088/1742-6596/426/1/012023

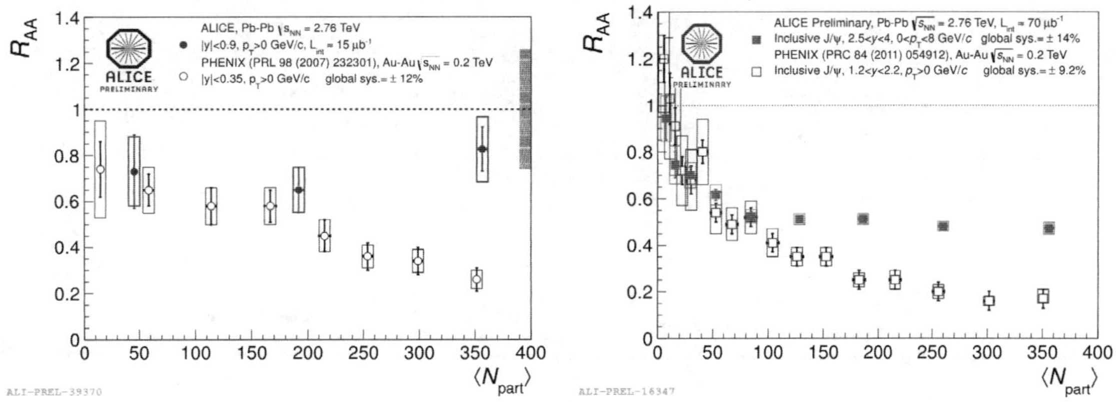

Figure 2. J/ψ nuclear modification factor $R_{\rm AA}$ as a function of the number of participant nucleons at mid-rapidity (left panel) and forward rapidity (right panel) compared with the measurements from PHENIX in Au-Au collisions at $\sqrt{s_{\rm NN}}$=200 GeV.

left panel of Fig.1). The contribution from the correlated e^+e^- continuum surviving the mixed event background subtraction are incorporated in the systematic uncertainties for the signal extraction. Finally, the raw J/ψ signal is obtained from bin counting in the range $2.92 - 3.16$ GeV/c^2. In the di-muon channel, thanks to the dedicated trigger system and better signal to background ratio, the raw J/ψ yields could be extracted using two different methods. One method employs fitting of the $\mu^+\mu^-$ distribution with a sum of a modified Crystal Ball (CB2) function and a variable width gaussian for background parameterization (right panel of Fig.1). Alternatively, the mixed event background was subtracted and the remaining distribution was fitted with the sum of a CB2 function for the signal and an exponential function to parameterize the remaining correlated background.

The raw J/ψ signal was corrected for trigger efficiency, acceptance and reconstruction efficiency using simulated Pb–Pb collisions employing the HIJING event generator enriched with primary J/ψ. The transport of the generated particles through the detector is done using GEANT3. Furthermore, in the case of the di-muon decay channel, the efficiencies were calculated by embedding simulated J/ψ into the raw data of real events. Details on the efficiencies and the systematic uncertainties can be found in [5, 6, 7].

3. Results and discussion

The nuclear modification factor is obtained using the J/ψ cross-section in pp collisions (see Eq.1) at $\sqrt{s}$ =2.76 TeV measured by ALICE [4]. The systematic uncertainty on the $R_{\rm AA}$ due to the statistical and systematic errors on the pp reference amounts to 26% for the di-electron channel. For the di-muon channel, the corresponding error, which is dominated by the systematic uncertainty, is 9% for the integrated results and 6% for the $p_{\rm T}$ or y dependent results.

Figure 2 shows our results on the $R_{\rm AA}$ as a function of the number of participant nucleons ($N_{\rm part}$) around mid-rapidity (left column) and at forward rapidity (right column). At mid-rapidity, the results show no $N_{\rm part}$ dependence within the experimental uncertainties and an $R_{\rm AA}$ value in the most central collisions of $0.83 \pm 0.09(\text{stat.}) \pm 0.26(\text{syst.})$ has been obtained. At forward rapidity, the results indicate a clear J/ψ suppression which is constant with centrality for $N_{\rm part} > 50$. In Figure 2 our results are compared with the ones from PHENIX measured in Au–Au collisions at $\sqrt{s_{\rm NN}}$=200 GeV [8] at mid-rapidity (left) and forward rapidity (right).

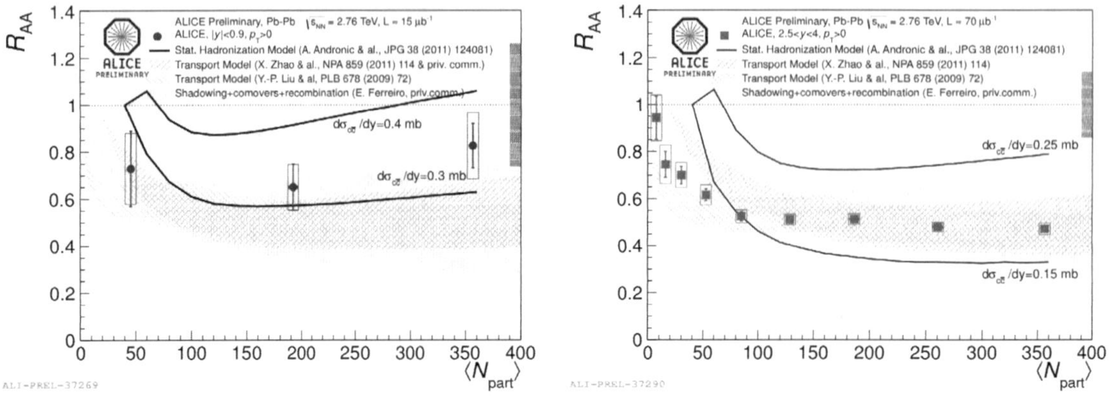

Figure 3. J/ψ nuclear modification factor R_{AA} as a function of the number of participant nucleons at mid-rapidity (left panel) and forward rapidity (right panel) compared with model calculations.

For the most central collisions, the J/ψ R_{AA} is significantly higher at the LHC at both mid- and forward rapidity. The R_{AA} measured by PHENIX shows a strong suppression in central collisions. Whether this observation is compatible with J/ψ suppression beyond the CNM effects, measured in d-Au collisions [10], is still under debate. Phenomenological calculations from [9] suggests that the observed nuclear suppression in Au–Au collisions may be explained solely with the CNM effects, which include nuclear shadowing and the breakup of correlated c$\bar{c}$ pairs in cold nuclear matter. At the LHC, it is expected that the high charm quark density enhances the formation probability of quarkonia during or at the end of the QGP evolution [11, 12]. Our results for central collisions suggest that this effect could have a significant contribution.

Figure 3 shows a comparison of our results with model calculations which take this effect into account. The hashed bands show the results from two transport models [13, 14] and from the comover interaction model [15]. In these models the fraction of J/ψ resulting from (re-)combined c$\bar{c}$ pairs from the medium is at most 50%, the rest being direct J/ψ produced during the hard partonic interactions. The solid lines show the prediction from the statistical hadronization model [16]. This model assumes that no charmonium is formed in the QGP phase and the charm quarks thermalize with the whole system. All the J/ψ and other charmed hadrons are then formed at the chemical freeze-out and their yields can be calculated based on the total charm cross section and the thermal parameters. All of the model calculations have uncertainties due to the unknown inclusive $c\bar{c}$ production cross section and the CNM effects. The CNM effects will be measured in p-Pb collisions scheduled for beginning of 2013.

Figure 4 shows the p_T (left) and the rapidity (right) dependence of the J/ψ R_{AA}. The p_T dependence is shown for both central (0-20%) and peripheral collisions (40-90%) in comparison with calculations from a transport model [13]. The results for central collisions show that R_{AA} is $\approx$0.6 at low p_T and decrease towards higher p_T. For the peripheral events, we observe no or little p_T dependence within the uncertainties. These results are in fair agreement with the model calculations from [13] with the exception of the peripheral collisions where the model seems to overestimate the data at low p_T. The rapidity dependence of R_{AA} for the most central (0-10%) collisions is shown in the right panel of Fig.4. We observe that the R_{AA} grows from forward to mid-rapidity. This behaviour was observed also at RHIC energy in the most central collisions. Although there are more effects which need to be taken into account, these results might suggest that the (re)generation mechanism becomes important at LHC since the

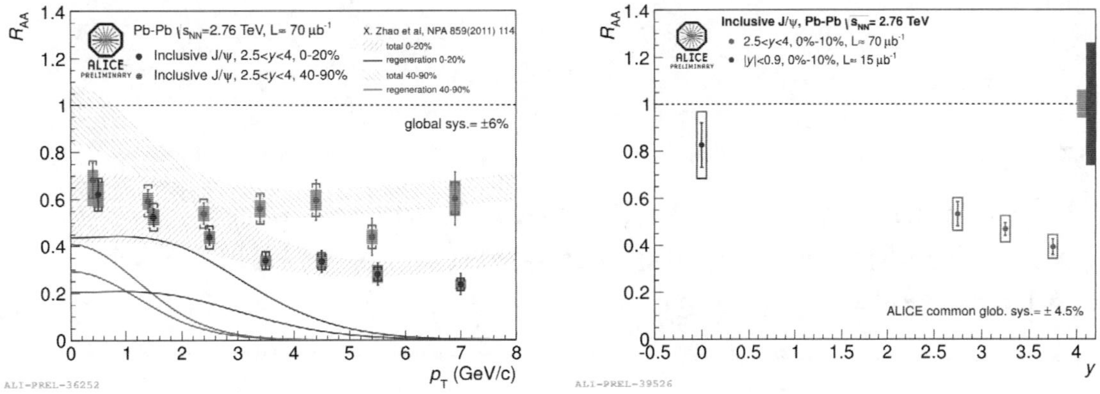

Figure 4. J/ψ nuclear modification factor R_{AA} as a function of transverse momentum at forward rapidities ($2.5 < y < 4$) (left) and as a function of rapidity (right).

charm density is expected to grow from forward to mid-rapidity.

4. Conclusions

To summarize, we reported on the latest ALICE measurements of the inclusive J/ψ nuclear modification factor as a function of centrality, p_{T} and y in Pb–Pb collisions at $\sqrt{s_{\mathrm{NN}}}$=2.76 TeV. In the most central collisions our integrated R_{AA} values are roughly 3 times higher than the results obtained in central Au–Au collisions by PHENIX collaboration. Model calculations which incorporate the (re)generation mechanism are in agreement with the data. Better knowledge of the total charm cross-section and the CNM effects will impose more stringent constraints to models. A comparison of the p_{T}dependence of the J/ψ R_{AA} in central collisions indicates that the (re)generation mechanism is significant at low transverse momentum. Finally, the J/ψ R_{AA} in central collisions is decreasing with increasing rapidity as was also observed at RHIC.

References

[1] Matsui T and Satz H 1986 *Phys.Lett.* B **178** 416.
[2] Aamodt K *et al.* (ALICE Collaboration) 2008 *JINST* **3** S08002
[3] Aamodt K *et al.* (ALICE Collaboration) 2011 *Phys.Lett.* B **704** 442
[4] Abelev B *et al.* (ALICE Collaboration) 2012 *Phys.Lett.* B **718** 295
[5] Wiechula J for the ALICE Collaboration arXiv: 1208.6566
[6] Arsene I for the ALICE Collaboration arXiv: 1210.5818
[7] Arnaldi R for the ALICE Collaboration arXiv: 1211.2578
[8] Adare A *et al.* (PHENIX Collaboration) 2007 *Phys.Rev.Lett.* **98** 232301
[9] Adare A *et al.* (PHENIX Collaboration) 2008 *Phys.Rev.* C **77** 024912
[10] Adare A *et al.* (PHENIX Collaboration) 2011 *Phys.Rev.Lett.* **107** 142301
[11] Braun-Munzinger P, Stachel J 2000 *Phys.Lett.* B **490** 196
[12] Andronic A, Braun-Munzinger P, Redlich K, Stachel J 2007 *Nucl.Phys.* A **789** 334
[13] Zhao X, Rapp R 2011 *Nucl.Phys.* A **859** 114 and private communication
[14] Liu Y P, Qu Z, Xu N, Zhuang P 2009 *Phys.Lett.* B **678** 72 and private communication
[15] Ferreiro E, arXiv:1210.3209 and private communication
[16] Andronic A, Braun-Munzinger P, Redlich K, Stachel J 2011 *J.Phys.* G **38** 124081

Journal of Physics: Conference Series **426** (2013) 012024 doi:10.1088/1742-6596/426/1/012024

J/ψ reconstruction in the di-muon decay channel with the CBM experiment for p+Au collisions at 30 GeV laboratory beam energy

V Akishina[1,2], **I Kisel**[2,3,4]
[1] Goethe Frankfurt University, Frankfurt am Main, Germany
[2] Laboratory of Information Technologies, Joint Institute for Nuclear Research, Dubna, Russia
[3] Frankfurt Institute for Advanced Studies, Frankfurt am Main, Germany
[4] GSI Helmholtzzentrum für Schwerionenforschung GmbH, Darmstadt, Germany

E-mail: v.akishina@gsi.de

Abstract. The study of p+Au collisions at 30 GeV laboratory beam energy is considered as a part of the CBM research program and will be performed in the first phase of FAIR with a start version of the CBM detector at SIS-100. Results of J/ψ reconstruction feasibility studies in di-muon channel in proton-nuclear collisions at SIS 100 energies are presented. The simulated detector concept includes STS and MuCh detectors. The study was performed for a wide range of MuCh detector inefficiency. Also the influence of MuCh stations misalignment on the charmonium reconstruction efficiency was studied. The algorithm allows to reconstruct J/ψ with an efficiency of 31.3 % and the signal to the background ratio of 24.4.

1. Introduction

The Compressed Baryonic Matter (CBM) experiment at the Facility for Antiproton and Ion Research (FAIR) in Darmstadt is designed for the detection of bulk and rare probes, and will benefit from the high-intensity heavy-ion beams provided by the FAIR accelerators. The CBM experiment is aimed to measure rare diagnostic probes never observed before at FAIR energies, and thus has a unique opportunity for the study of the baryonic matter at extremely high densities and moderate temperatures in the laboratory. In this region the QCD (Quantum chromodynamics) phase diagram is only little explored [1].

Measurements at SIS100 (Schwerionensynchrotron) energies will focus on the investigation of the properties of resonance matter in the vicinity of the phase boundary, and, therefore, will provide important information on this transition region of the QCD phase diagram. The measurement of charmonium is one of the key goals of the CBM experiment, since J/ψ mesons are expected to be sensitive to the conditions inside the early fireball [2]. The main difficulty lies in the extremely low multiplicity near J/ψ production threshold at SIS100 energies.

This paper presents results of feasibility studies of charmonium reconstruction in the di-muon channel in p+Au collisions at 30 GeV laboratory beam energy with the standard CBM detector setup. The effect of inefficiencies and layer misalignment in the MuCh detector was studied

2. Experimental setup

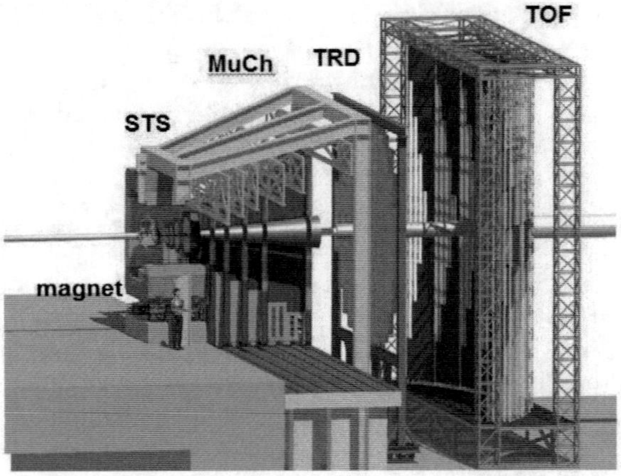

Figure 1. The setup of the Compressed Baryonic Matter experiment with the muon detection system

The CBM detector system layout is shown in Fig. 1 with the setup for muon identification. It consists of a Silicon Tracking System (STS) inside a superconducting dipole magnet as the primary tracking device. STS is followed by the muon system consisting of alternating detector and absorber layers.
The experimental challenge for muon measurements in heavy-ion collisions at FAIR energies is to identify low-momentum muons in an environment of high particle densities. The CBM concept is to track the particles through a hadron absorber system, and to perform a momentum-dependent muon identification. This concept is realized by segmenting the hadron absorber in several layers, and placing triplets of tracking detector planes in the gaps between the absorber layers. The absorber/detector system is placed downstream of STS which determines the particle momentum. The system has to be as compact as possible in order to reduce meson decays into muons. The actual design of the muon detector system consists of 6 hadron absorber layers (iron plates of thickness 20 cm, 20 cm, 20 cm, 30 cm, 35 cm, 100 cm) and 18 gaseous tracking chambers located in triplets behind each iron slab. The total absorber length in the current design amounts to 2.25 m of iron. The main difficulty for the muon chambers and for the track reconstruction algorithms is the huge particle density of up to 1 hit/cm^2 per event in the first detector layers after 20 cm of iron. Therefore, the detector development concentrates on the design of fast and highly granulated gaseous detectors based on GEM (Gas Electron Multiplier) technology. In order to provide continuous tracking, a TRD (Transition Radiation Detector) is installed after MuCh. A TOF (Time of Flight) wall completes the setup [3]

3. Feasibility studies

In order to study the feasibility of charmonium reconstruction in the di-muon channel simulations were performed for p+Au collision at 30 GeV laboratory beam energy within the simulation framework CBMROOT [4], which allows full event simulation and reconstruction. Decay muons from J/ψ were simulated with the PLUTO generator [5], where the multiplicity of charmonium production was taken from the predictions of the HSD model [6]. The background p+Au events were produced with UrQMD [7]. Signal embedded into background was transported through the standard CBM detectors using the transport code GEANT3 [8].
The simulated events were reconstructed using different software routines. Track reconstruction in the STS is performed using the L1 reconstruction package [9], which is based on a cellular automaton.

Journal of Physics: Conference Series **426** (2013) 012024 doi:10.1088/1742-6596/426/1/012024

Tracks reconstruction in the MuCh detector is done with the LIT-tracking package [10]. It uses tracks, which were found in the STS, as initial seeds. The LIT-tracking algorithm is based on the track following and the Kalman filter methods. In order to increase the statistics the background was simulated using the event mixing technique [11].

The J/ψ-meson reconstruction algorithm includes several steps: muon candidate tracks with more than 15 out of possible 18 hits in the MuCh detector are selected for charmonium search. The next step is to suppress the major source of background: muons from pions and kaons decays. In order to reject these soft muons a cut on the transverse momentum pt> 1 GeV/c of the reconstructed tracks was applied. On the next step the remaining positively charged tracks were combined with negatively charged tracks using the KFParticle package [12]; a good quality geometrical vertex ($\chi 2$ /ndf< 3 σ) was required to suppress combinatorial background.

An invariant mass spectra after the full analysis is shown in Fig. 2. The study indicates that charmonium can be easily identified above the combinatorial background, which is dominated by muons from weak pion decays. The signal yield is determined by fitting a Gaussian peak on top of a polynomial representing the combinatorial background. For the calculation of the signal-to-background ratio, the distribution was integrated in a $\pm 2\ \sigma$ window around the peak. The resulting value of S/B ratio is 24,4 and the J/ψ reconstruction efficiency is 31,3 %.

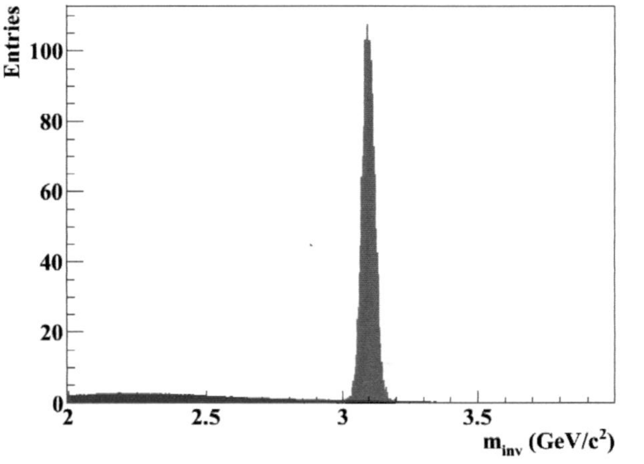

Figure 2. The reconstructed invariant mass spectrum for J/ψ candidates for p+Au collisions at 30 GeV laboratory beam energy

4. MuCh detector inefficiency and misalignment influence on the reconstruction results

For full exploitation of high resolution position sensitive detectors, it is crucial to determine the detector location and orientation with a precision better than their intrinsic resolution. It is a very demanding task to assemble large numbers of detector units in a large and complex detector system to this high precision. Since in real experiments detectors are always misaligned to some extent, we have studied the influence of MuCh detector layers misalignment on the charmonium reconstruction efficiency. Each station layer was shifted in both directions in the detector layer plane by a random value obtained with a uniform distribution with a width of misalignment in order to simulate the misalignment for the track reconstruction step. The muon track and charmonium reconstruction efficiency were studied for the misalignment up to 2 cm. The resulting dependencies are shown in Fig. 3. The muon track reconstruction efficiency decreases from 94,5 % in case of perfect alignment to 76,0 % in case of 1.0 cm misalignment. The J/ψ reconstruction efficiency decreases from 31,3 % to 21,5 %.

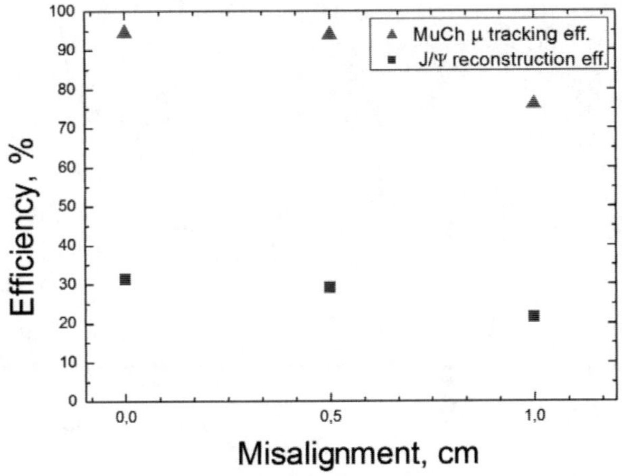

Figure 3. The charmonium reconstruction efficiency and muon track reconstruction efficiency dependencies on the misalignment of the MuCh station layers

Also the influence of the detector inefficiency on reconstruction results was studied. In order to implement it in the simulation, before tracking stage each fired digit was rejected with a certain probability. The results are shown in Fig. 4. The muon track reconstruction efficiency decreases from 94,5 % in case of perfect digits to 75,5 % in case of 90 % digit efficiency. The J/ψ reconstruction efficiency decreases from 31,3 % to 18,8 %.

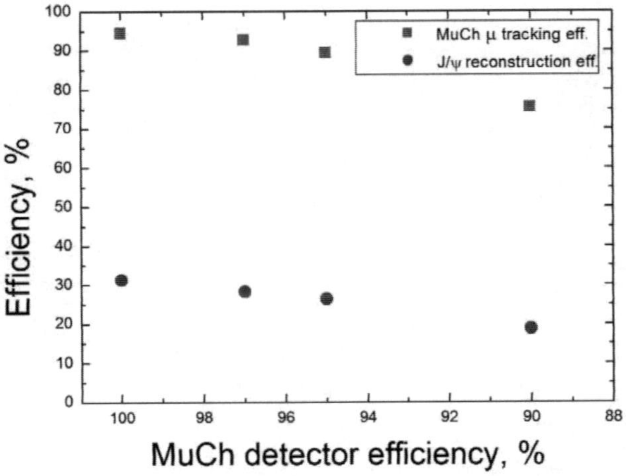

Figure 4. The charmonium reconstruction efficiency muon and track reconstruction efficiency dependencies on the MuCh detector digit efficiency

5. Conclusion

The feasibility study of J/ψ reconstruction in the di-muon channel for proton-nucleus collisions at 30 GeV laboratory beam energy shows that the CBM experimental setup allows to reconstruct charmonium with high efficiency (31,3 %). The influence of the MuCh detector inefficiency and station layer misalignment on the track and charmonium reconstruction has been studied. The study shows that tracking and particle reconstruction is stable up to 0,5 cm of misalignment and 95 % of MuCh detector efficiency.

References

[1] Friman B et al. *The CBM Physics Book*, Lect. Notes in Phys. **814** (2011) 838

[2] Matsui T and Satz H, *Phys. Lett.* **178** (1986) 416
[3] Hoehne C, *J. Phys .G: Nucl. Part. Phys.* **35** (2008) 104160
[4] Chattopadhyay S *J. Phys .G: Nucl. Part. Phys.* **35** (2008) 104027
[5] I. Frohlich et al., *PoS ACAT2007* (2007) 076
[6] Cassing W et al., *Nucl. Phys.* **A 691** (2001) 753
[7] Bass S A et al. , *Prog. Part. Nucl. Phys.* **41** (1998) 255
[8] Brun R et al., *GEANT User Guide*, CERN/DD/EE84-1 (1986)
[9] Kisel I, *Nucl. Instr. and Meth.* **A566** (2006) 85-88
[10] Lebedev A, Ososkov G, *CBM note QCD_CBM_SOFT-note-2008-002* (2008) 8
[11] Kopylov G, *Phys. Lett.* **B 50** (1974) 472
[12] Gorbunov S, Kisel I, *CBM-SOFT-note-2007-003* (2007)

The Ring Imaging Cherenkov detector for the CBM-Experiment

Christina Dritsa for the CBM collaboration
Justus-Liebig-Universität, Giessen
E-mail: Christina.A.Dritsa@exp2.physik.uni-giessen.de

Abstract. The mission of the future Compressed Baryonic Matter experiment, to be operated at the FAIR facility at GSI, Darmstadt, is the exploration of the properties of super-dense nuclear matter. In particular, the search for in-medium modifications of hadrons, the study of the transition from dense hadronic matter to quark-gluon matter, and the location of a critical endpoint in the phase diagram of strongly interacting matter are the principal physics goals of CBM. Detailed measurements of di-leptons and open charm probes are hoped to shed light to the existence of such effects.

The Ring Imaging Cherenkov detector of CBM aims at a clean and efficient electron identification. It is foreseen to have CO_2 as radiator gas and to be equipped with Hamamatsu multi-anode photomultiplier tubes. Herewith we present selected results of R&D studies and beam test measurements of the RICH prototype performed in October 2011 at the CERN/PS with negatively charged particles.

1. Introduction

The Compressed Baryonic Matter experiment (CBM) [1] will be one of the core experiments of the future FAIR facility at GSI, Darmstadt. CBM will study nuclear matter in the regime of highest net-baryon densities with various rare probes, among them low-mass vector mesons and charmonium. Low-mass vector mesons decay in leptons which have the particular property to leave the hot and dense fireball without undergoing any strong interactions. Therefore, their study is hoped to provide information on in-medium properties of vector mesons, as well as on charmonium production and propagation in hot and dense matter. Reconstructing these particles requires a set of high performance detectors. Among them is the Ring Imaging Cherenkov detector (RICH).

2. The RICH detector

The operation principle of RICH detectors is based on the fact that charged particles passing through a dielectric medium with a speed higher than the phase velocity of light in that medium emit electromagnetic radiation (cherenkov radiation). By measuring the angle of emission of the cherenkov radiation the identification of charged particles is achieved. The CBM-RICH is designed to provide electron identification for the CBM experiment in the momentum range of electrons from low-mass vector-meson decays, i.e. for $p \leq 8$ GeV/c. The detector will cover the full azimuthal angle and polar angles from 2.5° to 25°. For a sufficiently pure electron sample, a pion suppression factor in the order of 500 - 1000 is required for the RICH alone. The electron identification in the RICH is complemented by several layers of TRDs. A combined pion suppression of 10^4 will be achieved. Main challenges in the design of the RICH detector are set by the fact that it will be operated in a high track density environment: up to 1000 charged primary particles per event at interaction rates up to 10 MHz are expected in the detector acceptance for central Au+Au collisions if operated at the highest energies.

In order to cope with the high interaction rates, a vertically separated RICH detector is planned with CO_2 as radiator gas and MAPMTs with fast, self-triggered readout electronics as photodetector.

Given the high particle fluxes expected, a high electron-to-pion separation is needed. Assuming that pions can be separated from electrons up to 90% of the maximum Cherenkov opening angle, the choice of CO_2 is motivated by the need to separate electrons and pions with momenta up to $p \cong 10$ GeV/c. In CO_2 the transmittance of photons is limited by absorption to wavelengths $\lambda \geq 175$ nm.

Spherical glass mirrors with an Al+MgF_2 coating will be used to reflect the Cherenkov light cones on the 2.4 m^2 photodetector plane. Simulation studies have shown that the material budget of the mirrors is not crucial, therefore a thickness of 6 mm is tolerable. Several types of mirrors have been tested in the laboratory in terms of surface homogeneity and reflectivity (see Section 2.1). The photodetector plane is foreseen to be equipped with Hamamatsu H8500 multi-anode photomultiplier tubes (55000 channels). These are delivered from the factory with two different windows: borosilicate or UV enhanced window. The former exhibits an absorption edge at $\lambda \approx 270$ nm, the latter at $\lambda \approx 210$ nm. These limits are somewhat above the absorption limit for CO_2 (175 nm). Therefore, the option to use wavelength-shifting (WLS) films on top of the MAPMT windows in order to match the CO_2 absorption limit is under investigation. The usage of WLS films is based on their property to absorb deep UV photons and to re-emit them with longer wavelengths, where the transparency of the MAPMT window and the quantum efficiency of the photocathode is higher. Typical materials used for WLS are p-terphenyl (PT) and tetraphenyl-butadiene (TPB).

In order to cope with the high interaction rates (up to 10 MHz) required for the J/Ψ measurements in CBM, a fast, self-triggered readout electronics is foreseen. The data will be delivered to a large PC farm where they will be processed to allow for partial on-line event reconstruction.

2.1. Mirror tests

The Cherenkov photons are emitted in the form of light cones and a large fraction of the wavelengths lies in the UV region. These cones are focused with the use of spherical glass mirrors, forming rings on the photodetector plane. Detailed measurements were performed in cooperation with CERN for mirrors from different manufacturers: (i) FLABEG GmbH, Furth im Wald, Germany, (ii) Compas, Czech Republic, and (iii) JLO-Olomouc, Czech Republic. All mirrors are covered with an Al layer for best reflection of the UV photons and a MgF_2 layer to ensure protection from generation of aluminum oxidation layer which absorbs UV-photons. Table 1 summarizes all the important parameters of the tested mirrors: the radius of curvature r_0, the thickness of the glass substrate d, the area of the quadratic mirrors A, the thickness of the Al-and MgF_2 layer, $d(\mathrm{Al})$ and $d(\mathrm{MgF_2})$ (if measured), the results of D_0-measurements[1], as well as selected reflectivity data. Figure 1 summarizes the results on reflectivity measurements as a function of wavelength for all mirrors tested. The Flabeg mirrors (red rectangles) show the best reflectivity (≥85%) for wavelengths above 250 nm. However, this performance drops significantly below 80% for smaller wavelengths. The best compromise, i.e. very good reflectivity

[1] D_0 is defined as the diameter of the circle containing 95% of the total light intensity reflected by the mirrors when being fully illuminated.

Table 1. Parameters of the mirrors tested in the laboratory. Not all quantities are measured (indicated by n.a.). For details see text.

Producer	r_0 [mm]	d [mm]	A [cm^2]	d(Al) [nm]	$d(MgF_2)$ [nm]	D_0 [mm]	R(200 nm) [%]	R(300 nm) [%]
Flabeg	3200	6	40×40	55	120	diffuse	78	90
Compas	3000	6	20×20	n.a.	n.a.	≤ 3	63	76
JLO	3000	6	40×40	110	110	≤ 2	82	85

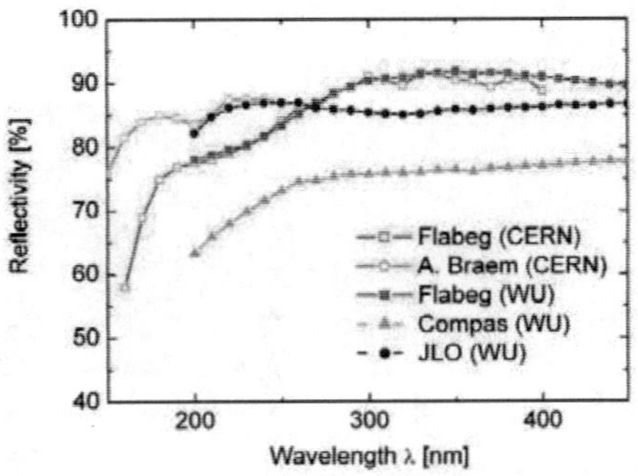

Figure 1. Reflectivity measurements, performed at CERN or at University of Wuppertal (WU), as a function of wavelength for various mirrors. Additionally, the reflectivity of a mirror coated by Andre Braem (CERN) is shown for comparison.

for all wavelengths down to $\sim$ 200 nm, is given by the JLO-Olomouc mirror (blue circles) which also shows the best results in terms of D_0 (see Table 1). Thanks to these performances, this type of mirror is chosen to be used for the prototype measurements in beam.

2.2. The RICH-prototype

A CBM-RICH prototype has been built and tested in beam in 2011 at the CERN/PS, line T9 [2]. Figure 2 on the left, shows the technical drawing of the prototype and a photo can be seen on the right of the same figure. The radiator length is identical as for the final CBM-RICH, i.e. 1.7 m. The individual modules are large enough to allow for first studies of system integration issues. The Cherenkov light cone is projected on the photodetector by four high reflectivity UV spherical mirror tiles of 40×40 cm^2 with radius R=3 m. Each mirror is mounted on three points with holders including an actuator. This mounting scheme minimizes distortions and allows for mirror rotation and alignment. The complete mirror array is mounted on two frames which allow orienting the Cherenkov light on different areas of the photodector. The latter consists of an array of 4×4 multianode photomultiplier tubes (MAPMTs). The gas system is computer controlled and aims at a clean and dry CO_2 flow. The evaluation of the prototype in this beam test mainly focuses in three topics: a) study of system integration issues (mirror, support, photodetector, electronics), b) test the ring finding and ring reconstruction algorithms and, c) study the detector performance under realistic conditions, i.e. in terms of number of photons per ring, noise levels, electron-to-pion separation, sensitivity to gas impurities, sensitivity to mirror distortions and misalignment.

Journal of Physics: Conference Series **426** (2013) 012025 doi:10.1088/1742-6596/426/1/012025

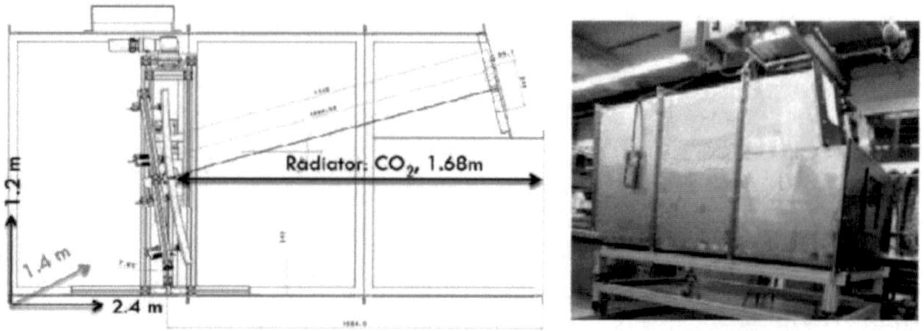

Figure 2. The technical drawing with dimensions (left) and a photo of the RICH prototype (right) are shown.

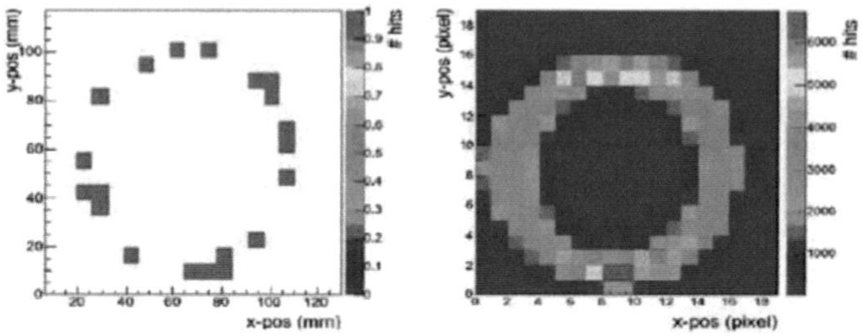

Figure 3. Rings from electrons of 6 GeV/c momentum. Left panel: Ring from single event containing about 20 hits. Right panel: Integrated rings from several thousand events.

3. Selected beam test results

A figure of merit for all RICH detectors is the ring hit multiplicity, i.e. the number of photons (hits) per ring. The latter was found between 19 and 22 photoelectrons without any correction for cross-talk. If correction for cross talk is applied, then the hit multiplicity varies between 16.7 and 19.4 photoelectrons, depending on the quantum efficiency of the MAPMTs. The WLS coated MAPMTs show enhanced hit multiplicities when compared to the non-coated ones. However, a detailed evaluation of this effect needs to take into account the efficiency of every MAPMT [3].

Figure 3 shows single and integrated rings from electrons of 6 GeV/c momentum (left to right) as obtained from the beam test. One also observes a high contrast between background noise and signal from electron rings. A very efficient shielding allowed reaching very low noise rates of the order of 10 Hz per channel (below 10 kHz for the whole photodetector).

The ring finding and fitting algorithms allowed fitting rings with ellipse or circle shapes. A 100% reconstruction efficiency was observed for rings with more than 8 hits (see Figure 4). Moreover, once a ring is found, it is succesfully fitted providing thus the ring parameters such as the number of hits per ring, the radius (when fit as circle) and the major and minor half-axis (when fit as an ellipse). In order to have an electron sample as clean as possible for the CBM measurements, a good electron-to-pion separation for momenta up to 8 GeV/c is one of the major requirements of the CBM-RICH. It was found that for particles with momenta of 8 GeV/c the electron-to-pion separation[2] is above $7\sigma_\pi$ which is an excellent performance. It should be stressed here that this measured performance is achieved under beam test conditions and does

[2] The electron-to-pion separation can be expressed in terms of standard deviations (σ_π) of the gauss fit of the distribution of pion ring radii.

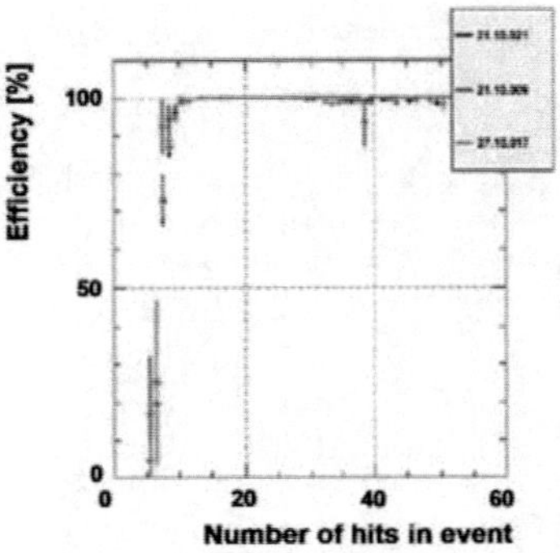

Figure 4. Ring finding efficiency as a function of the number of hits in the event. The efficiency is 100% for rings with more than 8 hits. The different colors correspond to data taken from different runs.

not include effects such as the track reconstruction and the track-ring matching efficiencies. Figure 5 shows the distribution of ring radii for negative particles of 8 GeV/c momentum. One sees a clear separation between pions and electrons based on their ring radius. These measured results match very well the corresponding simulation results for the prototype.

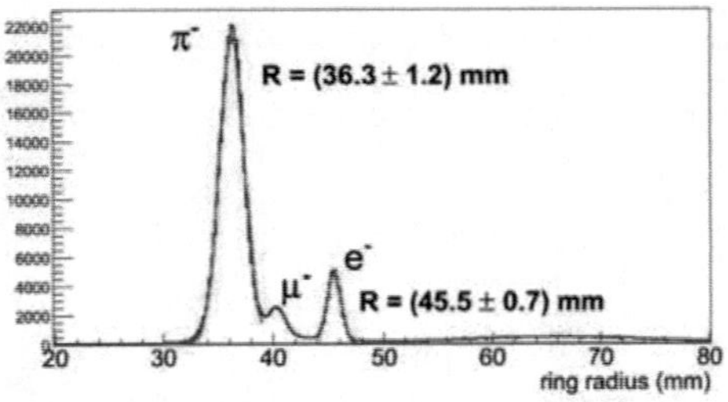

Figure 5. Distribution (blue line) of ring radii for negative pions, muons and electrons of 8 GeV/c momentum. A gauss fit is shown with the red line.

4. Summary

A RICH detector for CBM at FAIR is being developed for efficient and clean electron identification. The different R&D activities include the development of self-triggered readout electronics and the re-evaluation of wavelength-shifting films for enhanced quantum efficiency in the UV-range. For the envisaged glass mirrors, prototypes from various industrial providers have being evaluated in terms of mirror reflectivity and surface homogeneity tests. A real size RICH prototype has been built and tested at the CERN/PS in October 2011. This test revealed very promising results on the number of photoelectrons per ring, the effect of WLS coating films, the ring fitting and finding efficiency as well as the electron-to-pion separation capability of the RICH.

5. Acknowledgments

The authors would like to thank CERN for support in the WLS and mirror measurements, especially Dr. Christian Joram and Dr. Andre Braem. This work was supported by the Hessian

Journal of Physics: Conference Series **426** (2013) 012025 doi:10.1088/1742-6596/426/1/012025

LOEWE initiative through the Helmholtz International Center for FAIR (HIC for FAIR).

References

[1] C. Höhne, F. Rami, and P. Staszel, Nucl. Phys. News 310 16, 19 (2006).
[2] C. Bergmann et al., "Common CBM beam test of the RICH, TRD and TOF subsystems at the CERN PS T9 beamline", GSI Scientific Report 2011
[3] J.Kopfer, "In-beam test of a real-size CBM-RICH prototype at CERN PS" , GSI Scientific Report 2011

Journal of Physics: Conference Series **426** (2013) 012026 doi:10.1088/1742-6596/426/1/012026

Understanding the HADES dilepton invariant mass spectra with UrQMD simulations

C Behnke[1],Tetyana Galatyuk [2], Joachim Stroth [1]

[1]Goethe Universität Frankfurt, Max-von-Laue-Str. 1,60438 Frankfurt am Main
[2]Technische Universität Darmstadt, Schlossgartenstrasse 2 64289 Darmstadt

E-mail: C.Behnke@gsi.de

Abstract. Photons and lepton pairs emerging from decays of virtual photons are the most promising probes of dense hadronic matter. In the energy domain of 1 - 2 GeV per nucleon, HADES has measured electron pairs in C+C, Ar+KCl, p+p, d+p and p+Nb collisions. An experimentally constrained N+N reference spectrum was established. Moreover, for the first time at this energy the inclusive production cross sections for light vector mesons were extracted. This result allows putting tight constraints on vector meson production in heavy-ion collisions at beam energies of a few GeV per nucleon. In this contribution I will compare the HADES data to predictions from UrQMD microscopic transport model calculations and introduce an approach which will allow to separate in a transparent way the generation of the event background from the emission pattern of a physics observable under consideration.

1. Introduction

In order to transfer nuclear matter into the deconfined phase, one has to heat the system up or increase the baryon chemical potential. Heavy ion collisions are the only way to reach these conditions in the laboratory. Photons (γ) and leptons are penetrating the dense phase without strong interactions. By the decay of virtual photons (γ^*) lepton pairs are emitted during the whole evolution of the heavy ion collision. In the fist chance collision stage, at beam energies of a few Gev/u, the main source of lepton pairs is NN Bremsstrahlung. In the freeze out stage long living sources, which directly decay into lepton pairs or through Dalitz decays, are the dominant contributions. These are mainly π^0, η and ω mesons. The short lived sources, i.e. ρ mesons and baryonic resonances like Δ_{1232} and N^*, are the best tool to access the dense stage of the heavy ion collision, because they are produced and decaying there.

1.1. Heavy ion collisions and in medium modifications

It is predicted that the hadrons, which are embedded inside nuclei, change their properties [1]. In heavy ion collisions nucleons are excited into baryonic resonance states (Δ, N*). They decay by emitting mesons, which couple to a γ^*. The presence of baryons influences the vector meson spectral function in the medium. The ρ meson couples strongly to the 2π channel. Thus it can be seen as a broad $\pi + \pi$ resonance. Therefore, modifications of the ρ meson properties in hadronic matter can be linked to medium-dependent π propagation properties. Analog to the vacuum polarization of a photon [2] the ρ meson can create virtual π^+ and π^- pairs while propagating in the vacuum. With the presence of a medium the ρ meson self-energy becomes much more complex. Significant modifications of the ρ meson width are expected, since π couple strongly to δ-hole states in the medium. The direct coupling of the ρ meson to the baryonic resonances is also possible. To summarize, additional contributions to the ρ meson self-energy change the width of the ρ. Therefore the broadening of the ρ meson is a hint to the existence of the medium. A lepton pair enhancement above the trivial hadronic cocktail in the invariant mass region below the ρ meson pole mass can be interpreted as a broadening of the ρ meson due to the in-medium changes of its spectral function [3].

1.2. The HADES spectrometer

To study e^+e^- pairs in cold nuclear matter in heavy ion collisions as well as in elementary collisions, the High Acceptance DiElectron Spectrometer (HADES) is installed at the SIS 18 (GSI Helmholtzzentrum für Schwerionenforschung, Darmstadt). One of the goals was to validate the DLS pair excess and to solve the "DLS puzzle". It is build of 6 identical sectors, that covers a full azimuthal angle and polar angles from 18° to 85°. To study in a systematic way the origin of the e^+e^- pair yield and its dependence on the beam energy and system size a large amount of simulated data is needed. We have simulated C+C and Ar+KCl at beam energies already measured by HADES. In order to complete the systematics, HADES continued the investigations with larger collision systems and at the highest possible beam energies achievable at SIS18, Au+Au at 1.25 GeV/u in April 2012. This system have been simulated as well.

2. Dilepton pairs from heavy ion collisions simulated with UrQMD

Experimental results (C+C and Ar+KCl) will be compared to UrQMD transport model calculations. One would like to learn how different collision energies and different system sizes influence the distribution of invariant e^+e^- masses. The excess of electron pairs in C+C and Ar+KCl collisions was investigated by the HADES collaboration [4]. Figure 1 (a) shows the dependence of the integrated excess pair yield above the η in the invariant mass range $0.15 < M_{e^+e^-} < 0.55$ on the beam energy and on the system size. The results obtained from DLS and HADES are shown by red and black triangles (connected by dotted lines, representing arbitrary down-scaled pion excitation functions). The dilepton data is shown together with the π^0 and η multiplicities obtained with the TAPS photon calorimeter [5, 6]. Figure 1 (b) shows the excess systematics in the UrQMD simulations. The excess pairs yield was defined as a sum of all contributions (Δ_{1232}, ρ) besides the η contribution in the mass range $0.15 < M_{e^+e^-} < 0.55$ GeV/c^2. Please note, that the π^0 multiplicity obtained within the UrQMD model is consistent with the values measured by the TAPS collaboration. The HADES collaboration shows that the HADES and DLS excess pair yield from C+C collisions follow a remarkably similar trend with increasing beam energy like the mean π^0 multiplicity measured by TAPS. This is also the case for the UrQMD simulations. However, this is not longer true when going to the Ar+KCl system. At a given bombarding energy the excess pair yield scales with the number of participating nuclei stronger than the π^0 production. The multiplicity of the excess pair measured in C+C and Ar+KCl differs by a factor of 5, while only by factor of 2.7 for the π^0 multiplicities (see Fig. 1, a). A similar scaling is found in UrQMD, however the dependence of the excess yield on the system size is found to be not as strong as measured by HADES, i.e. just a factor of 3 (see Fig. 1, b). This might be interpreted as a fingerprint of in-medium effects related to multi-step collisions, with baryonic resonances playing an important role.

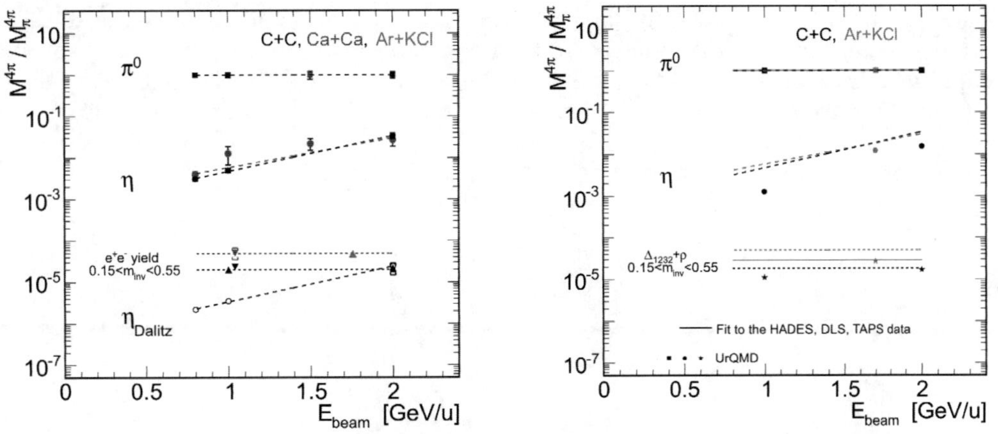

(a) Experimental results (b) Simulation results

Figure 1. Multiplicity of the pair yield above the η Dalitz in the mass range $0.15 < M_{e^+e^-} < 0.55$ GeV/c^2 as observed by DLS (open triangles) and HADES (full triangles) as function of beam energy and system size. More explanation see text.

2.1. Scaling of the dilepton pair yield with the system size

To study the scaling of the dilepton pair yield with the system size C+C, Ar+KCl, Ag+Ag, and Au+Au collisions at the same beam energy, i.e. 1.25 GeV/u have been simulated. Figure 2 shows the dependence of meson and baryon multiplicities as a function of the system size.

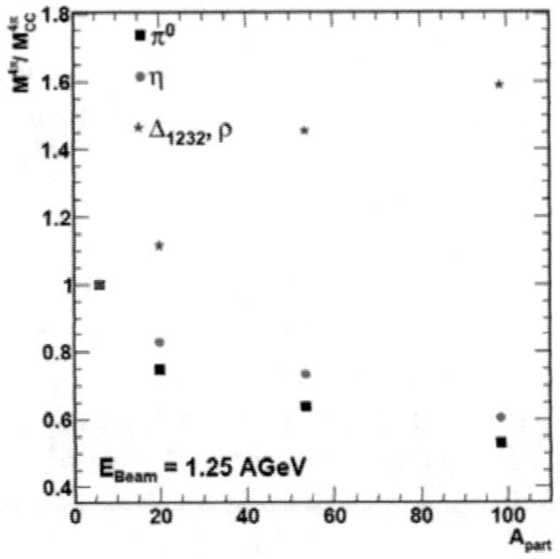

Figure 2. Dilepton mutiplicities extracted from C+C, Ar+KCl, Ag+Ag and Au+Au (per A_{part}) collisions at the same beam energy simulated with UrQMD

The multiplicities of e^+e^- pairs from π^0, η, Δ_{1232} and ρ decays were extracted for all simulated systems. Further the π^0, η, Δ_{1232} and ρ multiplicities are normalized to the corresponding one extracted from C+C collisions and normalised to A_{part}.
The multiplicities of the π^0 and η are found to be reduced when A_{part} increases. This might be explained by the time-integrated cross section for the Δ_{1232} reabsorption process ($N\Delta \to NN$), which is larger in Au+Au collisions compared to C+C collisions. Consequently, the total number of π^0 observed in the final state is up to 50% less in Au+Au collisions. The η mesons multiplicities (red dots) are decreasing with the system size by nearly 40%. In contrast, the excess pair yield, which is represented by the Δ_{1232} and the ρ in the mass region from 0.15 to 0.55 GeV/c^2, is rising with system size. From C+C the Au+Au excess is enhanced by around 60%.

3. Modeling Au+Au collisions

The ρ and Δ_{1232} multiplicities extracted from UrQMD simulations in the M_{ee} region from 0.15 - 0.55 GeV/c^2 shows a weak dependence on the beam energy but a stronger scaling with the system size. Therefore the HADES collaboration measured Au+Au collisions at 1.23 GeV/u in

Journal of Physics: Conference Series **426** (2013) 012026 doi:10.1088/1742-6596/426/1/012026

April 2012. For this system the time and the emission density evolution of the different particles is shown in Fig. 3. The figure depicts at which time and from which density of the collision the dileptons are emitted. The dileptons from π^0 meson decays (Fig. 3, (a)) are emitted at low densities and mainly at late stages of the collision. The Δ_{1232} decay (Fig. 3, (b)) is distributed over the whole time evolution and prominent at all density regions. But dileptons from this decay originate from the dense phase (ρ/ρ_0 >1.5) and a certain time region from 7-14 fm. The time evolution of the ρ meson is similar to that of the Δ_{1232}[7].

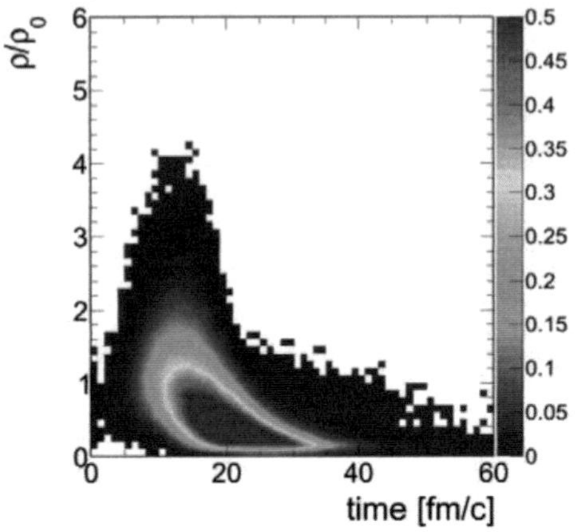

(a) π^0

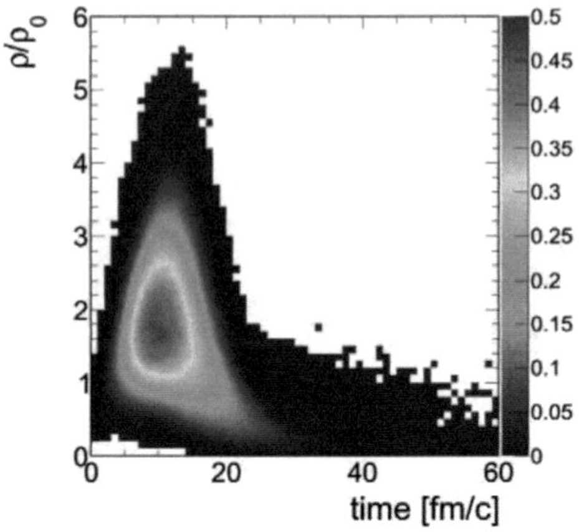

(b) Δ_{1232}

Figure 3. Emission density versus time from dilepton sources from Au+Au 1.25 GeV/u, simulated with UrQMD.

The evolution for a given meson is divided into 9 equal time steps. For each time step, the m_t spectra of the particles are analyzed with a χ^2 minimization method [7], and for each time step a density profile is extracted. This time dependence of effective slope parameters and emission density is shown in Fig. 4. For each hadron the temperature (black stars) and the density of the maximum decay probability of the dilepton sources (red squares) for the given time step is shown. Dileptons from π^0 decay (Fig. 4,(a)) are emitted at temperatures from 35 to 45 MeV for T_1 and 70 to 80 MeV for T_2. Both temperatures show a slight rise to later stages of the collision. Note that T_1 is dominated by resonance decays, so only for T_2 the expression "temperature" is fully correct. However, the late stages suffer from statistics especially for the higher temperature (T_2). The decreasing of temperature T_2 could be a hint to freeze out effects. The dileptons from π^0 decays are emitted from dilute stages. The maximum decay rate is reached at 1.2 times normal nuclear matter density in early stages of the collision. Δ_{1232} Dalitz decays inside the dense phase, with ρ/ρ_0 up to 1.9 times normal nuclear matter density (Fig. 4,(b)). Temperatures from 48 to 78 MeV, with small errors are extracted [7].

4. Summary

In Au+Au collisions at 1.25 GeV/u a large e^+e^- pair yield comes from the emission density region which is a factor of 2 larger compared to normal nuclear matter density. Since UrQMD uses only vacuum spectral functions the predictions for e^+e^- radiation from high density stages of the heavy ion collision could be more selected. In the SPS energy regime a thermal model like from [8] is able to describe the measured NA60 data [9], which also show clear medium effects. A description of heavy ion collisions with a thermal model seems to be a good method to model

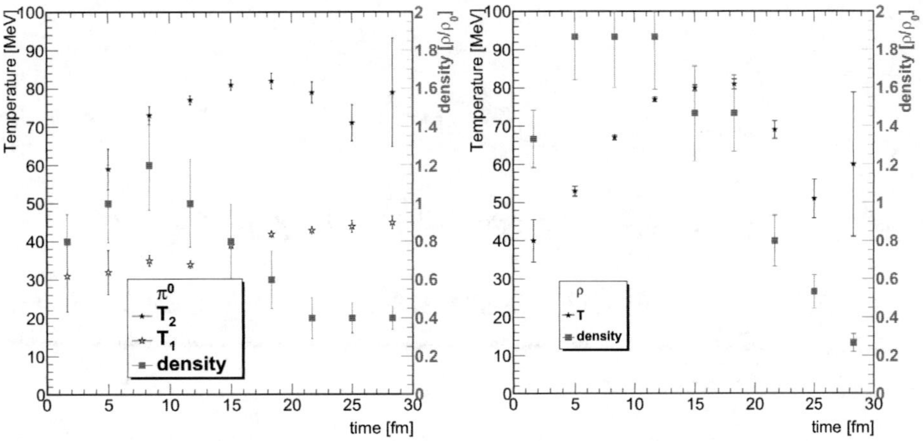

(a) π^0 (b) Δ_{1232}

Figure 4. Emission temperature and emission density versus time from dilepton sources from Au+Au 1.25 GeV/u.

e^+e^- production from the dense stage of the heavy ion collision. The extracted temperatures and densities can be used as an input to a thermal model. The combination of UrQMD and an thermal model is expected to give a good approach for describing dilepton radiation from the dense stage of a heavy ion collision at a wide energy range.

Acknowledgments

This work has been supported by BMBF (06 FY 9100 I), HIC for FAIR, EMMI, GSI, HGS-Hire and H-QM .

References

[1] J. Wambach, Meson properties in dense hadronic matter, Nuclear Physics A 699.

[2] I. Levine, et al., Measurement of the electromagnetic coupling at large momentum transfer., Phys. Rev. Lett. 78 (3) (1997) 424–427.

[3] E. L. Bratkovskaya, C. M. Ko, Low-mass dileptons and dropping rho meson mass, Phys. Lett. B445 (1999) 265–270.

[4] G. Agakishiev, et al., Dielectron production in Ar+KCl collisions at 1.76A GeV, Phys.Rev. C84 (2011) 014902.

[5] R. Holzmann, Production of p and light mesons in carbon-induced relativistic heavy-ion collisions, Zeitschrift für Physik A Hadrons and Nuclei 359 65–73.

[6] R. Averbeck, R. Holzmann, V. Metag, R. S. Simon, Neutral pions and eta mesons as probes of the hadronic fireball in nucleus-nucleus collisions around 1 AGeV, Phys. Rev. C 67 (2003) 024903.

[7] C. Behnke, Development of an event generator for dilepton production in heavy ion collisions in the SIS energy regime (unpublished master thesis) (2012).

[8] R. Rapp, J. Wambach, H. van Hees, The Chiral Restoration Transition of QCD and Low Mass Dileptons, Nucl. Phys.arXiv:0901.3289.

[9] S. Damjanovic, et al., NA60 results on the rho spectral function in In-In collisions, Nucl.Phys. A783 327–334.

Journal of Physics: Conference Series **426** (2013) 012027 doi:10.1088/1742-6596/426/1/012027

Recent results from NA61/SHINE and NA49

Anar Rustamov for the NA61/SHINE and NA49 Collaborations
Institut für Kernphysik, Goethe-Universität Frankfurt, Frankfurt am Main, Germany
E-mail: a.rustamov@cern.ch

Abstract. Preliminary results on particle spectra and fluctuations in inelastic p+p interactions measured by NA61/SHINE at the CERN SPS are presented and its future ion program is discussed. We also show results on the excitation functions of event-by-event particle ratio fluctuations in central Pb+Pb collisions from NA49 using a novel approach. We found a dependence of the results on phase space coverage which appears to explain the reported difference between measurements of NA49 and those of STAR in central Au+Au collisions.

1. Introduction

As the fundamental theory of strong interactions, Quantum Chromodynamics (QCD), is asymptotically free, the created matter at high temperature and/or high density may be dominated by the state of quasi-free quarks and gluons referred to as the Quark-Gluon Plasma (QGP). By colliding heavy-ions at high energies one hopes to heat and/or compress the matter to energy densities at which the production of the QGP begins. Indeed, lattice QCD calculations predict this new phase at high temperatures. The motivation of investigating relativistic heavy ion collisions is the experimental study of hadronic matter under extreme conditions. A wealth of ideas in the past few decades have been proposed to explore the phase structure of strongly interacting matter. Among other possible probes event-by-event fluctuations of different observables may be sensitive to the transitions between hadronic and partonic phases. In particular, the location of the critical point may be signalled by a characteristic pattern in the energy and system size dependence of the measured fluctuation signals.

The SPS Heavy Ion and Neutrino Experiment (NA61/SHINE) is operating since 2007 in the North Area of the CERN Super Proton Synchrotron (SPS). NA61/SHINE uses the apparatus of the NA49 [1] experiment with numerous upgrades [2]. Its large acceptance and excellent particle identification capabilities [1, 2] make possible a rich physics program, in particular the ion program which is the main subject of this contribution. The ultimate goal is to explore the phase diagram of strongly interacting matter using collisions of different systems from p+p, through p+A, to A+A collisions at projectile momenta of 13A, 20A, 30A, 40A, 80A and 158A GeV/c. The energy scan of p+p reactions was completed in 2009-2011 and preliminary results will be presented. Data on Be+Be collisions at three top beam momenta and at 13A GeV/c were recorded in 2011 and 2012. In 2013 Be+Be data at the remaining energies will be collected. The energy scans of p+Pb, Ar+Ca and Xe+La collisions are foreseen until 2016. In addition a beam energy scan of Pb+Pb collisions is planned. The explored energy range probes an important region in the phase diagram of strongly interacting matter. Indeed, the NA49 collaboration reported signals for the onset of deconfinement encoded in non-monotonic behavior of excitation functions of several hadronic observables [3]. These observations have been recently confirmed by the RHIC and LHC data [4]. Moreover, the critical point at the location predicted by several theory groups [5] can be probed at SPS energies.

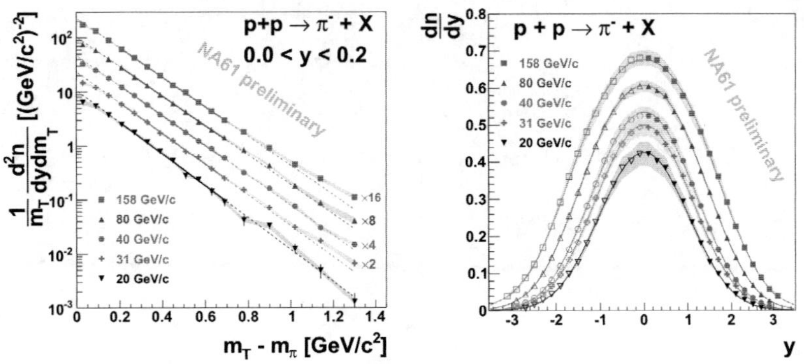

Figure 1. (Color online) Left: Transverse-mass spectra at mid-rapidity for negatively charged pions fitted with an exponential function in the $0.2 < m_T - m_\pi < 0.7$ GeV/c^2 interval. Right: Rapidity spectra of negatively charged pions fitted with the sum of two Gaussian functions. Results refer to inelastic p+p interactions at beam momenta of 20-158 GeV/c.

2. Inclusive particle spectra in p+p interactions

Results from the energy scan of p+p interactions will serve as an important reference for interpretation of the p+A and A+A data at corresponding beam energies collected with the same apparatus. Figure 1 (left) presents the transverse mass (m_T) spectra at mid-rapidity for negatively charged pions in inelastic p+p collisions at 20, 31, 40, 80 and 158 GeV/c [6]. The results are corrected for reconstruction inefficiencies, acceptance, and feed-down using Monte Carlo simulations. The m_T spectra are extrapolated by fitting the distributions with an exponential function. Figure 1 (right) depicts the m_T integrated rapidity spectra. The latter allow to obtain the mean number of negatively charged pions in full phase space from which one can calculate the total number of pions using isospin coefficients. The NA61/SHINE pion multiplicities are in good agreement with the world data [6].

3. Event by Event fluctuations in Pb+Pb and p+p collisions.

A motivation for our work presented in this section is to shed light on reported differences between the results from STAR and NA49 on fluctuations of identified hadrons [7, 8, 9]. The studied measure of dynamical particle ratio fluctuations $\nu_{dyn}[A, B]$ is defined as [10]:

$$\nu_{dyn}[A, B] = \frac{\langle A(A-1)\rangle}{\langle A\rangle^2} + \frac{\langle B(B-1)\rangle}{\langle B\rangle^2} - 2\frac{\langle AB\rangle}{\langle A\rangle\langle B\rangle}, \tag{1}$$

where A and B stand for multiplicities of different particle species.

As is evident from Eq.(1) all second moments of the multiplicity distributions of particles A and B have to be known in order to construct ν_{dyn}[A,B]. The standard approach is to identify these particles event-by-event. However, this approach is hampered by incomplete particle identification, which is taken care of by either selecting suitable phase space regions or by applying a fitting procedure event-by-event. The latter typically introduces artificial correlations which are usually corrected for by the event mixing technique.

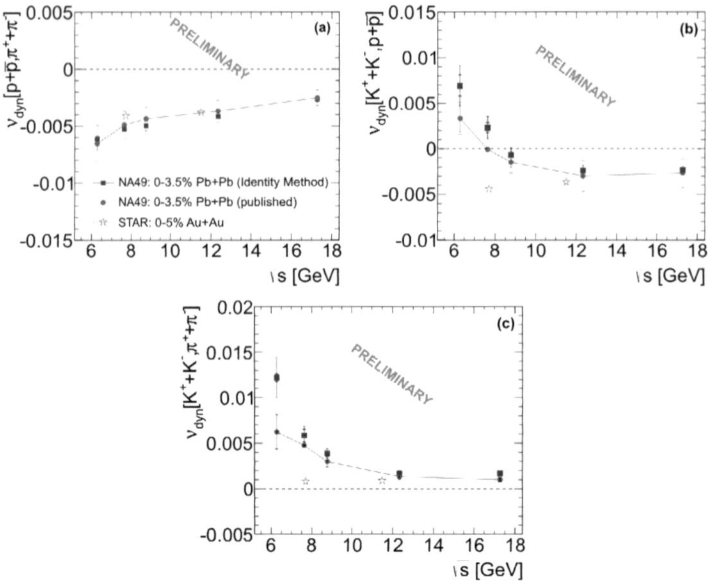

Figure 2. (Color Online) Energy dependence of (a) $\nu_{dyn}[p+\bar{p}, \pi^{+}+\pi^{-}]$, (b) $\nu_{dyn}[K^{+}+K^{-}, p+\bar{p}]$ and (c) $\nu_{dyn}[K^{+}+K^{-}, \pi^{+}+\pi^{-}]$. Results from the Identity Method for central Pb+Pb data of NA49 are shown by blue squares. Published NA49 results are indicated by circles. Stars represent results of the STAR collaboration for central Au+Au collisions.

In this context a novel approach, called *Identity Method* [11, 12] was developed for fluctuation analysis. It uses a probabilistic approach which avoids the event-by-event fitting as well as has a rigorous mathematical derivation. Furthermore, there is no need for corrections based on event mixing anymore. Results using the *Identity Method* are presented in Fig. 2 for central Pb+Pb data of NA49 and compared to published NA49 and STAR results from its Beam Energy Scan (BES) program [9]. The energy dependence of $\nu_{dyn}[p+\bar{p}, \pi^{+}+\pi^{-}]$ is consistent with the NA49 publication. Moreover the increasing trend of the excitation functions of $\nu_{dyn}[K^{+}+K^{-}, p+\bar{p}]$ and $\nu_{dyn}[K^{+}+K^{-}, \pi^{+}+\pi^{-}]$ is confirmed by this analysis. It was found in Ref. [13] that ν_{dyn} exhibits an intrinsic dependence on the multiplicities of accepted particles. Published NA49 results are essentially performed at forward rapidities, whereas the STAR acceptance covers the mid-rapidity region without the low p_T range. As an example we illustrate in Fig. 3 the phase space coverage of pions, kaons and protons at different projectile energies in the acceptance of NA49. In the same figure the acceptance of the STAR apparatus at corresponding center-of-mass energies are presented with colored lines. We, therefore, studied the acceptance dependence of ν_{dyn} by performing the analysis in different acceptance bins going from forward rapidity to mid-rapidity. Technically different acceptance bins were selected by applying several upper momentum cuts for each reconstructed track and calculating the corresponding maximum rapidity of protons at p_T=0. Thereafter we will call this quantity a proton rapidity cut. The dependence of obtained results on the proton rapidity cut is illustrated in Fig. 4. At $20A$ and $30A$ GeV/c $\nu_{dyn}[K^{+}+K^{-}, p+\bar{p}]$ and $\nu_{dyn}[K^{+}+K^{-}, \pi^{+}+\pi^{-}]$ show a strong acceptance dependence, and eventually hit the STAR point in a particular acceptance bin. Interestingly the acceptance dependence weakens above $30A$ GeV/c where no difference was observed with STAR. It is also remarkable that $\nu_{dyn}[p+\bar{p}, \pi^{+}+\pi^{-}]$ shows little acceptance dependence. This detailed study of fluctuation results in different acceptance bins appears to explain the difference between the

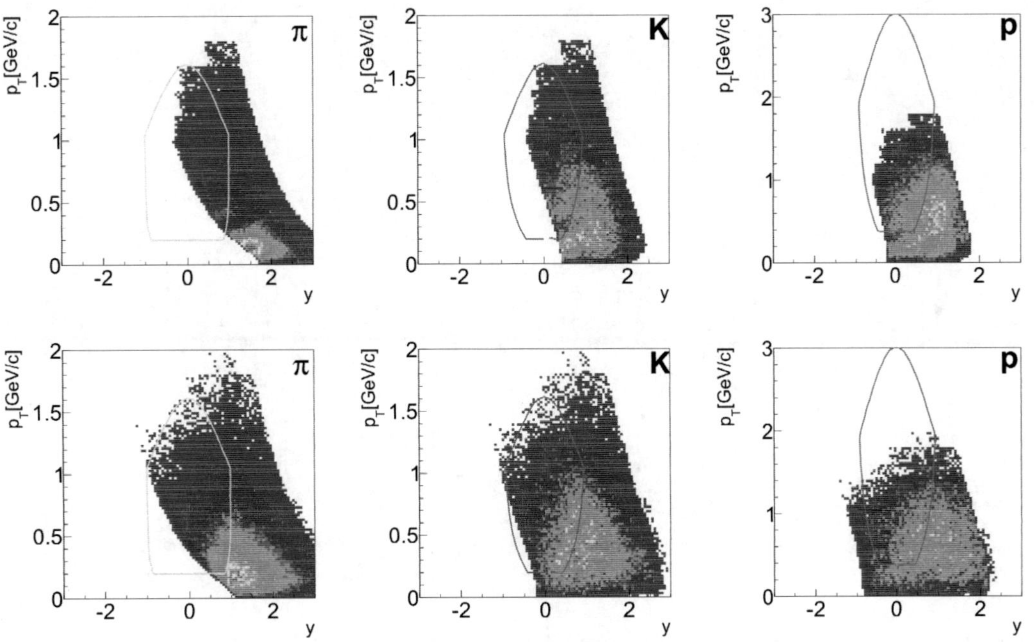

Figure 3. (Color Online) The phase space coverage of pions, kaons and protons in the acceptance of the NA49 experiment for the Pb+Pb collisions at 30A GeV/c (upper panel) and 158A GeV/c (lower panel) used for the event-by-event fluctuations presented in Fig. 2. The acceptance of the STAR apparatus at corresponding center-of-mass energies are depicted with colored lines.

STAR BES and NA49 measurements.

As mentioned in the introduction, the obtained fluctuation signals from A+A collisions should be compared systematically to the baseline obtained from analysis of p+p interactions at corresponding energies. It is however important to properly take into account trivial differences between A+A and p+p collisions e.g., in the size of the colliding systems. In thermodynamics the quantities which are proportional to the volume of the system are called extensive quantities. For example, the mean number of particles in a relativistic gas within the grand canonical ensemble is an extensive quantity. The ratio of two extensive quantities does not depend on the system volume and is referred to as an intensive quantity. The studied fluctuation measure ν_{dyn} is even more complicated because it is inversely proportional to the system volume or to the number of wounded nucleons within the Wounded Nucleon Model. On the other hand the scaled variance, $(\langle N^2 \rangle - \langle N \rangle^2)/\langle N \rangle$, is an intensive quantity. It is therefore more intuitive to get a baseline from p+p interactions for the scaled variance. The scaled variance of multiplicity distributions, obtained via the *Identity Method* is presented in Fig. 5 for the p+p data of NA61/SHINE (filled circles) and compared to results from the EPOS model (solid lines) [14]. These are the first released results on fluctuations of identified hadrons in p+p interactions. Therefore, a semi-quantitative agreement with the predictions of the EPOS model is remarkable. More work is needed to interpret remaining differences between the data and the model.

An additional complication in the experimental study of fluctuations in A+A collisions are unavoidable volume fluctuations from event to event. These additional sources of fluctuations

Journal of Physics: Conference Series **426** (2013) 012027 doi:10.1088/1742-6596/426/1/012027

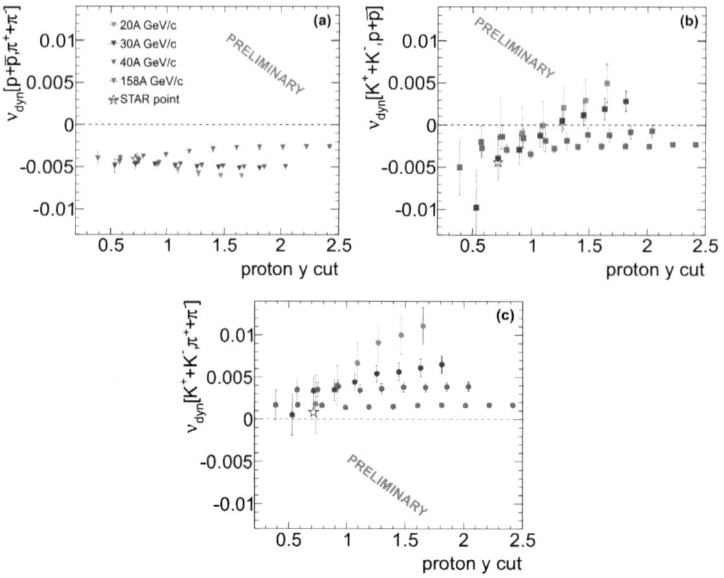

Figure 4. (Color Online) Acceptance dependence of (a) $\nu_{dyn}[p+\bar{p},\pi^{+}+\pi^{-}]$, (b) $\nu_{dyn}[K^{+}+K^{-},p+\bar{p}]$ and (c) $\nu_{dyn}[K^{+}+K^{-},\pi^{+}+\pi^{-}]$ in central Pb+Pb collisions of NA49 (triangles, squares, dots). Stars show measurements of the STAR collaboration. Results are plotted versus the maximum of the proton rapidity at p_T=0.

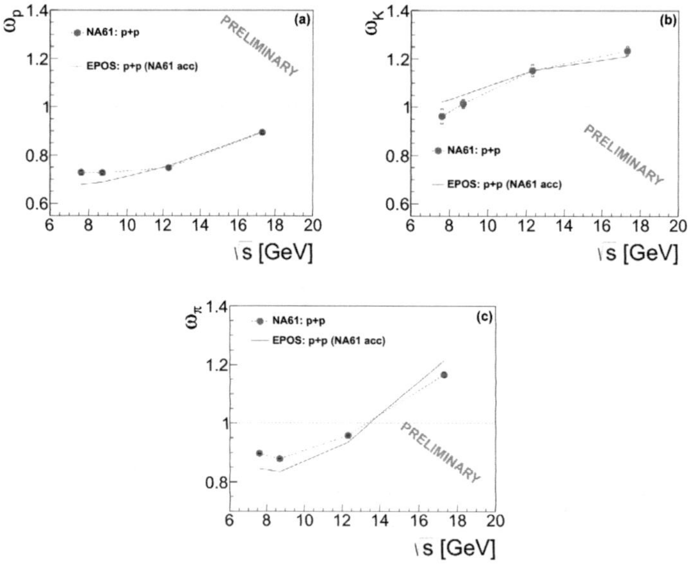

Figure 5. (Color Online) Scaled variance for (a) $p+\bar{p}$, (b) $K^{+}+K^{-}$ and (c) $\pi^{+}+\pi^{-}$ multiplicity distributions for p+p collisions. The NA61/SHINE results are shown by filled circles and compared to EPOS model calculations depicted by a solid blue curve.

may mask the fluctuations of interest. For example, the scaled variance is sensitive to the volume fluctuations. A set of so-called *strongly intensive* quantities, which within a grand canonical ensemble depend neither on the system volume nor on its fluctuations were proposed in Ref [15]. The experimental study of these quantities is ongoing.

4. Summary

We presented preliminary results from NA61/SHINE on inclusive particle production in inelastic p+p collisions and discussed its future ion program. Event-by-event fluctuations of particle ratios were analyzed using a novel approach, the *Identity Method*. Results on excitation functions of ν_{dyn} in central Pb+Pb collisions confirm an increasing trend towards lower energies, reported earlier by NA49. The detailed study of ν_{dyn} reveals a strong acceptance dependence at low energies for the $\nu_{dyn}[p, K]$ and $\nu_{dyn}[\pi, K]$. We conclude that the different energy dependence of ν_{dyn} measured by NA49 and STAR (BES program for central Au+Au collisions) is due to the different phase space coverage. Moreover, using the *Identity Method*, the scaled variance of the multiplicity distributions of protons, kaons and pions were reconstructed in inelastic p+p collisions of NA61/SHINE. These will serve as an important reference for A+A collisions.

4.1. Acknowledgments

The support by the German Research Foundation (DFG grant GA 1480/2.1) is gratefully acknowledged.

References

[1] S. Afanasiev et al. (NA49 Collab.), Nucl. Instrum. Meth. A430, 210 (1999)
[2] N. Antoniou et al. (NA61/SHINE Collab.), CERN-SPSC-2006-034, CERN-SPSC-P-330 (2006)
[3] C. Alt et al. (NA49 Collab.), Phys. Rev. C77, 024903 (2008)
[4] A. Rustamov, Central Eur. J. Phys. 10, 1267-1270 (2012), arXiv:1201.4520v1 [nucl-ex] (2012)
[5] Z. Fodor and S. D. Katz, JHEP 0404, 050 (2004)
[6] S. Pulawski, QM2012, poster session
[7] C. Alt et al. (NA49 Collab.), Phys. Rev. C79, 044910 (2009)
[8] T. Anticic et al. (NA49 Collab.) Phys. Rev. C83, 061902(R) (2011)
[9] T. J Tarnowsky (STAR Collab.) J. Phys. G: Nucl. Part. Phys. 38 124054 (2011)
[10] C. Pruneau, S. Gavin, S. Voloshin, PRC 66, 044904 (2002)
[11] M. Gazdzicki et al., Phys. Rev. C83, 054907 (2011); M. I. Gorenstein, Phys. Rev. C84, 024902 (2011)
[12] A. Rustamov, M. I. Gorenstein, Phys. Rev. C86, 044906 (2012)
[13] V. Koch and T. Schuster, Phys. Rev. C81, 034910 (2010)
[14] M. Máckowiak, QM2012, poster session.
[15] M. I. Gorenstein and M. Gazdzicki, Phys.Rev. C84, 014904 (2011)

Journal of Physics: Conference Series **426** (2013) 012028 doi:10.1088/1742-6596/426/1/012028

Probing early-time correlations in heavy ion collisions

Sean Gavin[1] **and George Moschelli**[2]

[1] Department of Physics and Astronomy, Wayne State University, 666 W Hancock, Detroit, MI, 48202, USA

[2] Frankfurt Institute for Advanced Studies, Johann Wolfgang Goethe University, Ruth-Moufang-Str. 1, 60438 Frankfurt am Main, Germany

E-mail: [1] sean@physics.wayne.edu [2] moschelli@fias.uni-frankfurt.de

Abstract. We investigate the common influence of correlations arising from initial state parton density fluctuations on measured multiplicity and momentum fluctuations as well as flow fluctuations. We calculate both a universal correlation scale in an initial stage Glasma flux tube picture and the modification to these correlations from later stage hydrodynamic flow. We find quantitative agreement with measurements over a range of collision systems and energies.

1. Introduction

In the first instants of every heavy ion collision event a unique system is produced. This happens because of the random orientation of nucleons before the collision and the finite probability for nucleon-nucleon or partonic interaction at the instant of the collision. The irregular nature of these first moments leads to inhomogeneous or "lumpy" energy/parton density deposits in the collision volume and the distribution of these lumps defines a distinct event geometry. Assuming that the subsequent evolution of the system progresses from the initial shape in a deterministic way, fluctuations in the initial geometry could explain the origin of odd harmonic flow coefficient measurements such as v_3. Just as measurements of harmonic flow coefficients of all orders, v_n, can provide insight into the initial geometrical distribution of lumps, fluctuations of v_n can provide information about the lumps themselves.

We argue that partons emerging from the same source or lump are more likely to share the same phase space throughout the collision lifetime and this correlation simultaneously gives rise to observed multiplicity and transverse momentum fluctuations as well as fluctuations of harmonic flow coefficients [1, 2]. In Sec. 2, we define these observables and demonstrate that they are related by a common model independent momentum correlation function. In Sec. 3, we consider contributions to this correlation function from the flow modification of initial state spatial correlations of partons originating from common density lumps. In this formulation parton density lumps emerge from the dissolution of Color Glass Condensate (CGC) -Glasma flux tubes. We find the strength of this correlation is determined by the size, number, and parton multiplicity of the lumps and sets a common scale that governs most of the centrality and center of mass collision energy dependence [3, 4]. Finally, we summarize in Sec. 4.

2. Correlations and Fluctuations

Fluctuation observables are often measured utilizing two-particle correlation methods. The pair distribution of hadrons with momenta $\mathbf{p}_1$ and $\mathbf{p}_2$ is $\rho_2(\mathbf{p}_1, \mathbf{p}_2) = dN/d\mathbf{p}_1 d\mathbf{p}_2$ and in the absence of correlations, it factorizes into the square of the singles distribution, $\rho_2(\mathbf{p}_1, \mathbf{p}_2) \to \rho_1(\mathbf{p}_1)\rho_1(\mathbf{p}_2)$ where $\rho_1(\mathbf{p}) = dN/d\mathbf{p}$. In this case, fluctuation measurements would depend only on event by event differences in $\rho_1(\mathbf{p})$. Alternatively, if correlations do exist, then fluctuation observables fundamentally depend on their behavior. In this work, we consider contributions to pairwise correlations only from the existence of parton density lumps, specifically, the correlation induced between particles that originate at the same transverse position. We remark, however, that the formulae in this section apply in general to any two-particle correlation in the system.

In the presence of correlations, we can write the pair distribution as $\rho_2(\mathbf{p}_1, \mathbf{p}_2) = \rho_1(\mathbf{p}_1)\rho_1(\mathbf{p}_2)+r(\mathbf{p}_1, \mathbf{p}_2)$, where $r(\mathbf{p}_1, \mathbf{p}_2)$ represents the genuine two-particle correlation function [1, 2, 5, 6]. We can isolate these correlations by writing

$$r(\mathbf{p}_1, \mathbf{p}_2) = \rho_2(\mathbf{p}_1, \mathbf{p}_2) - \rho_1(\mathbf{p}_1)\rho_1(\mathbf{p}_2). \tag{1}$$

Integrating over all momenta yields the "correlation strength" defined as

$$\mathcal{R} \equiv \frac{1}{\langle N\rangle^2}\int r(\mathbf{p}_1, \mathbf{p}_2)d\mathbf{p}_1 d\mathbf{p}_2 = \frac{\langle N(N-1)\rangle - \langle N\rangle^2}{\langle N\rangle^2} = \frac{Var(N) - \langle N\rangle}{\langle N\rangle^2}. \tag{2}$$

This quantity distinguishes multiplicity fluctuations as non-Poissonian particle production since Poissonian distributions have the property that $Var(N) = \langle N\rangle$. Consequently, non-zero values of $\mathcal{R}$ indicate a correlation in the production of hadrons. As suggested by the name, the correlation strength, $\mathcal{R}$, sets the scale of all fluctuation measurements related to two-particle correlations.

Dynamic fluctuations of transverse momentum are defined using the covariance

$$\langle \delta p_{t1}\delta p_{t2}\rangle = \frac{\langle \sum_{i\neq j}\delta p_{ti}\delta p_{tj}\rangle}{\langle N(N-1)\rangle} = \int d\mathbf{p}_1 d\mathbf{p}_2 \frac{r(\mathbf{p}_1, \mathbf{p}_2)}{\langle N(N-1)\rangle}\delta p_{t1}\delta p_{t2}, \tag{3}$$

where $\delta p_{ti} = p_{ti} - \langle p_t\rangle$ and the average transverse momentum is $\langle p_t\rangle = \langle P_t\rangle/\langle N\rangle$ for $P_t = \sum_i p_{ti}$ the total transverse momentum in an event [7, 8, 9, 10]. This quantity vanishes when particles i and j are uncorrelated. Equation (3) sums the product of differences in particle momenta from the average for each hadron pair. Pairs in which both particles have higher than average momentum add to $\langle \delta p_{t1}\delta p_{t2}\rangle$. Lower-than-average pairs also add to the covariance, while high/low pairs subtract from it. In *global* equilibrium $\langle \delta p_{t1}\delta p_{t2}\rangle \equiv 0$. Positive measurements of $\langle \delta p_{t1}\delta p_{t2}\rangle$ could be explained if, for example, particle production is dominated by jets, or a collectively expanding medium is comprised of spots of *locally* equilibrated matter. In the next section we investigate the latter case and take the locally equilibrated spots of fluid to be the results of initial state parton density lumps.

The correlation strength, $\mathcal{R}$, sets the scale of (3) in the sense that the same (correlated) pairs are counted in (2), only here they are weighted by their momenta. Significantly, both observables count correlated pairs independent of their direction and therefore do not depend on the asymmetry of the system. However, because of the momentum weights, (3) is sensitive to the *average* expansion and temperature.

The relative angles of correlated pairs is quantified by harmonic flow correlation and fluctuation measurements. These measurements reflect both global correlations based on the shape of the collision as well as correlations represented by (1). The global anisotropy of the collision can be characterized by the moments $\langle v_n\rangle = \langle \cos n(\phi - \psi_{RP})\rangle$, where $\langle ...\rangle$ represents an average over particles and events and ψ_{RP} is the reaction plane angle defined by the plane spanned by the impact parameter $\mathbf{b}$ and the beam direction. Measurements utilizing two-particle correlations do not require knowledge of the reaction plane and are defined as

$$v_n\{2\}^2 = \frac{\langle \sum_{i\neq j}\cos n(\phi_i - \phi_j)\rangle}{\langle N(N-1)\rangle} = \int \frac{\rho_2(\mathbf{p}_1, \mathbf{p}_2)}{\langle N(N-1)\rangle}\cos n(\phi_1 - \phi_2)\, d\mathbf{p}_1 d\mathbf{p}_2 \tag{4}$$

[5, 6, 11, 12, 13]. Cast in terms of the pair distribution, we use (1), to write (4) as $v_n\{2\}^2 = \langle v_n\rangle^2 + 2\sigma_n^2$, where the flow coefficient relative to the reaction plane is $\langle v_n\rangle = \int \rho_1(\mathbf{p})\cos n(\phi - \Psi_{RP})d\mathbf{p}$ comes from the single particle distribution and the factor of two in the fluctuation term σ is conventional [14]. Assuming the four-particle harmonic flow correlation measurement is approximately $v_n\{4\} \approx \langle v_n\rangle$, as in Ref. [14], we have

$$\sigma_n^2 = \frac{v_n\{2\}^2 - v_n\{4\}^2}{2} = \int d\mathbf{p}_1 d\mathbf{p}_2 \frac{r(\mathbf{p}_1, \mathbf{p}_2)}{2\langle N(N-1)\rangle} \cos n\Delta\phi, \tag{5}$$

where $\Delta\phi = \phi_1 - \phi_2$ is the relative azimuthal angle between the pair. This observable quantifies the contribution of genuine two-particle correlations to the harmonic flow coefficients (4). Although this contribution is often regarded as "non-flow", anisotropic flow modifies the relative angles of correlated pairs emerging from common density sources. Given the assumption that a density lump produces a locally thermalized fluid element, the push on this element from flow completely determines the $\Delta\phi$ distribution of partons in that cell. We discuss this idea further in the next section.

3. Glasma Correlations

We examine correlations arising only from particles emerging from a common source and their modifications from the transverse expansion of the system. We take the sources to be initial state density lumps which are consequences of the dissolution of CGC-Glasma flux tubes. In general one can describe two-particle spatial correlations with $c(\mathbf{x}_1, \mathbf{x}_2) = n_2(\mathbf{x}_1, \mathbf{x}_2) - n_1(\mathbf{x}_1)n_1(\mathbf{x}_2)$ Here $\mathbf{x}_1$ and $\mathbf{x}_2$ are transverse spatial coordinates. If the pair distribution, $n_2(\mathbf{x}_1, \mathbf{x}_2)$, contains no correlations, it factorizes into the square of the singles distribution, $n_1(\mathbf{x}_i)$ and $c(\mathbf{x}_1, \mathbf{x}_2)$ vanishes. In our flux tube lump picture we write

$$c(\mathbf{x}_1, \mathbf{x}_2) = \langle N\rangle^2 \mathcal{R}\,\delta(\mathbf{r}_t)\rho_{FT}(\mathbf{R}_t). \tag{6}$$

The delta function in relative transverse elliptical coordinates $\mathbf{r}_t = \mathbf{r}_{t1} - \mathbf{r}_{t2}$ enforces the condition that correlated partons originate from the same flux tube source, presuming the flux tube transverse size is small compared to that of the collision area. The flux tube probability distribution, written in average coordinates $\mathbf{R}_t = (\mathbf{r}_{t1} + \mathbf{r}_{t2})/2$, is $\rho_{FT}(\mathbf{R}_t) \approx (2/\pi R_A^2)\,(1 - R_t^2/R_A^2)$. Here the shape of ρ_{FT} resembles the nuclear thickness function and πR_A^2 is the area of the overlap region. The distribution ρ_{FT} represents an average over all possible shapes, which we have taken to be a simple ellipse.

The correlation strength, $\mathcal{R}$, governs the magnitude of correlations in (6). To compute its value we implement a CGC-Glasma framework in which the initial moments of a collision event is comprised of a system of longitudinally long color flux tubes each with transverse size $\sim Q_s^{-2}$. Here Q_s is the scale of momentum transfer in collisions of dense partonic systems that signals the onset of gluon saturation. For recent review see [15]. Each event produces a random number of K flux tubes that is proportional to the transverse area R_A^2 divided by the area per flux tube, $\sim Q_s^{-2}$ [16]. K fluctuates from event to event with average $\langle K\rangle$, and we calculate in Ref. [3] that $\mathcal{R} \propto \langle K\rangle^{-1}$. Each Glasma flux tube yields an average multiplicity of $\sim \alpha_s^{-1}(Q_s)$ gluons and as in Ref. [16], the number of gluons in a rapidity interval Δy is then $\langle N\rangle = (dN/dy)\Delta y \sim {\alpha_s}^{-1}(Q_s)\langle K\rangle$. Combining these two results we find the Glasma correlation scale, $\mathcal{R}dN/dy = \kappa\alpha_s^{-1}(Q_s^2)$ which is dimensionless and depends only on the saturation scale, Q_s^2 [1, 17]. Measurements of the ridge at various beam energies, target masses, and centralities fix the dimensionless coefficient κ and are in accord with the leading-order dependence [3, 4]. As a comparison, the correlation strength is equivalent to the inverse of the width parameter of negative binomial multiplicity distributions [18] as measured by the PHENIX collaboration [19]. The comparisons are shown in Fig. 1 for various beam energies.

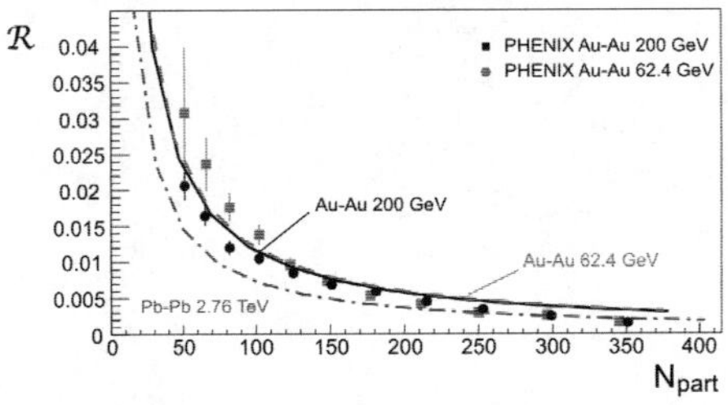

Figure 1. Calculated $\mathcal{R}$ compared to PHENIX data from [19]

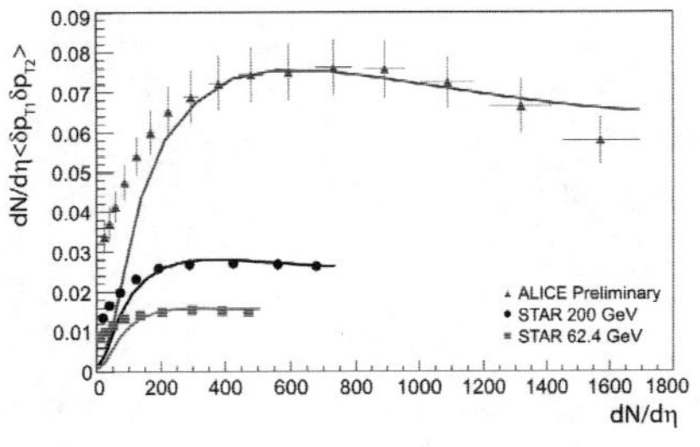

Figure 2. Transverse momentum fluctuations. Data from [10, 20].

Correlations (6) are modified by the transverse expansion of the system. Depending on its initial position a parton receives a transverse boost from flow; partons originating near the center get a small push while those near the periphery get a larger push. Additionally, the geometry of the collision influences this boost since, depending on the shape, some directions have faster expansions than others. The key is that, on average, partons originating at the same position experience the same boost. Their transverse momenta is therefore enhanced in a common way; pairs experiencing a larger transverse boost end up with more columnated final transverse momenta. In this manner, correlations in coordinate space (6) result in correlations in momentum space (1). In this spirit we describe (1) as

$$r(\mathbf{p}_1, \mathbf{p}_2) = \int c(\mathbf{x}_1, \mathbf{x}_2) f(\mathbf{x}_1, \mathbf{p}_1) f(\mathbf{x}_2, \mathbf{p}_2) d\Gamma_1 d\Gamma_2. \tag{7}$$

Here (6) links blast-wave descriptions of the Boltzmann phase-space densities $f(\mathbf{x}, \mathbf{p})$ for two particles with Cooper-Frye freeze out surfaces $d\Gamma$ [1, 2].

We are now in a position to calculate (3) and (5) using (7). In Fig. 2 we compare $dN/dy\langle\delta p_{t1}\delta p_{t2}\rangle$ to data from the STAR and ALICE collaborations [10, 20]. In Fig. 3 we compare flow fluctuations in terms of the observable $\sigma_{v_n}/\langle v_n\rangle = \left((v_n\{2\}^2 - v_n\{4\}^2)/(v_n\{2\}^2 + v_n\{4\}^2)\right)^{1/2}$ which resembles the so-called "coefficient of variation" defined as the standard deviation divided by the mean. Care should be taken here since the definition of σ_n^2, Eq.(5), is not strictly the variance. In both cases the closest agreement with data occurs in central collisions, the region where the Glasma description is most valid. If flow fluctuations in this region arise only from fluctuations in the shape of the collision area, then their governing correlation function would predict $\langle\delta p_{t1}\delta p_{t2}\rangle = 0$ and have almost no dependence on collision energy. The energy dependence of $\mathcal{R}$ is dominated by that of the saturation scale Q_s and we note that the most significant result here is the simultaneous agreement of all mentioned observables with changes in collision system and center of mass energy.

Lastly in Fig. 4 we show v_3 fluctuations from Glasma. We further emphasize that the energy dependence in Fig. 4 is in good accord with data, again supporting the Glasma scaling with Q_s^2. To use (5) to calculate $v_3\{2\}$, we must come to grips with the fact that our symmetric blast wave parametrization assumes $v_3\{4\} = 0$. While this seems to be the case for the STAR measurements, ALICE has measured a non-zero $v_3\{4\}$ for Pb+Pb collisions at 2.76 TeV. Therefore, we prefer to report only our calculation of σ_3^2.

4. Summary

In this work we study the connection between correlations and fluctuations of multiplicity, momentum, and harmonic flow coefficients. We argue in Sec. 2 that given the existence of

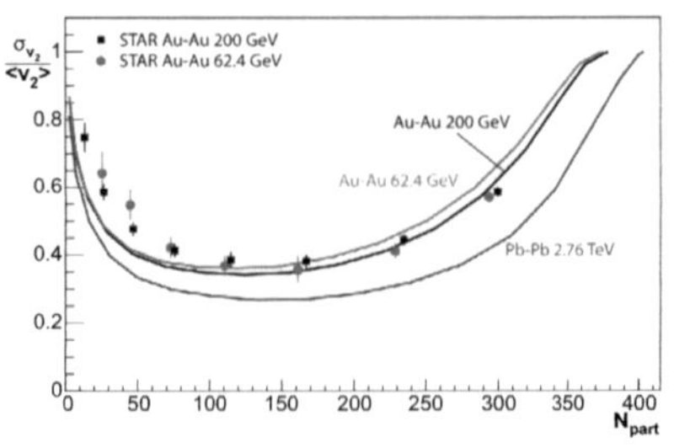

Figure 3. Elliptic flow fluctuations as the coefficient of variation compared to STAR data [21].

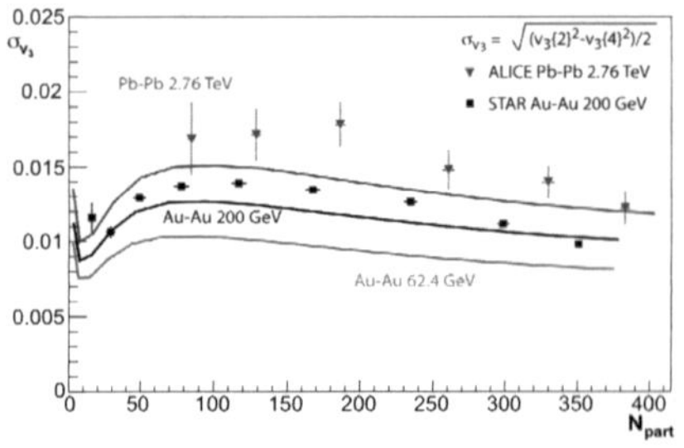

Figure 4. Triangular flow fluctuations. Data computed from $v_3\{2\}$ and $v_3\{4\}$ measurements [22, 23].

genuine two-particle correlations in the hadron pair distribution, (1), multiplicity fluctuations, momentum fluctuations, and flow fluctuations are all governed by a common scale $\mathcal{R}$. This is the first calculation to link these observables. We then consider contributions to the genuine correlations due to fluctuating or lumpy initial parton distributions where correlated partons emerge from the same source/lump. We take the origin of initial state parton density lumps from Glasma flux tubes and present our model calculations in Sec. 3. The reasonable agreement of our calculations with the collision energy dependence of the measurements supports saturation based descriptions of the initial state. Further improvements of this method will come with implementation of event-by-event hydrodynamic calculations which inherently embody the dynamics of collisions that our current expansion model only approximates.

Acknowledgments
This work was supported in part by the U.S. NSF grant PHY-0855369 (SG) and The Alliance Program of the Helmholtz Association (HA216/EMMI) (GM)

References

[1] Gavin S and Moschelli G 2012 *Phys.Rev.* **C85** 014905 (*Preprint* 1107.3317)
[2] Gavin S and Moschelli G 2012 (*Preprint* 1205.1218)
[3] Gavin S, McLerran L and Moschelli G 2009 *Phys.Rev.* **C79** 051902 (*Preprint* 0806.4718)
[4] Moschelli G and Gavin S 2010 *Nucl.Phys.* **A836** 43–58 (*Preprint* 0910.3590)
[5] Borghini N, Dinh P M and Ollitrault J Y 2001 *Phys. Rev.* **C63** 054906 (*Preprint* nucl-th/0007063)
[6] Borghini N, Dinh P M and Ollitrault J Y 2001 *Phys.Rev.* **C64** 054901 (*Preprint* nucl-th/0105040)
[7] Voloshin S, Koch V and Ritter H 1999 *Phys.Rev.* **C60** 024901 (*Preprint* nucl-th/9903060)
[8] Voloshin S A (STAR) 2001 591–596 (*Preprint* nucl-ex/0109006)
[9] Adamova D *et al.* (CERES) 2003 *Nucl.Phys.* **A727** 97–119 (*Preprint* nucl-ex/0305002)
[10] Adams J *et al.* (STAR) 2005 *PHRVA,C72,044902.2005* **C72** 044902 (*Preprint* nucl-ex/0504031)
[11] Bilandzic A, Snellings R and Voloshin S 2011 *Phys.Rev.* **C83** 044913 (*Preprint* 1010.0233)
[12] Voloshin S A, Poskanzer A M and Snellings R 2008 (*Preprint* 0809.2949)
[13] Sorensen P 2009 (*Preprint* 0905.0174)
[14] Voloshin S A, Poskanzer A M, Tang A and Wang G 2008 *Phys.Lett.* **B659** 537–541 (*Preprint* 0708.0800)
[15] Gelis F 2012 (*Preprint* 1211.3327)
[16] Kharzeev D and Nardi M 2001 *Phys. Lett.* **B507** 121–128 (*Preprint* nucl-th/0012025)
[17] Gavin S and Moschelli G 2012 *J.Phys.Conf.Ser.* **389** 012038 (*Preprint* 1208.1741)
[18] Gelis F, Lappi T and McLerran L 2009 *Nucl. Phys.* **A828** 149–160 (*Preprint* 0905.3234)
[19] Adare A *et al.* (PHENIX) 2008 *Phys. Rev.* **C78** 044902 (*Preprint* 0805.1521)
[20] Heckel S (ALICE) 2011 (*Preprint* 1107.4327)
[21] Pruthi N K (The STAR Collaboration) 2011 (*Preprint* 1111.5637)
[22] Sorensen P (STAR Collaboration) 2011 *J.Phys.G* **G38** 124029 (*Preprint* 1110.0737)
[23] Aamodt K *et al.* (ALICE Collaboration) 2011 *Phys.Rev.Lett.* **107** 032301 (*Preprint* 1105.3865)

FAIRNESS 2012: FAIR NExt Generation of ScientistS 2012 IOP Publishing
Journal of Physics: Conference Series **426** (2013) 012029 doi:10.1088/1742-6596/426/1/012029

The Micro-Vertex-Detector for the $\overline{\mathsf{P}}$ANDA experiment

Laura Zotti on behalf of the PANDA-MVD Group
Università degli Studi di Torino & INFN Torino, Via P. Giuria 1, 10125 Torino, Italy
E-mail: zotti@to.infn.it

Abstract. $\overline{\mathsf{P}}$ANDA is a fixed target experiment that will be carried out at the future FAIR facility. $\overline{\mathsf{P}}$ANDA will provide an excellent tool to address fundamental question in the field of hadronic physics, with a physic program that extends from the investigation of QCD (providing insight in the mechanisms of mass generation and confinement) to the test of fundamental symmetries.
The Micro-Vertex-Detector located in the innermost part of the central tracking system will be composed by hybrid pixel and double-sided micro-strip silicon detectors. The Micro-Vertex-Detector will play an important role for the $\overline{\mathsf{P}}$ANDA physics goals. The possibility to reconstruct the secondary vertices and the applicability of a precise D meson tagging is essential for the spectroscopy in the open charm sector and the charmonium mass region. To this aim the Micro-Vertex-Detector features a spatial resolution better than 100μm, a time resolution better than 20ns, a limited material budget, and a high data rate capability in a triggerless environment.
An overview of the Micro-Vertex-Detector related to the physics goals will be presented.

1. Introduction

$\overline{\mathsf{P}}$ANDA is one of the new experiments that will take place at the facility FAIR under construction at GSI. It is a fixed target experiment where antiproton-proton and antiproton-nuclei annihilations will be studied in a fixed target setup. The main feature is the antiproton beam, that will be provided by the High Energy Storage Ring (HESR) with an unprecedented intensity and quality. The antiproton momentum will be in the range from 1.5 up to 15GeV/c. The physics program foresees studies in the charmonium region (such as spectroscopy, gluonic excitation and in-medium modification), in hypernuclei field and CP-violation, see [1].

Charmonium Spectroscopy

One of the physics topics of $\overline{\mathsf{P}}$ANDA is the hadron spectroscopy, in particular in the charm sector. The charmonium mass spectrum can be divided into two part [2]:

- the charmonium spectrum below the open charm threshold, which is both experimentally and theoretically established. Here we find 8 states, all experimentally confirmed, and with the conventional potential models in good agreement.
- the charmonium spectrum above the open charm threshold. Here the situation is rather incomplete, and far away to be fully understood. In fact only four charmonia have been confirmed, meanwhile a lot of new states, such as the X(3872) or the Y(4260), have been discovered at the B-factories that do not seem to fit in the theoretical models.

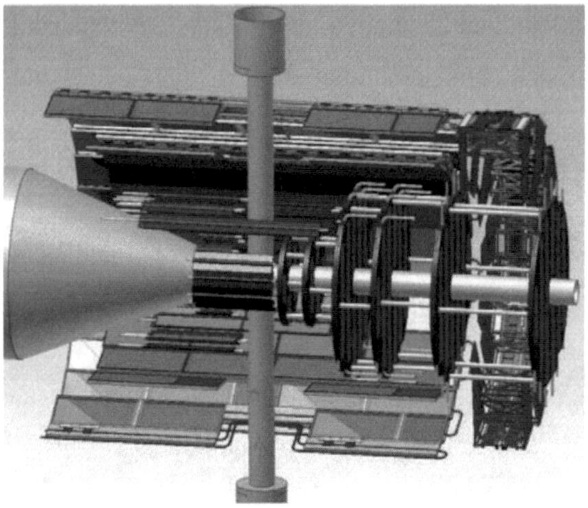

Figure 1. Layout of the $\overline{\text{P}}$ANDA MVD

$\overline{\text{P}}$ANDA represents a unique tool to improve both the statistics and the precision of existing data, and to further explore the charm quark sector. The unprecedented energy definition of the antiproton beam will allow to perform precise mass scans. In fact the running and future planned experiments will not provide precise energy scan and, moreover, they will be limited by the directly accessible quantum number. The possibility to perform such a physic program is strongly related to the capability to reconstruct with high precision the D meson decay, exactly to this aim the MVD has been designed.

2. The Micro Vertex Detector

The Micro Vertex Detector (MVD) is located in the target spectrometer, and it is the detector closest to the interaction point. It will extend for +/-23cm with respect to the interaction point, and the radius ranges from 2.2 up to 13.5cm. Two types of detectors will be implemented: hybrid pixels and double-sided micro-strips. The first one will be placed in the 6 forward disks and in the two internal barrels, while the last one will be placed in the two external barrels and as external radial component of the last two disks, see Fig. 1. The detector will have almost 10^6 pixel readout channels and around $2{\cdot}10^5$ double-sided readout channels [3].

2.1. Requirements and Features

Since the main aim of the MVD is to allow the reconstruction of the secondary vertices of D mesons, the detector should guarantee a spatial resolution better than 100µm. Moreover, the MVD will impose important constraints on the tracking algorithm leading to an overall momentum resolution of the order of 2%. Since the $\overline{\text{P}}$ANDA data acquisition will be triggerless and the nominal interaction rate will exceed $2{\cdot}10^7$ annihilations per second, the time resolution of the detector should be better than 20ns. All the materials have to be chosen both to be radiation hardness (the expected radiation damage level, in the case of $\bar{p}$-p annihilations at 15GeV/c, will be strongly forward peaked and has been estimated to be of the order of $10^{14}\text{n}_{\text{eq}}\text{cm}^{-2}$ for 10 years of data taking and a 50% of duty cycle) and to not affect the material budget (the full MVD is below 10% X/X_0, in unit of radiation length). The readout electronics should be able to work in a triggerless environment, to handle a high data rate and to perform energy loss measurements in order to contribute to the global PID.

Journal of Physics: Conference Series **426** (2013) 012029 doi:10.1088/1742-6596/426/1/012029

2.2. Hybrid Pixel Detector

The hybrid pixels are composed of a thin segmented silicon sensor connected via the bump bonding technique to a custom readout chip. The sensors will be made of epitaxial silicon wafer, grown on a Czochralski substrate (Cz). Each pixel will have an area of 100x100μm^2, and an active thickness of 100μm. This last value is a reasonable compromise between the sensor radiation tolerance and the possibility to perform energy loss measurements and to have a good signal to noise ratio.

The pixel readout electronics has to provide for each hit a simultaneous position, time and energy loss measurement, and it has also to digitize this information at the pixel level in order to transmit only digital information. The main constraint in the design of such a readout is the pixel geometry. The ASIC for the $\overline{\text{P}}$ANDA hybrid pixel has been designed in the 130nm CMOS technology, and it is called Topix. The chip has to be able to work with a high system clock of 155.52MHz (time resolution of ~6ns), which is essential in order to perform a continuous digitization of the data in a triggerless environment. Energy loss measurement will be performed by the Time over Threshold technique (ToT) [4].

2.3. Double-Sided Micro-Strip

The use of double-sided strip detectors offers the capability of precise point reconstruction with the advantage of less detector channels compared to pixel detectors, and to a low material budget setup. However, ambiguities occur in case of multiple particles crossing the detector in the same time frame. Therefore the $\overline{\text{P}}$ANDA MVD utilizes silicon strip detectors at the outer layers of the MVD where the particle flux is reduced in contrast to the inner layers. The starting material for the strip detector will be a mono-crystalline floating zone silicon wafer. Three type of geometry will be used: the rectangular and quadratic sensors, with a tilt angle of 90°, will be used in the barrel part, while the trapezoidal shape sensor will be placed in the external ring of the last forward disks, with a tilt angle of 15°. The pitch will be 130μm for the barrel strip and 67.5μm for the forward strip, in order to recover the worsened resolution coming from a smaller stereo angle.

The possibility to use a modified version of Topix, for the strip readout is under study [3].

3. MonteCarlo Simulation

In order to understand the environment where the detector should work and to foresee its performance several MonteCarlo simulations have been performed, using the $\overline{\text{P}}$ANDA framework [5]. The accuracy of the prediction will be proportional to the detailed description of the detector. To this aim a converter has been developed to transform the Computer Aided Drawing (CAD) model into a root-type geometry suited for the simulation. Taking advantage of this tool the developed MVD geometry contains approximately 70.500 individual volumes [6].

3.1. Rate Study

The evaluation of the expected count rates in different detector regions is crucial for the developments of the readout electronics, moreover the integrated numbers coming from all the readout chips will allow the estimation of the overall data load to be handled by the MVD. The simulation results show that the maximum rate will be strongly correlated to the physical topology of the annihilations. In the case of antiproton-proton annihilations at 15GeV/c the maximum will be in the forward part, i.e. the disk part, meanwhile in the case of antiproton-nuclei annihilation the maximum has to be expected in the barrel part. The maximum expected data load in the full MVD is of the order of 38Gb/s.

3.2. Physic Benchmark

A typical signature common for several of the channels within the scope of PANDA is for example: $\bar{p}p \rightarrow \psi(3770) \rightarrow D^+D^- \rightarrow K^-\pi^+\pi^+K^+\pi^-\pi^-$. In fact, this reaction features secondary vertices with a short decay length (312µm for the charged D mesons) and a relatively large number of ejectiles to be reconstructed in a exclusive analysis. A first selection of the candidates is done requiring that the reconstructed D meson mass differs no more than 750 MeV/c from the nominal D mass, moreover a vertex fit is carried out and in case of multiple candidates the selection is done on the χ^2. The obtained vertex resolutions are 55µm in the xy-direction and 104µm in the z-direction, see [7]. The obtained mass resolution is 16MeV/c^2.

4. Prototype Test

The first beam test of the hybrid pixel detectors have been carried out at the COSY synchrotron of the Forschungszentrum Jülich with 2.7GeV/c protons. The first assembly prototypes with an area of 2x3.2mm^2, composed by the ToPix3 readout chip and a dedicated epitaxial silicon sensor, with 640 pixels each, of 100x100µm^2 area, and 100µm thickness. These sensors were obtained without removing the most of the Cz substrate on which the epitaxial layers have grown. Actually, each sensor was composed by 100µm epitaxial thickness and 525µm of Cz. Because of the low resistivity of the Cz (10-20 mΩ·cm), the sensor backside was not aluminized and a direct ohmic bias connection was realized.

The basilar bench includes the testing board hosting the assembly and a Xilinx evaluation board equipped with a Virtex 6FPGA. With this bench, a pixel tracking station of 4 assembly has been realized, for a total length of almost 20cm, and a distance between two sensor surfaces of 6cm.

The data acquisition was performed with a 50MHz clock, and the raw data contains the following information: 12bit leading and trailing edge (for ToT measurements) and the address of the pixel (column and row). A 50MHz clock was used for the time and ToT measurement. Since the acquisition was performed in a triggerless mode, a really precise time information is needed, so a Time-Stamp is obtained summing the leading edge and a 32bit End of Counter. In this way all the hit have a time information of 44bit and this allow to identify the tracks looking at the correspondence of the Time-Stamp. This procedure of Time-Stamp matching consists in finding the hit of the different planes with a time-stamp equal or different by $\pm$ 2 clock cycles (this range assures to take into account the TimeWalk effect).

In the left part of Fig. 2 a picture of the time information and of the Time-Stamp matching procedure is shown. On the right part of Fig. 2 the ToT distribution of the four planes is shown. The ToT distributions have been fitted with the convolution of a Gaussian with a Landau function. This convolution is useful in order to reproduce either the Gaussian response of the detector either the Landau shape of the energy loss. The most probable values (MP) are compatible within the error (Gsigma), the variations are due to the chip-to-chip technology fluctuations. In order to have an information in terms of collected electrons a fine channel-to-channel calibration is in progress. The alignment results show that the best spatial resolutions (taking into account that the setup was 20% X/X_0) were reachable on the two internal planes, and they were of the order of 40µm both in the x and y directions. The first beam test has shown an overall good functionality both of the sensors and of the readout electronics.

Conclusion

The design of the MVD have been finalized, the MonteCarlo simulations show a good performance on the capability to reconstruct D mesons. The first prototypes of both pixel and strip (see [8]) have shown promising results.

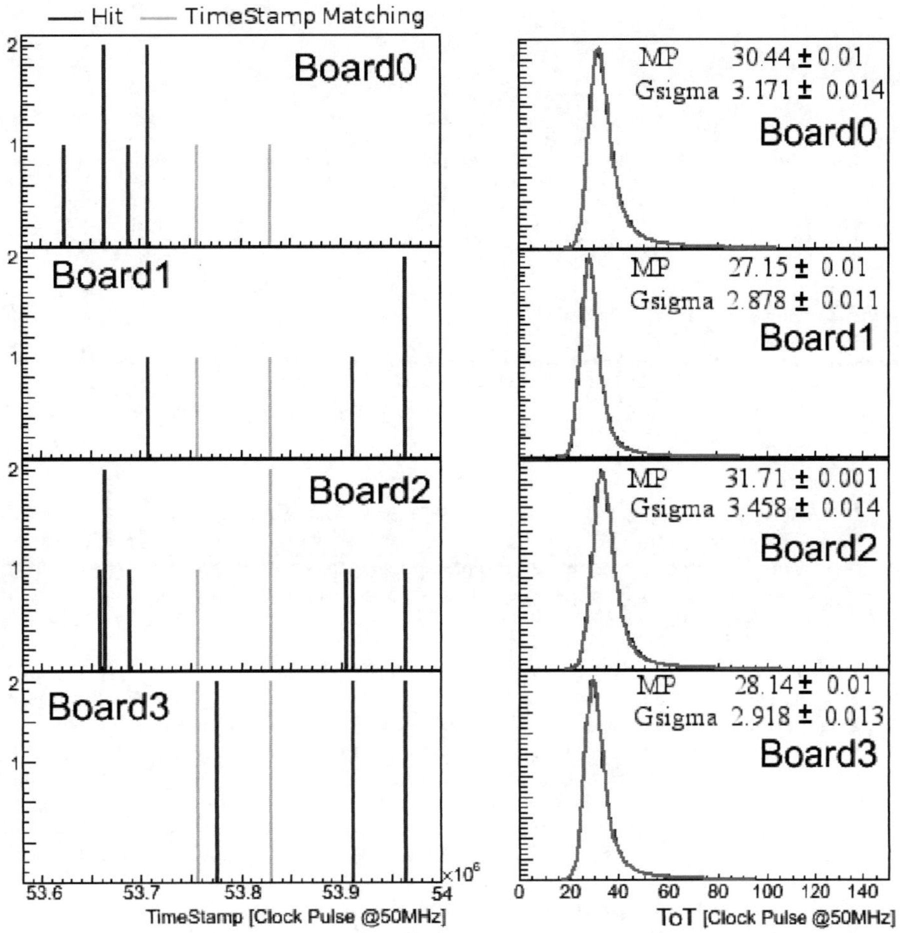

Figure 2. On the left an example of the hit time information on the different board is shown, where the Time-Stamp matching procedure is illustrated. On the right the four ToT distributions are shown, the MP and Gsigma are the value of the Gauss+Landau convoluted function used to fit the distribution.

Reference

[1] PANDA Collaboration, 2009, Physics Performance Report, *arXiv:hep-ex* 0903.3905v1
[2] D. Bettoni, 2012, *PoS(Bormio)* **004**
[3] PANDA Collaboration, 2012, MVD Technical Design Report, *arXiv:physics.ins-det* 1207.6581v2
[4] G. Mazza et al., 2012, *JINST*, **7** C02015
[5] S. Spataro, 2011, *J. Phys. Conf. Ser.* **331** 032031
[6] S. Bianco et al., 2011, *Nucl. Instr. Meth. A*, **654** 630633
[7] PANDA Collaboration, 2012, STT Technical Design Report, *arXiv:physics.ins-det* f1205.5441v2
[8] S. Bianco et al., 2011, *PoS(RD11)*, **025**

Journal of Physics: Conference Series **426** (2013) 012030 doi:10.1088/1742-6596/426/1/012030

High precision γ spectroscopy of $\Lambda\Lambda$-Hypernuclei at the PANDA experiment

A. Sanchez Lorente on behalf of the PANDA collaboration
Helmhotz Institute Mainz, Johannes Gutenberg-Universität, 55099 Mainz, DE
E-mail: a.sanchez@gsi.de

Abstract. Hypernuclear research will be one of the main topics addressed by the PANDA experiment at FAIR at Darmstadt (Germany). Thanks to the use of stored antiproton beams, copious production of double $\Lambda\Lambda$-Hypernuclei is expected at the PANDA experiment, which will enable high precision gamma spectroscopy of such nuclei for the first time. At PANDA excited states of Ξ^- hypernuclei will be used as a starting point for the formation of double $\Lambda\Lambda$-Hypernuclei. In order to predict the yield of particle-stable double hypernuclei a microcanonical decay model was developed. For the detection of these nuclei, a devoted hypernuclear detector setup is planned. This set-up consists, in addition to the general purpose of the PANDA set-up, of a primary nuclear target for the production of $\Xi^- + \bar{\Xi}$ pairs, a secondary active target for the hypernuclei formation and the identification of associated decay products and a germanium array detector to perform gamma spectroscopy. Moreover, one of the most challenging issues of this project is the fact that all detector systems need to operate in the presence of a high magnetic field and a large hadronic background. In these proceedings details concerning the identification procedure of double hypernuclei and the suppression of background will be presented. In addition, the current status of the activities related to the detector developments for this challenging programme will be briefly given.

1. Introduction

The simultaneous production and implementation of two Λ particles into a nucleus is intricate. There is a possibility to produce multi-strange hypernuclei in heavy ion collisions via coalescence [1, 2]. The first observation of antihypernuclei by the STAR collaboration impressively illustrates the potential of this method [3]. However, high resolution spectroscopy of excited states is not feasible. To produce double hypernuclei in a more 'controlled' way the conversion of a captured Ξ^- and a proton into two Λ particles can be used. This process releases, ignoring binding energy effects, only 28 MeV. For light nuclei there exists, therefore, a significant probability of the order of a few percent that both Λ hyperons are trapped in the same excited nuclear fragment [4, 5, 6, 7, 8, 9]. After an atomic cascade, the Ξ^- hyperon is eventually captured by a secondary target nucleus and converted via the $\Xi^- p \rightarrow \Lambda\Lambda$ reaction into two Λ hyperons. In a similar two-step process relatively low momentum Ξ^- can also be produced using antiproton beams in $\bar{p}p \rightarrow \Xi^-\overline{\Xi}^+$ or $\bar{p}n \rightarrow \Xi^-\bar{\Xi}$ reactions if this reactions happens in a complex nucleus where the produced Ξ^- can re-scatter[10, 11]. The advantage as compared to the kaon induced reaction is that antiprotons are stable and can be retained in a storage ring. This allows a rather high luminosity even with very thin primary targets. Because of the two-step mechanism, spectroscopic studies, based on two-body kinematics like in single hypernucleus production, cannot be performed. Spectroscopic information on double hypernuclei can therefore only be obtained via their decay products. The kinetic energies of weak decay products are sensitive to the binding energies of the two Λ hyperons. While the double pionic decay of light double hypernuclei can be used as an effective experimental filter to reduce the background [11] the unique identification of hypernuclei ground states only via their pionic decay is usually hampered by the limited resolution. Instead, γ-rays emitted via the sequential decay of excited double hypernuclei may provide precise information on the level structure.

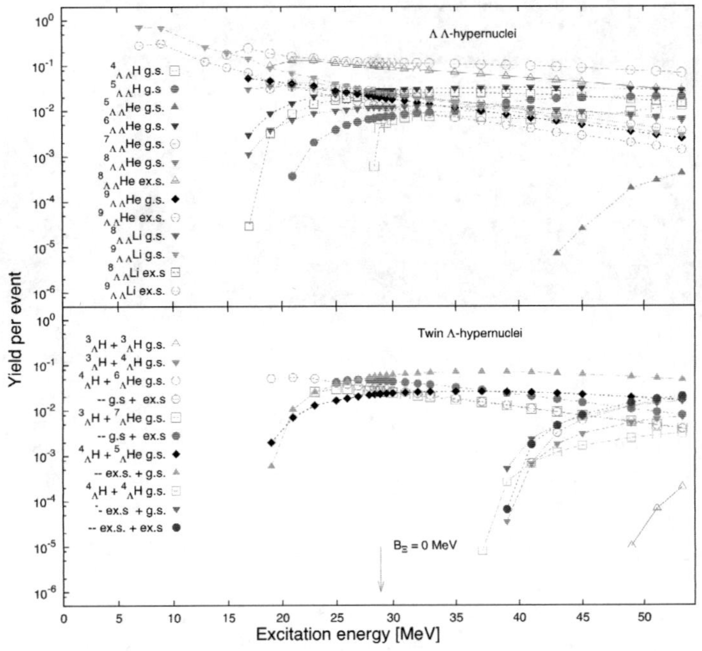

Figure 1. Relative production probability of double hypernuclei (top part) and single hypernuclei (lower part) for an excited $^{10}_{\Lambda\Lambda}Li$ nucleus as a function of its excitation energy.

2. Statistical decay of excited doubly Strange Nuclei

In order to limit the number of possible transitions and thus to increase the possible signal to background ratio, the experiment will focus on light secondary target nuclei. To describe this break-up process of the excited primary $\Lambda\Lambda$ nucleus and in order to estimate the population of individual excited states in double hypernuclei after the conversion of the Ξ^-, we have developed a statistical decay model which is reminiscent of the Fermi break-up model [12]. We assume that the nucleus decays simultaneously into cold or slightly excited fragments. In the case of conventional nuclear fragments, we adopt their experimental masses in ground states, and take into account their particle-stable excited states. For single hypernuclei, we use the experimental masses and all known excited states. For double hypernuclei we apply theoretically predicted masses and excited states [13].

In the model we consider all possible break-up channels, which satisfy the mass number, hyperon number, charge, energy and momentum conservation, and take into account the competition between these channels. Since the excitation energy of the initially produced double hypernuclei is not exactly known, we performed the calculations as a function of the binding energy B_{Ξ} of the captured Ξ^-.

As an example Fig.1 shows excitation function of the relative production probability of double hypernulei (top part) and single hypernuclei (lower part) for a primary ${}^{10}_{\Lambda\Lambda}$Li. Since the conversion of the Ξ^- is expected to take place close to B_{Ξ}=0 MeV the production of excited double hypernuclei is predicted to dominate in the PANDA experiment [14].

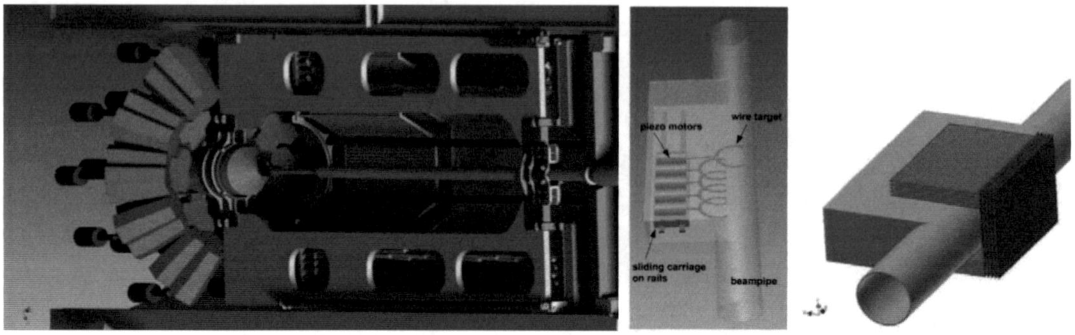

Figure 2. Schematic view of the hypernuclear setup in PANDA. The picture on the left shows a side view of the hypernuclear experimental setup. The two pictures on the right show two different views of the mechanism which helds the target at a constant interaction rate by an automatic feedback system which continuously adjusts the wire positions via stepping motors. The target mechanism is mounted inside the vertex vessel, surrounded by the moveable parts of the secondary active target.

3. Integration inside the PANDA spectrometer

A big challenge to be solved in order to place the hypernuclear detector setup, is the limited space available at the entrance of the PANDA spectrometer. Therefore, some modifications of the inner part of the former spectrometer is needed. In particular, due to the fact that double Λ-Hypernuclei formation proceeds in a two-steps process, a dedicated target system consisting of a primary and secondary interaction point with an associated beam line is required (see Fig. 2). Since this target system has to be placed outside the interaction region of the PANDA spectrometer, detectors such as the Micro Vertex Detector (MVD) and Backward EndCap Calorimeter will be removed to avoid unnecessary radiation damage. In addition, the central frame dedicated to hold the beam line and the central tracker detectors will also be accordingly adapted to the hypernuclear setup.

3.1. Primary target

The main role of the primary target will be the production of low momentum Ξ hyperons. Although heavy targets are more efficient for re-scattering of the produced primary particles and hence for the emission of low momentum Ξ hyperons, light targets will be preferred in order to avoid a high hadronic background into backward axial angles, where the germanium detectors are located. Furthermore, Coulomb scattering in heavy primary targets leads to significant losses of antiprotons. Additionally, it is required that the luminosity of the $\bar{p}$-beam remains as constant as possible. In addition, high interaction rates will be avoided by choosing an appropriate fraction of the beam halo onto the target, what can be achieved by an monitoring mechanism where the beam as well as the target can be steered till the desired interaction rate is reached (see Fig.2). As a result, the use of thin carbon micro-ribbons [15] seems to be the best candidate to allow a stable operation over a sufficiently long measuring period.

3.2. Secondary active target

The purpose of the active volume of the secondary target is the tracking of charged particles generated during the first and the secondary interaction. As a consequence, its geometry is based on a compact structure where silicon micro-strip detectors layers are in direct contact with absorber material [16]. In analogy to the germanium detectors array, this device has to be able to operate in extreme conditions such as a large hadronic environment, since it is close to the interaction point. Furthermore, the material budget on the detector volume must be kept low. The feasibility of such a device has recently been studied in Mainz, by evaluating the influence of putting layers of absorber material directly on a silicon sensor [16]. Results have not shown any significant change on the preamplifier signal caused by the vicinity of an boron layer [16]. The sensor utilized is a double sided micro-strip detector with dimensions of $2{\times}2\,\mathrm{cm}^2$, a strip pitch of $50\,\mu$m, punch-through biasing and AC coupled contact pads. The sensor is mounted on a L-shaped PCBoard displayed in Fig.3 and is bonded to a APV25-S1 front-end chip which features 128 channels. In order to avoid a huge load of hadronic flux on the readout electronics, the use of Ultra-thin Al-Polyimide micro-cables is foreseen. The reason for that, is the routing of analog signals from silicon sensors to the readout system outside the interaction region and as a result to decrease the amount of material budget on the detecting volume. A prototype of

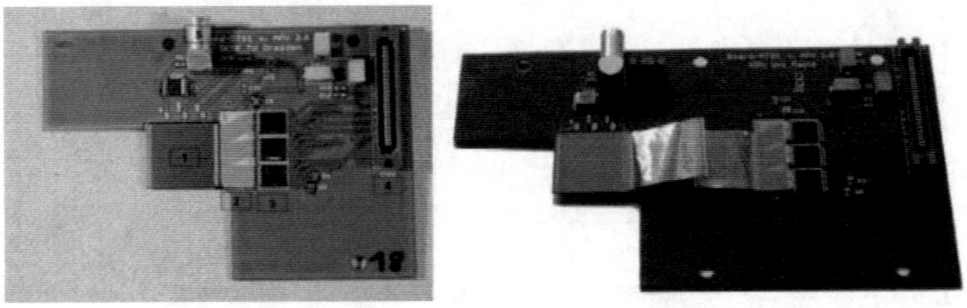

Figure 3. Overview of the old and new PCB prototypes. The two pictures above show a photograph of a sensor module equipped to perform a double sided readout of the sensor(1) consisting of a PCB. Main components of each board are:(2) pitch adapter;(3) APV25 front-ends; (4) high density connector. The detector bias voltage (5) is supplied via one board. The picture on the right panel shows a modified board to enable the connexion to a flexible ultra-thin cable.

such as cables has already been manufactured in SE SRTIIE, Kharkov, Ukraine. These cables were made on the basis of adhesiveless aluminium-polyimide foiled dielectrics. In addition, the PCB used for testing the analog signals, has to be modified accordingly (see Fig.3). The cable will be connected to the chip via bonding wires and to the readout electronics via a fine pitch connector. Further activities dedicated to investigate the effect of the cable length on the signal transmission are in progress. In addition, the performance of a holding structure for the whole device, is being studied. The mechanics of such a device has to be optimized considering thermal and mechanical properties of the holding structure.

4. The HPGe cluster Array

In order to increase the detection efficiency needed for a high resolution γ-Spectroscopy, the high-purity germanium (HPGe) detectors array has to be placed as close as possible to the interaction point and cover a wide solid angle. Due to the limited space of the PANDA spectrometer, the

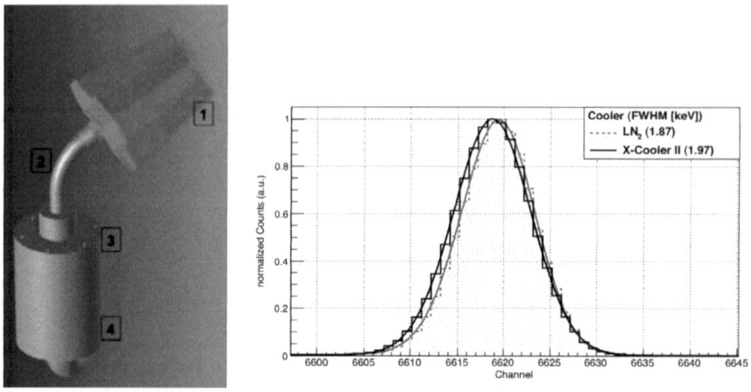

Figure 4. HPGe triple cluster array assembled to a X-Cooler device. The encapsulated n-type coaxial HPGe crystals (EUROBALL) are arranged in a triangular form. The free space behind the crystals is foreseen for electronics. The connection of the crystal vessel to the cold head of the cooler is flexible to enable the placement of each cluster at backward axial angles. (1) encapsulated crystals, (2) flexible neck, (3) X- Cooler cold head, (4) X-Cooler Cable. The figure on the right shows the energy spectra corresponding to the 1,332 keV line of ^{60}Co for two different cooling devices. The dashed spectrum corresponds to the case where the germanium crystal has been cooled electro-mechanically showing a slightly worsening of the energy resolution.

arrangement of the germanium detector array will only be possible at backward axial angles. That means for instance, that the operation of these detectors will have to withstand a large flux of hadronic background and a high magnetic field [17], which can influence the good energy resolution ($\sim$ 3 keV at the 1,332 MeV line of Co^{60}) of these detectors. A possible solution for the space limitation has been to replace the standard cooling system, based on liquid nitrogen dewars, by a mechanical cooling device [18]. Fig. 4 shows a prototype for a triple germanium cluster array cooled by an electromechanical device. After installation of each of the encapsulated germanium crystals in the cryostat or vacuum vessel, the system has to reach optimum vacuum conditions to be properly operated.

The cooling efficiency of these devices has been successfully tested for three encapsulated germanium crystals without observing any additional worsening of the energy resolution [19]. Further investigations are currently taking place in Mainz at a dedicated test station. The scope of these studies is to evaluate to what extent the energy resolution of a germanium detector, cooled electro-mechanically, can be influenced. For this reason, an ORTEC GEM-75205P device and analog readout electronics has been used. The energy resolution of such a device has been measured with a ^{60}Co source considering two different cooling devices, namely a standard liquid nitrogen dewar and a X-Cooler device. For the case of a standard cooling system, the energy resolution was found to be 1.86 KeV for 1.332 keV line [1]. The one achieved by the X-cooler device has been 1.97 keV, which seems to be consistent. Fig. 4 shows the energy spectra corresponding to the 1.332 keV line of ^{60}Co, obtained by considering the two cooling systems named above. The dashed spectrum corresponds to the case where the germanium crystal has been cooled electro-mechanically and although the energy resolution for this case is slightly worse, one can see that the use of the X-Cooler as a cooling option seems to be nevertheless acceptable. Further information about temperature effects and their impact on the spectroscopy properties as well as the performance of the detector and the X-Cooler assembly is in progress. In addition, activities concerning the use of digital electronics to evaluate pile-up effects, and radiation damages in a

[1] The energy resolution provided by Ortec is about 2.05 keV

Journal of Physics: Conference Series **426** (2013) 012030 doi:10.1088/1742-6596/426/1/012030

high flux hadronic environment are also under preparation [20].

References

[1] Kerman A K and Weiss M S 1973 *Phys. Rev.* C 8, **408**.
[2] Botvina A S and Pochodzalla J 2007 *Phys. Rev.* C **76** 024909
[3] The STAR Collaboration Science 2010 328 **58-62**.
[4] Yamamoto Y Sano M and Wakai M 1994 *Theor. Phys. Suppl.* **117** 265.
[5] Yamada T and Ikeda K 1997 *Phys. Rev.* C **56** 3216.
[6] Yamamoto Y Wakai M Motoba T and Fukuda T 1997 *Nucl. Phys.* A **625** 107.
[7] Hirata Y *et al.* 1998 *Nucl. Phys.* A **639**, 389c.
[8] Hirata Y *et al.* 1999 *Prog. Theor. Phys.* **102** 89.
[9] Aoki S *et al.* 2009 *Nucl. Phys.* A **828** 191.
[10] Ferro F *et al.* 2007 *Nucl. Phys.* A **789** 209
[11] Physics Performance Report for PANDA: Strong Interaction Studies with Antiprotons, `arXiv:0903.3905v1`.
[12] Fermi E 1950 *Prog. Theor. Phys.* **5** 570.
[13] Hiyama E *et al.* 2002 *Phys. Rev. C* **66** 024007.
[14] Sanchez Lorente A, Botvina A S and Pochodzalla J 2011 *Phys. Lett.* B **697** 222-228.
[15] `http://www-panda.gsi.de/db/papersDB/PC19-050217\_panda\_tpr.pdf`
[16] Bleser S, PhD.Thesis, under preparation.
[17] Sanchez Lorente A *et al.* 2007 *Nucl. Inst. Meth.* A **573** 410 (2007).
[18] ORTEC. X-Cooler$^{\mathbf{TM}}$ II- Mechanical Cooler for HPGe detectors user manual.
[19] Wolf M Kojouharova J Kojouharov I *et al.* Proceedings COMSOL Users Conference, Grenoble, 2007.
[20] Steinen M Master Thesis.

Journal of Physics: Conference Series **426** (2013) 012031 doi:10.1088/1742-6596/426/1/012031

Monte-Carlo simulation of lepton pairs production in "$p\bar{p} \rightarrow \mu^+\mu^- + X$" events at $E_{beam} = 5$ GeV

Anna Skachkova

Joint Institute for Nuclear Research, Joliot-Curie 6, 141980 Dubna, Moscow region, Russia

E-mail: Anna.Skachkova@cern.ch

Abstract. Muon pairs production via the quark-antiquark annihilation process in collisions of antiproton beam with $E_{beam} = 5$ GeV (which corresponds to the center-of-mass energy of the $p\bar{p}$ system $E_{cm} = 3.363$ GeV) with the proton target in PANDA experiment is modeled by using of PYTHIA6.4 generator. The considered quark level subprocess goes through the production of virtual photon which converts into lepton pair ($q\bar{q} \rightarrow \gamma^* \rightarrow \mu^+\mu^-$). Quark-antiquark annihilation process of hadron-hadron collision may provide an interesting information about the quark dynamics inside the hadron. The measurement of the total transverse momentum of a lepton pair as a whole system may provide an important information about the intrinsic transverse momentum k_T that appears due to the Fermi motion of quarks inside the nucleon. The distributions of different kinematical variables of final state muons are presented. The problems due to the presence of fake leptons that appear from meson decays, as well as due to the background caused by minimum bias events are also discussed. The set of cuts which allows to separate the signal events with lepton pairs from the backgrounds is proposed.

Introduction

Intermediate energy experiment PANDA ($E_{beam} \leq 15$ GeV) may play an important role because it allows to study the energy range where the perturbative methods of QCD (pQCD) come into interplay with a rich physics of resonance production. A detailed and high-precision experimental study at PANDA may allow to discriminate between a large variety of existing nonperturbative approaches and models that already exist or are under development now. Dilepton events may serve as a powerful tool to get out the information about the parton distribution functions (PDFs) in hadrons [1]. The plans to study this process are included into the LoI [3], TPR [4] and "Physics book" [5] of PANDA experiment at HESR (High-Energy Storage Ring at FAIR). This study may provide an interesting information about quark dynamics inside the nucleon [2].

Observations and Interpretation

This work is a continuation of the work [1], [2] done to study the lepton pairs production in the processes $p\bar{p} \rightarrow \mu^+\mu^- + X$ at the beam energy $E_{beam} = 14$ GeV. Here we consider the same process but in the case of beam energy E_{beam}= 5 GeV (which corresponds to the center-of-mass energy of the $p\bar{p}$ system $E_{cm} = 3.363$ GeV) which is the lowest possible for simulation of this kind of process with PYTHIA6.4 [6]. The main goal of this work for the moment is an estimation of the leptons kinematical distributions and search for the cuts criteria for signal and background separation.

We use PYTHIA6.4 to generate 100000 $p\bar{p} \to \mu^+\mu^-$ events which include the $2 \to 2$ quark level $q\bar{q} \to \gamma^* \to \mu^+\mu^-$ subprocess. The simulation is done starting from the assumption of the ideal muon detection system and electromagnetic calorimeter covering full phase space without taking into account the influence of magnetic field. We consider the case where both initial-state radiation (ISR) and final-state radiation (FSR) are switched on simultaneously by choosing the corresponding values of PYTHIA parameters. Also we have used the CTEQ3L parametrization of parton distributions and the default value of the parameter, which allows one to take into account the primordial k_T-effect (i.e. quarks Fermi motion).

The study of kinematical characteristics of lepton pair as a whole system was also done.

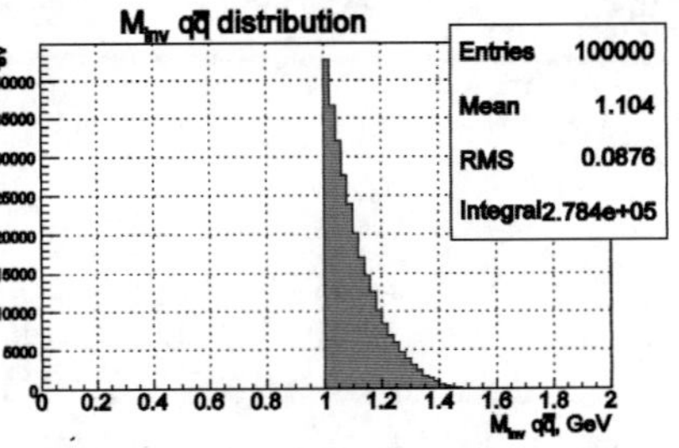

Figure 1. The invariant mass of initial-state quark-antiquark system $M_{inv}^{q\bar{q}}$.

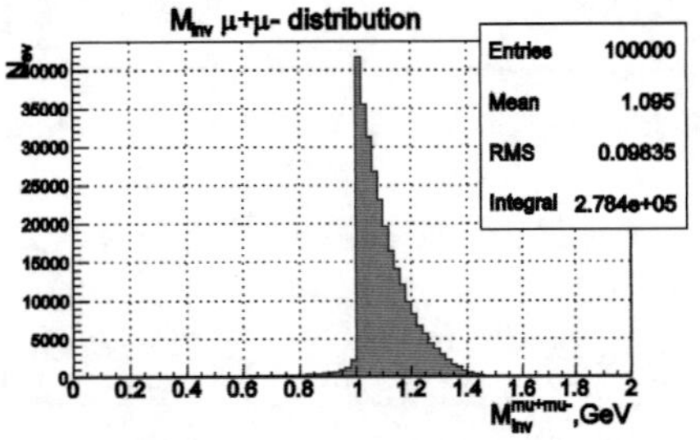

Figure 2. The invariant mass of final-state $\mu^+\mu^-$ pair $M_{inv}^{\mu^+\mu^-}$.

Figs.1 and 2 show, correspondingly, the distributions of the invariant masses of initial-state quark-antiquark pairs $M_{inv}^{q\bar{q}} = \sqrt{(P^{q\bar{q}})^2}$ and the invariant mass of the final-state $\mu^+\mu^-$ pair $M_{inv}^{\mu^+\mu^-} = \sqrt{(P^{\mu^+\mu^-})^2} = Q$, where $Q^2 = q^2 = (P_\alpha^{\mu^+} + P_\alpha^{\mu^-})^2$. Both invariant mass distributions look rather similar. It is shown in Fig.2 that the spectrum of the invariant mass of the lepton pair decreases rather fast and vanishes at $M_{inv}^{l+l-} = 1.45$ GeV. The square of the invariant mass $(M_{inv}^{l^+l^-})^2 = Q^2 = q^2 = (P^{l^+} + P^{l^-})^2$ has the meaning of the square of the momentum transferred from the hadronic system of initial quark-antiquark pair to the electromagnetic system of the final state lepton pair. From these figures it can be seen that it can have the values in the region $Q^2 \leq 2.1\ GeV^2$.

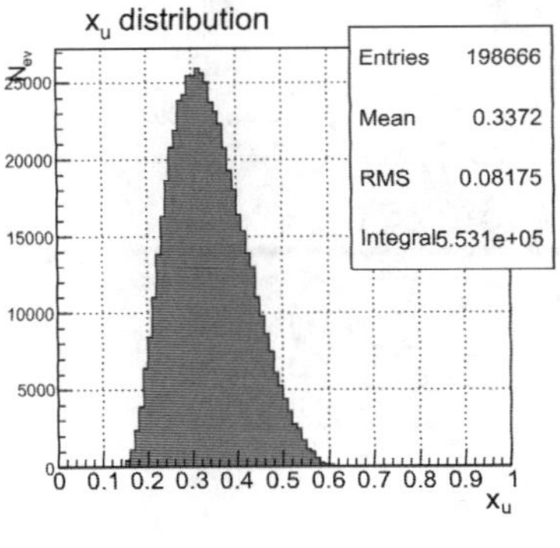

Figure 3. Distribution of Bjorken x_u.

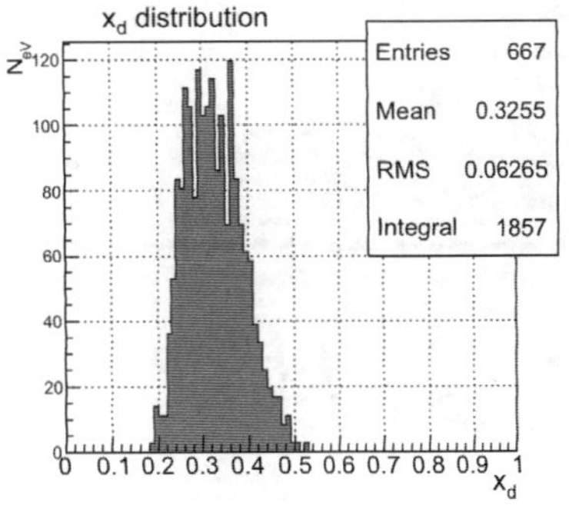

Figure 4. Distribution of Bjorken x_d.

Figs.3 and 4 contain the distributions of up- and down-quarks at $E_{beam} = 5$ GeV, respectively. They demonstrate that the contribution of up-quarks dominates over the down quarks, while both of them cover practically the same region of x's: $0.15 \leq x \leq 0.6$. Let us emphasize that the measurements in this region of positive ("time-like") $q^2 \equiv (P^{l^+} + P^{l^-})^2 = Q^2 \geq 0$ would be a good extension of studies planned to be done at JLab in the region of negative ("space-like") values of $q^2 = (P^l_{in} - P^l_{out})^2 \leq 0$.

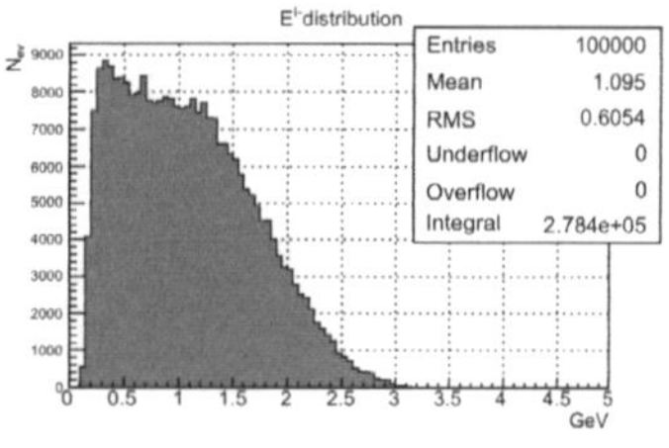

Figure 5. Muon's energy E_μ.

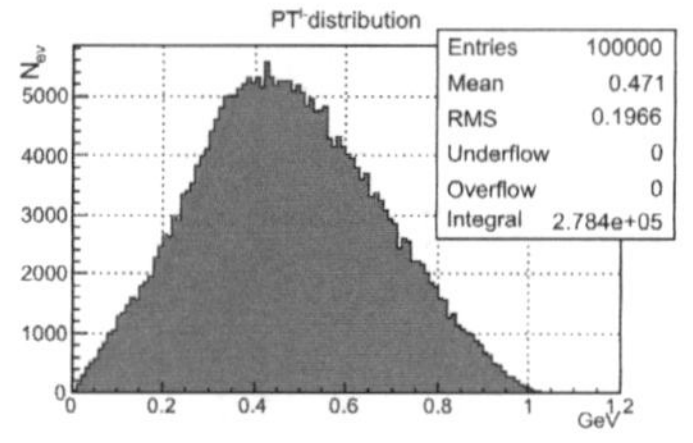

Figure 6. Muon's transverse momentum PT_μ.

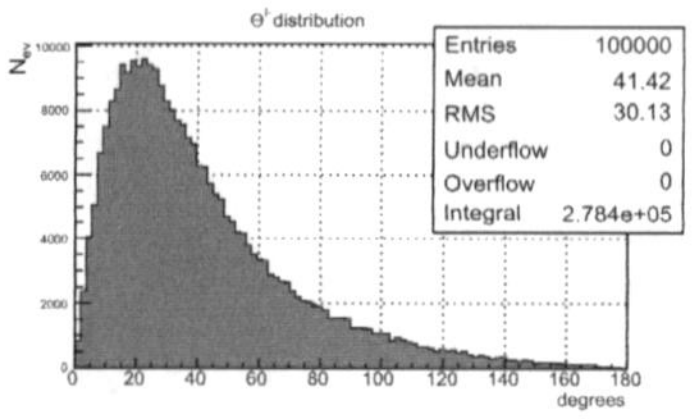

Figure 7. Muon's angle Θ_μ.

The distributions of the number of generated signal events N_{ev} versus the muon energy $E^{\mu^\pm}$, as well as versus the modulus of the transverse momentum $PT^{\mu^\pm}$ and of the polar angles $\Theta^{\mu^\pm}$, measured from the z-axis directed along the beam line, are given in the Figs.5-7. At PYTHIA level of simulation (for "ideal" detector) the distributions for positive and negative leptons are identical. One can see from Fig.5 that the most part of leptons energy is contained in the interval $0 < E^{\mu^\pm} < 3$ GeV. Its spectrum has a mean value $< E^{\mu^\pm} > = 1.09$ GeV and a peak at $E^{\mu^\pm}_{peak} \approx 0.3$ GeV. The $PT^{\mu^\pm}$ spectrum (Fig.6) has an analogous peak at $PT^{\mu^\pm} \approx 0.4$ GeV. The main part of $PT^{\mu^\pm}$ spectrum is confined within a rather narrow interval $0 < PT^{\mu^\pm} < 1$ GeV. The number of events spectrum versus the polar angle $\Theta^{\mu^\pm}$ (Fig.8) has a peak around $\Theta^{\mu^\pm} \approx 20^o$ and the mean value $< \Theta^{\mu^\pm} >= 41.4^o$. One sees that while 98% of signal leptons fly in the forward direction $\Theta^{\mu^\pm} < 90^o$, there is still 2% of them which fly into the back hemisphere $\Theta^{\mu^\pm} > 90^o$.

The next Figures include the correlation plots done for the signal leptons having the largest energy in the lepton pair, μ_{fast}, and for the leptons having a smaller energy in the pair, μ_{slow}. We shall call them, correspondingly, as "fast" and "slow" muons.

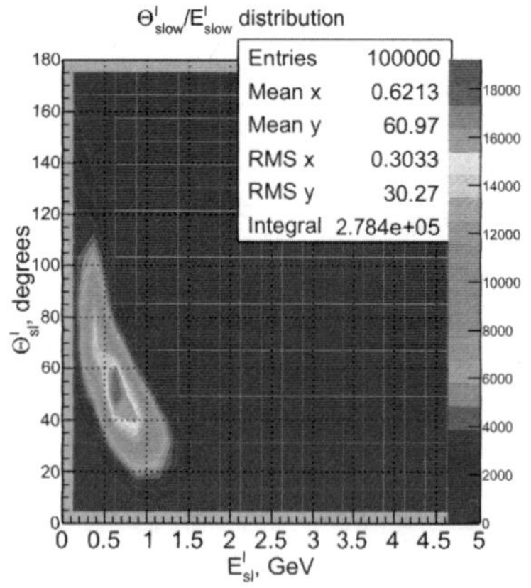

Figure 8. $\Theta^\mu_{slow} - E^\mu_{slow}$ correlation.

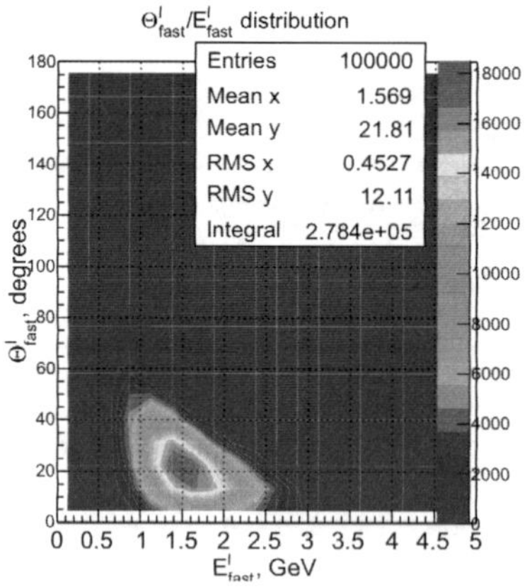

Figure 9. $\Theta^\mu_{fast} - E^\mu_{fast}$ correlation.

Figures 8-9 contain two projections of 3-Dimensional "Angle-Energy" correlation plots for slow $\Theta^\mu_{slow} - E^\mu_{slow}$ (left plot) and fast $\Theta^\mu_{fast} - E^\mu_{fast}$ (right plot) muons in signal pairs onto $\Theta - E$ planes. This allows to show the boundary contours of regions with different density of number of events. The the color code on Zaxis in each correlation plot shows the correspondence between the contour color and the density of events (N_{ev} per year).

After a discussion of individual lepton distributions let us turn to the distributions that characterize the produced pair of leptons as a whole system. The next Figures 10-11 show the Energy-Energy $E^{\mu}_{slow} - E^{\mu}_{fast}$ (left plot) and Angle-Angle $\Theta^{\mu}_{slow} - \Theta^{\mu}_{fast}$ (right plot) correlations.

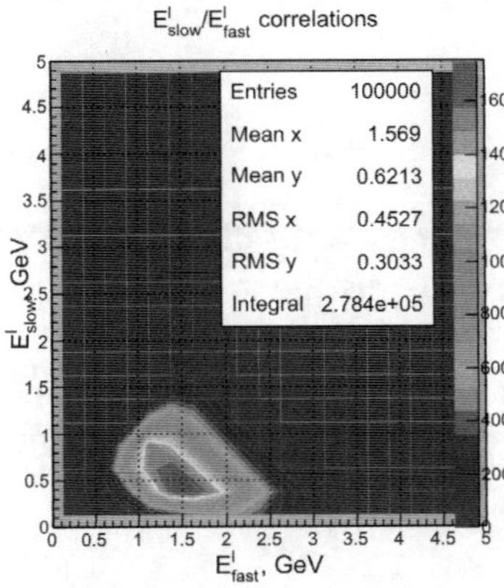

Figure 10. $E^{\mu}_{slow} - E^{\mu}_{fast}$ correlation.

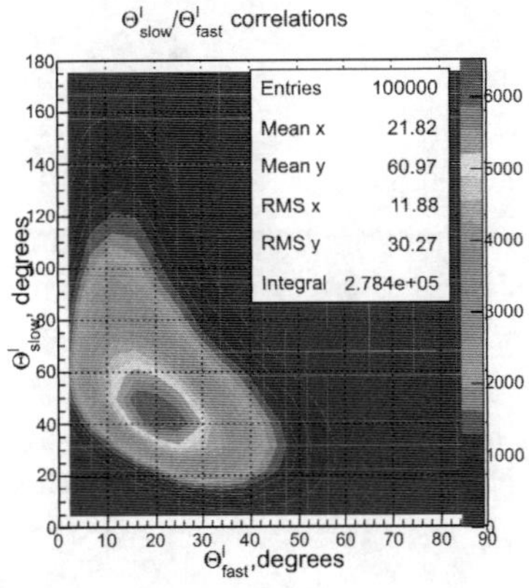

Figure 11. $\Theta^{\mu}_{sl} - \Theta^{\mu}_{slow}$ correlation.

Backgrounds

The signal events, defined by the $q\bar{q} \to \mu^+\mu^-$ subprocess, also may contain up to 8 hadrons in the final state. Fortunately, their number is essentialy restricted by the upper limit on the PANDA beam energy. This circumstance may simplify greatly the identification of final state particles and the physical analysis due to reduction of the phase space and therefore to the reduction of the number of hadrons and other particles which may be produced in event directly or in the decay cascades of other hadrons. These hadrons may decay within the detector volume and thus produce the background leptons which may fake the signal leptons produced in a signal annihilation subprocess.

The first kind of background contains the muons which may be produced additionally to a signal $\mu^+\mu^-$ pair in the signal events due to hadron decays. According to simulation by PYTHIA the fraction of signal processes, which include fake muons, is about 1.2%. Criteria for separation of such background were proposed in [1],[3]. The choice of the events with two muons, having: a) different charges, b) $E^{\mu} > 0.2\ GeV, PT^{\mu} > 0.2\ GeV$ and c) the production vertex is close to the interaction point, allows to get rid of events having the fake decay muons by the loss of 15.8% of signal events.

The main background comes from minimum-bias events (i.e. low PT scattering, single diffractive processes etc.), which total cross section is more than $7 \cdot 10^8$ times higher than the cross section of the signal event.

Comparison of figures of background muon distribution with their analogs for the signal muons shows that background muons are less energetic that the signal ones. The average value of the signal muons energy $< E^{\mu} >= 0.97$ GeV corresponds to such a point of the background muons spectrum where the contributions of the last ones is practically absent. Analogically the average value of PT-distribution of signal muons $< PT^{\mu}_{sig} > = 0.33$ GeV corresponds to such a point where PT^{μ}_{dec} spectrum disappears. From the muon production vertex distributions one can see that the largest part of the background muons are produced far away from the interaction point and the vertices of their production are distributed over the detector volume.

Journal of Physics: Conference Series **426** (2013) 012031 doi:10.1088/1742-6596/426/1/012031

From distribution of the number of muons per background event one can see that in a case of beam energy of $E_{beam} = 5$ GeV there can be up to 6 muons per event (while for $E_{beam} = 14$ GeV this number can reach 7 muons per event).

To get rid of minimum-bias events contribution the following set of five cuts (see also [1],[3]) was used:

- 1. We choose the events with 2 muons in the final state with $E^{\mu}_{dec} > 0.2\ GeV$, $PT^{\mu}_{dec} > 0.2\ GeV$;
- 2. These muons have the opposite charges;
- 3. Their production vertex is close to the interaction point;
- 4. Their invariant mass $M_{inv}(\mu^+\mu^-) > 0.9\ GeV$;
- 5. Muons are isolated, i.e. the summarized energy of charged particles within the cone around the muon momentum, having the radius which can cover about three cells of PANDA calorimeter, is less than 0.5 GeV.

The table below shows the dependence of signal to background ratio (S/B) on the application of the choosen five cuts for two cases of using the lower bound for muon transverse momentum, i.e. one with $PT^{\mu} > 0.2\ GeV$ and the other one with $PT^{\mu} > 0.5\ GeV$. Efficiency shows the realtion $N^{cut_I}_{bkg-events}/N^{cut_{I-1}}_{bkg-events}$, thus the less is efficiency the more effective is the cut.

Table 1. Influence of the proposed cut for S/B separation.

N of cut	S/B ratio for $PT^{\mu} > 0.2$ GeV	Efficiency	S/B ratio for $PT^{\mu} > 0.5$ GeV	Efficiency
1	$1.02 \cdot 10^{-6}$	$1.47 \cdot 10^{-3}$	$2.37 \cdot 10^{-6}$	$4.22 \cdot 10^{-4}$
2	$1.13 \cdot 10^{-6}$	0.906	$2.57 \cdot 10^{-6}$	0.921
3	$3.70 \cdot 10^{-4}$	$3.04 \cdot 10^{-3}$	$6.09 \cdot 10^{-4}$	$4.21 \cdot 10^{-3}$
4	$6.58 \cdot 10^{-2}$	0.056	$7.04 \cdot 10^{-3}$	0.086
5	1.5	0.004	2	0.003

These results of backgrouns study are very preliminary because they are done for the case of "ideal" detector. Nevertheless they are rather promising and give the confidence in a good signal and background separation. The numbers of S/B ratio may be increased by adjustment the value of each of the cuts parameters. More precise study would be done using the PANDARoot simulation.

Acknowledgments

This work was supported by Helmholz association by the grant IK-RU-002.

References

[1] Skachkova A N, Skachkov N B and Alexeev G D Monte-Carlo simulation of lepton pair production in "p pbar → l+l- + X" events at $E_{beam} = 14$ GeV arXiv: hep-ph/0506139v2;
[2] *http://www-panda.gsi.de/archive/public/panda_loi.pdf*
[3] *http://www-panda.gsi.de/archive/public/panda_tpr.pdf*
[4] PANDA collab. Erni W et al Physics Performance Report for PANDA: Strong Interaction Studies with Antiprotons 2007 arXiv: hep-ex/0903.3905
[5] Skachkova A N and Skachkov N B On lepton pair production in proton-antiproton collisions at intermediate energies 2009 *Phys.Part.Nucl.Lett* **6** N4(153) 504-18.
[6] Sjöstrand T, Mrenna S and Skands P 2006 *JHEP* **05**(2006) 026 hep-ph/0603175

Journal of Physics: Conference Series **426** (2013) 012032 doi:10.1088/1742-6596/426/1/012032

Heavy quark transport at RHIC and LHC

Thomas Lang[1,2], Hendrik van Hees[1,2], Jan Steinheimer[3], Yu-Peng Yan[4,5], Marcus Bleicher[1,2]

[1] Frankfurt Institute for Advanced Studies (FIAS),Ruth-Moufang-Str. 1, 60438 Frankfurt am Main, Germany

[2] Institut für Theoretische Physik, Johann Wolfgang Goethe-Universität, Max-von-Laue-Str. 1, 60438 Frankfurt am Main, Germany

[3] Lawrence Berkeley National Laboratory, 1 Cyclotron Road, Berkeley, CA 94720, USA

[4] School of Physics, Institute of Science, Suranaree University of Technology, Nakhon Ratchasima 30000, Thailand

[5] Thailand Center of Excellence in Physics (ThEP), Commission on Higher Education, Bangkok 10400, Thailand

Abstract. We calculate the heavy quark evolution in heavy ion collisions and show results for the elliptic flow v_2 as well as the nuclear modification factor R_{AA} at RHIC and LHC energies. For the calculation we implement a Langevin approach for the transport of heavy quarks in the UrQMD (hydrodynamics + Boltzmann) hybrid model. As drag and diffusion coefficients we use a Resonance approach for elastic heavy-quark scattering and assume a decoupling temperature of the charm quarks from the hot medium of 130 MeV. At RHIC energies we use a coalescence approach at the decoupling temperature for the hadronization of the heavy quarks to D-mesons and B-mesons and a sub-following decay to heavy-flavor electrons using PYTHIA. At LHC we use an additional fragmentation mechanism to account for the higher transverse momenta reached at higher collision energies.

1. Introduction

Heavy quarks are an ideal probe for the QGP. They are produced in the primordial hard collisions of the nuclear reaction and therefore probe the created medium during its entire evolution process. When the system cools down they hadronize, and their decay products can finally be detected. Therefore, heavy-quark observables provide new insights into the interaction processes within the hot and dense medium. Two of the most interesting observables are the elliptic flow, v_2, and the nuclear modification factor, R_{AA}, of open-heavy-flavor mesons and their decay products. The measured large elliptic flow, v_2, of open-heavy-flavor mesons and the electrons from heavy-flavor decays underline that heavy quarks take part in the collective motion of the bulk medium, consisting of light quarks and gluons. The nuclear modification factor shows a large suppression of the open-heavy-flavor particles' spectra at high transverse momenta (p_T) compared to the findings in pp collisions. This also supports a high degree of thermalization of the heavy quarks with the bulk medium.

In this letter we explore the medium modification of heavy-flavor p_T spectra, using a hybrid model, consisting of the Ultra-relativistic Quantum Molecular Dynamics (UrQMD) model [1, 2] and a full (3+1)-dimensional ideal hydrodynamical model [3, 4] to simulate the bulk medium. The heavy-quark propagation in the medium is described by a relativistic Langevin approach [5]. Similar studies have recently been performed in a thermal fireball model with a combined coalescence-fragmentation approach [6, 7, 8, 9, 10, 11, 5], in an ideal hydrodynamics model with a lattice-QCD EoS [12, 13], in a model from Kolb and Heinz [14], in the BAMPS model [15, 16], the MARTINI model [17] as well as in further studies and model comparisons [18, 19, 20, 21, 22].

2. Description of the model

The UrQMD hybrid model has been developed to combine the advantages of transport theory and (ideal) fluid dynamics [23]. It uses initial conditions, generated by the UrQMD model [24, 25], for a full (3+1) dimensional ideal fluid dynamical evolution, including the explicit propagation of the baryon current. After a Cooper-Frye transition back to the transport description, the freeze out of the system is treated dynamically within the UrQMD approach. The hybrid model has been successfully applied to describe particle yields and transverse dynamics from AGS to LHC energies [23, 26, 27, 28, 29] and is therefore a reliable model for the flowing background medium.

The diffusion of a "heavy particles" in a medium consisting of "light particles" can be described with help of a Fokker-Planck equation [31, 32, 18, 31, 32, 33, 37, 6, 35, 36] as an approximation of the collision term of the corresponding Boltzmann equation. It can be mapped into an equivalent stochastic Langevin equation, suitable for numerical simulations.

The drag and diffusion coefficients for the heavy-quark propagation within this framework are taken from a Resonance approach [33], where the existence of D-mesons and B-mesons in the QGP phase is assumed, as well as a T-Matrix approach [6] in which quark-antiquark potentials are used for the calculation of the coefficients in the QGP.

The initial production of charm quarks in our approach is based on a Glauber approach. For the realization of the initial collision dynamics we use the UrQMD model. We perform a first UrQMD run excluding interactions between the colliding nuclei and save the nucleon-nucleon collision space-time coordinates. These coordinates are used in a second, full UrQMD run as possible production coordinates for the charm quarks.

As momentum distribution for the initially produced charm quarks at $\sqrt{s_{NN}} = 200$ GeV we use

$$\frac{dN}{2\pi p_T dp_T} = \frac{C \cdot \left(1 + A_1 \cdot p_T^2\right)^2}{\left(1 + A_2 \cdot p_T^2\right)^{A_3}}, \tag{1}$$

with $A_1 = 2.0/\mathrm{GeV}^2$, $A_2 = 0.1471/\mathrm{GeV}^2$ and $A_3 = 21.0$ and for bottom quarks

$$\frac{dN}{2\pi p_T dp_T} = \frac{C}{\left(1 + A_1 \cdot p_T^2\right)^{A_2}}, \tag{2}$$

with $A_1 = 0.0173/\mathrm{GeV}^2$ and $A_2 = 5.04$. These distributions are taken from [37, 6]. The p_T distribution for charm quarks at 2.76 TeV is obtained from a fit to PYTHIA calculations.

$$\frac{dN}{2\pi p_T dp_T} = \frac{C}{(1 + A_1 \cdot p_T^2)^{A_2}} \tag{3}$$

with the coefficients $A_1 = 0.379/\mathrm{GeV}^2$ and $A_2 = 5.881$. C is an arbitrary normalization constant. with the unit $1/\mathrm{GeV}^2$.

Starting with these distributions as initial conditions we propagate the heavy quarks at each hydro-timestep. We use the UrQMD/hydro's cell velocities, the cell temperature, the size of the time-step, and the γ-factor for the calculation of the momentum transfer, propagating all quarks independently. Our approach provides us only with the heavy-quark distributions. Since heavy quarks cannot be measured directly in experiments we include a hadronization mechanism for D-mesons and B-mesons, via the use of a quark-coalescence mechanism. To implement this coalescence we perform our Langevin calculation until the decoupling temperature is reached. Subsequently we add the momenta of light quarks to those of the heavy quarks.

3. Results

First we performed our calculations in Au+Au collisions at $\sqrt{s_{NN}} = 200$ GeV in a centrality range of 20%-40%. To compare our results to the single-electron spectra measured by PHENIX we use PYTHIA for the decay of the heavy quarks to heavy-flavor electrons and apply a rapidity cut of $|y| < 0.35$. Fig. 1 (left) shows our results for the elliptic flow v_2. For a decoupling temperature of 130 MeV we obtain a reasonable agreement with the experimental data except for low p_T bins. Here a depletion effect can be seen. In case of a developed elliptic flow particles in x direction have a higher velocity than in y direction. Consequently there is a depletion of particles with high v_x in the low p_T region and thus a smaller elliptic flow. This effect is more important for heavier particles and a larger radial flow [38, 39].

In Fig. 1 (right) the nuclear modification factor R_{AA} for the heavy-flavor decay electrons is depicted.

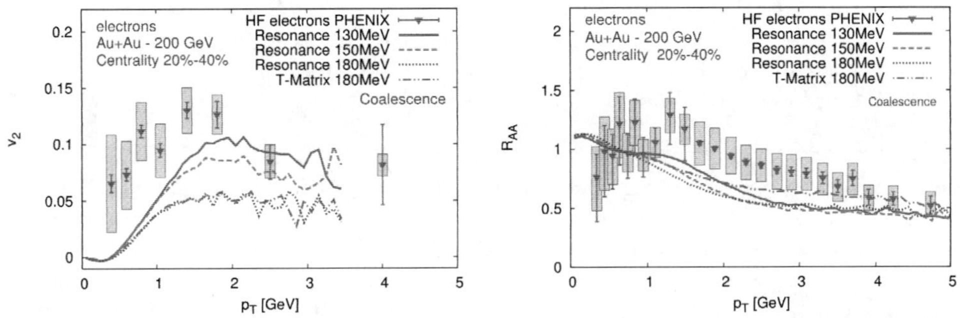

Figure 1. (Color online) Elliptic flow v_2 (left) and nuclear modification factor R_{AA} (right) of electrons from heavy quark decays in Au+Au collisions at $\sqrt{s_{NN}} = 200$ GeV using a coalescence mechanism. We use a rapidity cut of $|y| < 0.35$. For a decoupling temperature of 130 MeV we get a reasonable agreement to data [40]

Also here we obtain a good agreement with the data, especially in case of using the T-Matrix coefficients or a low decoupling temperature.

Now we performed the same calculations, but in Pb+Pb collisions at $\sqrt{s}_{NN} = 2.76$ TeV in a centrality range of 30%-50%. The analysis is done in a rapidity cut of $|y| < 0.35$ in line with the ALICE data. Here we made use of the coalescence mechanism with a decoupling temperature of 130 MeV only since we achieved the best results using this configuration at RHIC energies. In the ALICE experiment D-mesons are measured. Therefore we do not need to perform the decay to electrons this time. Fig. 2 (left) depicts our results for the elliptic flow compared to ALICE measurements. Additionally, apart form the calculation using the coalescence mechanism, also a calculation using a fragmentation mechanism is shown, since fragmentation might get more important at higher p_T bins, as measured at LHC. As fragmentation mechanism we used the Peterson fragmentation [42].

$$D_Q^H(z) = \frac{N}{z[1-(1/z)-\epsilon_Q/(1-z)]^2},$$

Here N is a normalization constant, z the relative-momentum fraction obtained in the fragmentation of the charm quarks and $\epsilon_Q = 0.05$. Both v_2 calculations are in agreement with the ALICE data set. Using the fragmentation function a sharper rise of the elliptic flow at low p_T is reached, while at medium p_T the flow using the coalescence approach is stronger. At high p_T both hadronization mechanisms lead to similar results.

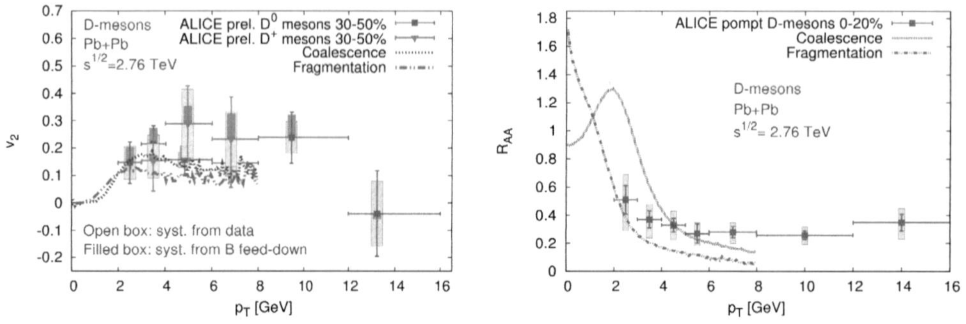

Figure 2. (Color online) Left: Flow v_2 of D-mesons in Pb+Pb collisions at $\sqrt{s_{NN}} = 2.76\,\text{TeV}$ compared to data from the ALICE experiment. (Talk by Z. Conesa del Valle at QM 2012, data not published yet.) A rapidity cut of $|y| < 0.35$ is employed. Right: R_{AA} of D-mesons in Pb+Pb collisions at $\sqrt{s_{NN}} = 2.76$ TeV compared to experimental data from ALICE [41]. A rapidity cut of $|y| < 0.35$ is employed.

A complementary view on the drag and diffusion coefficients is provided by the nuclear suppression factor R_{AA}. Figure 2 (right) shows the calculated nuclear modification factor R_{AA} of D-mesons compared to the experimental ALICE measurements [41]. In line with the experimental data the simulation is done for a more central bin of $\sigma/\sigma_{to} = 0\%$-20%. In case of the coalescence approach we find a maximum of the R_{AA} at about 2 GeV followed by a sharp decline to an R_{AA} of about 0.2 at high p_T. The "peak-like" structure of the R_{AA} is due to the coalescence of low p_T heavy quarks that are pushed to higher p_T due to the transverse momentum of the light quarks. The fragmentation approach leads to a different result at low p_T. A very sharp R_{AA} drop-off from low to high p_T is seen. At high p_T the two approaches nearly converge. Concerning the difference of the results using the fragmentation and coalescence mechanism new v_2 and R_{AA} measurements, especially at low p_T, would be very helpful to draw conclusions on the hadronization mechanism at LHC.

To summarize, we presented in this letter our results on the medium modification of heavy quarks at RHIC and LHC energies using the nuclear modification factor R_{AA} and the elliptic flow v_2 as observables. At RHIC energies we compared different sets for drag and diffusion coefficients and obtained the best agreement to experimental measurements if using a Resonance model with a decoupling temperature of 130 MeV. At LHC we compared a coalescence approach and a fragmentation approach as hadronization mechanism using the Resonance model at a decoupling temperature of 130 MeV. Both approaches describe the elliptic flow v_2 in pretty good agreement with the experimental data while for the R_{AA} a major disagreement between our models at low p_T can be seen that needs to be resolved by new measurements.

4. ACKNOWLEDGMENTS

We are grateful to the Center for Scientific Computing (CSC) at Frankfurt for providing computing resources. T. Lang gratefully acknowledges support from the Helmholtz Research School on Quark Matter Studies. This work is supported by the Hessian LOEWE initiative through the Helmholtz International Center for FAIR (HIC for FAIR). J. S. acknowledges a Feodor Lynen fellowship of the Alexander von Humboldt foundation. This work is supported by the Office of Nuclear Physics in the US Department of Energy's Office of Science under Contract No. DE-AC02-05CH11231.

[1] S. A. Bass *et al.*, Prog. Part. Nucl. Phys. **41** (1998) 255 [Prog. Part. Nucl. Phys. **41** (1998) 225] [arXiv:nucl-th/9803035].
[2] M. Bleicher *et al.*, J. Phys. **G25** (1999) 1859 [arXiv:hep-ph/9909407].
[3] D. H. Rischke, S. Bernard and J. A. Maruhn, Nucl. Phys. A **595** (1995) 346 [arXiv:nucl-th/9504018].
[4] D. H. Rischke, Y. Pursun and J. A. Maruhn, Nucl. Phys. A **595** (1995) 383 [Erratum-ibid. A **596** (1996) 717] [arXiv:nucl-th/9504021].
[5] R. Rapp and H. van Hees, arXiv:0903.1096 [hep-ph].
[6] H. van Hees, M. Mannarelli, V. Greco and R. Rapp, Phys. Rev. Lett. **100** (2008) 192301 [arXiv:0709.2884 [hep-ph]].
[7] H. van Hees, V. Greco and R. Rapp, arXiv:0706.4456 [hep-ph].
[8] V. Greco, H. van Hees and R. Rapp, arXiv:0709.4452 [hep-ph].
[9] H. van Hees, M. Mannarelli, V. Greco and R. Rapp, Eur. Phys. J. C **61** (2009) 799 [arXiv:0808.3710 [hep-ph]].
[10] R. Rapp, D. Cabrera, V. Greco, M. Mannarelli and H. van Hees, arXiv:0806.3341 [hep-ph].
[11] R. Rapp and H. van Hees, arXiv:0803.0901 [hep-ph].
[12] M. He, R. J. Fries and R. Rapp, arXiv:1204.4442 [nucl-th].
[13] M. He, R. J. Fries and R. Rapp, arXiv:1208.0256 [nucl-th].
[14] J. Aichelin, P. B. Gossiaux and T. Gousset, Acta Phys. Polon. B **43** (2012) 655 [arXiv:1201.4192 [nucl-th]].
[15] J. Uphoff, O. Fochler, Z. Xu and C. Greiner, Phys. Rev. C **84** (2011) 024908 [arXiv:1104.2295 [hep-ph]].
[16] J. Uphoff, O. Fochler, Z. Xu and C. Greiner, Phys. Lett. B **717** (2012) 430 [arXiv:1205.4945 [hep-ph]].
[17] C. Young, B. Schenke, S. Jeon and C. Gale, Phys. Rev. C **86** (2012) 034905 [arXiv:1111.0647 [nucl-th]].
[18] G. D. Moore and D. Teaney, Phys. Rev. C **71** (2005) 064904 [arXiv:hep-ph/0412346].
[19] I. Vitev, A. Adil and H. van Hees, J. Phys. **G34** (2007) S769 [arXiv:hep-ph/0701188].
[20] P. B. Gossiaux, J. Aichelin, T. Gousset and V. Guiho, J. Phys. **G37** (2010) 094019 [arXiv:1001.4166 [hep-ph]].
[21] P. B. Gossiaux, S. Vogel, H. van Hees, J. Aichelin, R. Rapp, M. He and M. Bluhm, arXiv:1102.1114 [hep-ph].
[22] P. Gossiaux, J. Aichelin and T. Gousset, Prog. Theor. Phys. Suppl. **193** (2012) 110 [arXiv:1201.4038 [hep-ph]].
[23] H. Petersen, J. Steinheimer, G. Burau, M. Bleicher and H. Stocker, Phys. Rev. C **78** (2008) 044901 [arXiv:0806.1695 [nucl-th]].
[24] S. A. Bass, A. Dumitru, M. Bleicher, L. Bravina, E. Zabrodin, H. Stoecker and W. Greiner, Phys. Rev. C **60** (1999) 021902 [arXiv:nucl-th/9902062].
[25] A. Dumitru, S. A. Bass, M. Bleicher, H. Stoecker and W. Greiner, Phys. Lett. B **460** (1999) 411 [arXiv:nucl-th/9901046].
[26] J. Steinheimer, M. Bleicher, H. Petersen, S. Schramm, H. Stocker and D. Zschiesche, Phys. Rev. C **77** (2008) 034901 [arXiv:0710.0332 [nucl-th]].
[27] J. Steinheimer, V. Dexheimer, H. Petersen, M. Bleicher, S. Schramm and H. Stoecker, Phys. Rev. C **81** (2010) 044913 [arXiv:0905.3099 [hep-ph]].
[28] H. Petersen, G. Y. Qin, S. A. Bass and B. Muller, Phys. Rev. C **82** (2010) 041901 [arXiv:1008.0625 [nucl-th]].
[29] H. Petersen, Phys. Rev. C **84** (2011) 034912 [arXiv:1105.1766 [nucl-th]].
[30] J. Steinheimer, S. Schramm and H. Stocker, Phys. Rev. C **84** (2011) 045208 [arXiv:1108.2596 [hep-ph]].
[31] B. Svetitsky, Phys. Rev. D **37** (1988) 2484.
[32] M. Golam Mustafa, D. Pal and D. Kumar Srivastava, Phys. Rev. C **57** (1998) 889 [Erratum-ibid. C **57** (1998) 3499] [nucl-th/9706001].
[33] H. van Hees and R. Rapp, Phys. Rev. C **71** (2005) 034907 [nucl-th/0412015].
[34] H. van Hees, V. Greco and R. Rapp, Phys. Rev. C **73** (2006) 034913 [nucl-th/0508055].
[35] P. B. Gossiaux and J. Aichelin, Phys. Rev. C **78** (2008) 014904 [arXiv:0802.2525 [hep-ph]].
[36] M. He, R. J. Fries and R. Rapp, Phys. Lett. B **701** (2011) 445 [arXiv:1103.6279 [nucl-th]].
[37] H. van Hees, V. Greco and R. Rapp, Phys. Rev. C **73** (2006) 034913 [arXiv:nucl-th/0508055].
[38] P. Huovinen, P. F. Kolb, U. W. Heinz, P. V. Ruuskanen and S. A. Voloshin, Phys. Lett. B **503** (2001) 58 [arXiv:hep-ph/0101136].
[39] D. Krieg and M. Bleicher, Eur. Phys. J. A **39** (2009) 1 [arXiv:0806.0736 [nucl-th]].
[40] A. Adare *et al.* [PHENIX Collaboration], Phys. Rev. C **84** (2011) 044905 [arXiv:1005.1627 [nucl-ex]].
[41] *et al.* [ALICE Collaboration], JHEP **1209** (2012) 112 [arXiv:1203.2160 [nucl-ex]].
[42] C. Peterson, D. Schlatter, I. Schmitt and P. M. Zerwas, Phys. Rev. D **27** (1983) 105.

Journal of Physics: Conference Series **426** (2013) 012033 doi:10.1088/1742-6596/426/1/012033

Dilepton production at SIS energies with the UrQMD model

Stephan Endres[1,2] **and Marcus Bleicher**[1,2]

[1]Institut für Theoretische Physik, Universität Frankfurt, Max-von-Laue-Straße 1, 60438 Frankfurt, Germany

[2]Frankfurt Institute for Advanced Studies, Ruth-Moufang-Straße 1, 60438 Frankfurt, Germany

E-mail: `endres@th.physik.uni-frankfurt.de`

Abstract. The production of lepton pairs in elementary and nucleus-nucleus collisions at SIS energies is studied within the Ultra-relativistic Quantum Molecular Dynamics transport approach. In comparison to previous calculations, we now included an improved treatment of the Δ_{1232} resonance. The resulting invariant mass and transverse momentum spectra are compared to available HADES data. In general, our results give a good description of the data. With regard to the observed overestimation of the ρ contribution, the different production channels and the respective production cross-sections are investigated. A problem remains the lack of experimental data for cross-sections and branching ratios of intermediate baryonic resonances, which are crucial for the ρ production. However, recent HADES cross-section measurements close to the ρ threshold might help to further constrain the model parameters.

1. Motivation

Dileptons are clean probes of hot and dense matter created in nuclear collisions. As they do not interact strongly, there is no interplay between them and the surrounding hadronic matter after their production. In consequence, they can provide insights to the physics of strongly interacting matter at high baryon densities and temperatures and deliver information on medium modifications of vector mesons and the restoration of chiral symmetry [1, 2]. As neutral vector mesons can directly decay into lepton pairs, one is able to deduce specific information concerning the mesons and their in-medium properties from the experimentally measured dilepton spectra.

Although several experimental and theoretical studies were conducted during the last years, it is still controversial how exactly the vector meson mass or width may change in hot nuclear matter. Proposed scenarios for in-medium effects are a dropping of the vector meson mass [3] and a broadening of the spectral function [4]. At low collision energies, dilepton measurements were conducted by the DLS experiment at BEVALAC [5]. The results showed an excess in the invariant mass spectra which could not completely be resolved by the proposed in-medium effects (the so-called "DLS puzzle"). Recently the measurements were confirmed in the same energy regime with higher precision and resolution by the HADES Collaboration [6, 7, 8, 9].

In these proceedings we present our ongoing work with a detailed study of lepton pair production in elementary and heavy-ion collisions at SIS energies. We use the Ultra-relativistic Quantum Molecular Dynamics (UrQMD) model for the calculations and compare our results with the experimental data of the HADES Collaboration.

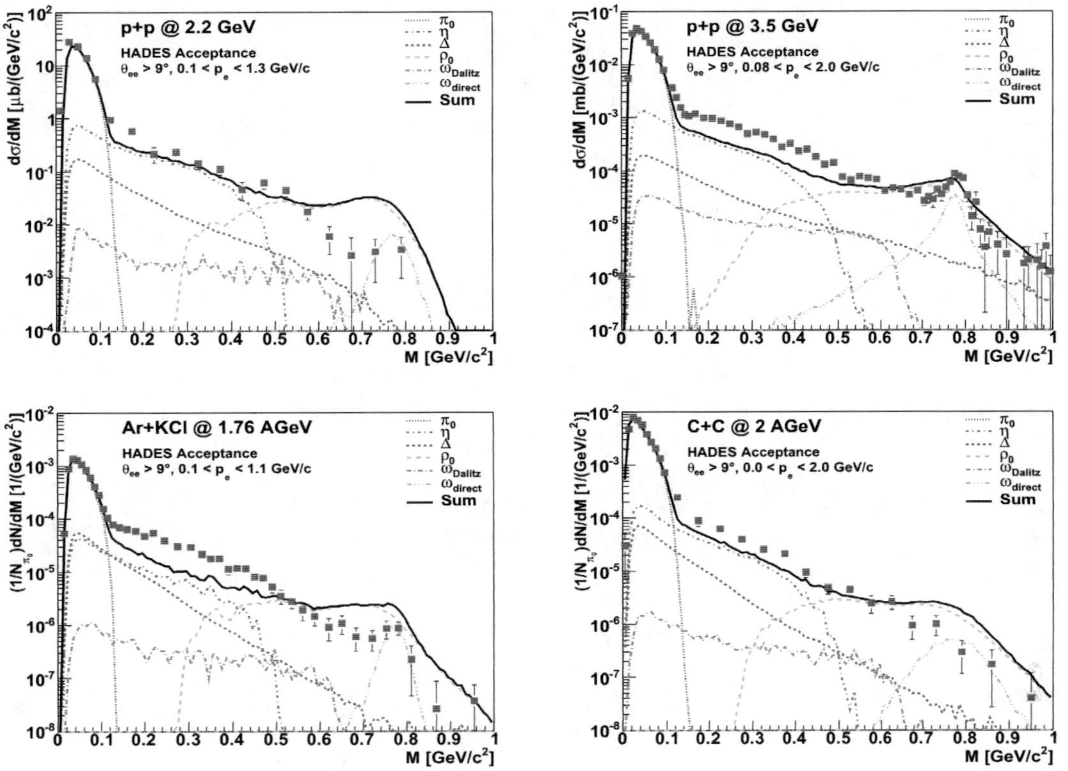

Figure 1. (Color Online.) Invariant mass spectra obtained with the UrQMD transport model for p+p and A+A collisions at SIS energies. The results are corrected for the HADES acceptance and compared with data from the HADES Collaboration [6, 7, 8, 9].

2. The Model

The microscopic Ultra-relativistic Quantum Molecular Dynamics model is a hadronic non-equilibrium transport approach based on the quantum molecular dynamics concept [10]. It includes all baryons and mesons with masses up to 2.2 GeV. The dynamics of reactions in a nuclear collision are described in terms of a n-particle phase-space distribution.

For lower beam energies, vector meson production in UrQMD takes place via two different reaction mechanisms: Either they are produced in a two step process that proceeds via the excitation of a heavier resonance and the subsequent decay (e.g. N+N $\rightarrow$ N+N* $\rightarrow$ N+N+ρ) or directly in collision processes, e.g $\pi + \pi \rightarrow \rho$. The probability for a resonance decaying into a certain channel is determined by the mass dependent branching ratio which we use in accordance with the values given by the Particle Data Group [11]. All further details concerning the general properties of the approach can be obtained from [12, 13, 14].

Regarding dilepton production, several channels are included in the calculations: Direct decays of ρ^0, ω and ϕ vector mesons, Dalitz decays of pseudoscalar mesons, like π^0, η, η', and of the ω vector meson respectively. Also the lowest Delta resonance, Δ_{1232}, contributes via Dalitz decay. The Dalitz decay for pseudoscalar and vector mesons is decomposed into a two-step process, first the corresponding decay into a virtual photon and then the subsequent decay of the photon via electromagnetic conversion. The form factors used in this description are obtained by the vector meson dominance model [15, 16]. For the Delta Dalitz decay we

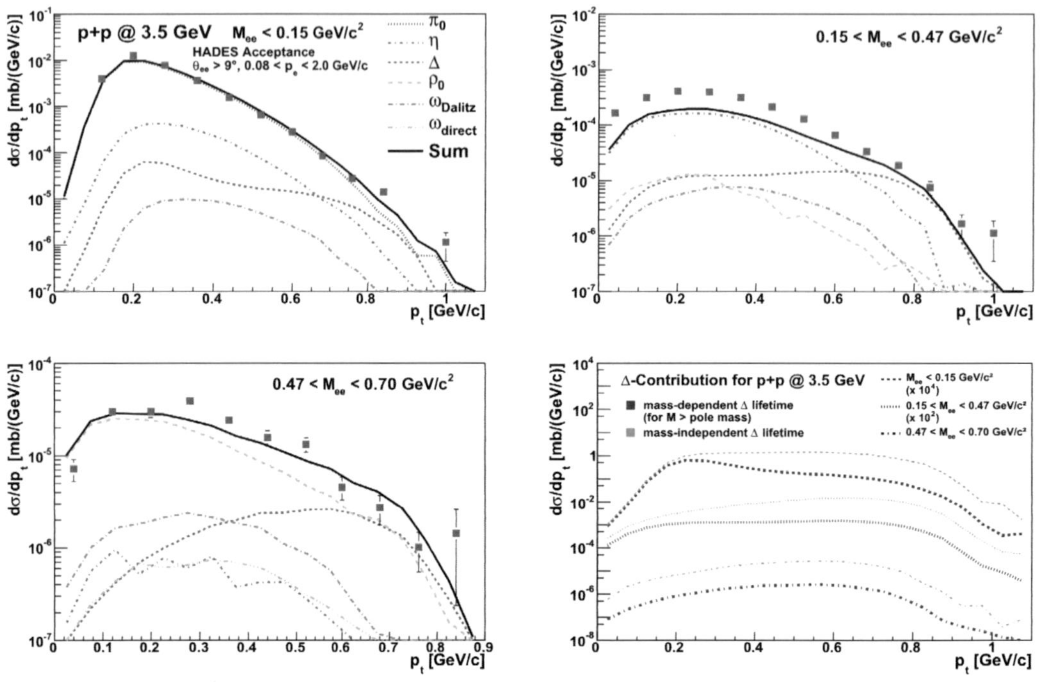

Figure 2. (Color Online.) Transverse momentum spectra obtained with the UrQMD transport model for p+p collisions at $E_{lab} = 3.5$ GeV. The results for the different mass regions are acceptance corrected and compared with data by the HADES Collaboration [9]. The plot at the bottom right shows the difference in the Delta contribution between the present calculations for mass-dependent Δ_{1232} lifetime and the mass-independent case, as used previously.

use the parametrization by Wolf [17] with a modified coupling to fit the radiative decay width [18]. To calculate the dilepton emission rates, we use the so called "shining method", i.e. that a resonance steadily emits dileptons over its lifetime. This method increases the statistics and also includes resonances that do not decay but are reabsorbed in the medium [19].

In contrast to previous dilepton calculations with UrQMD (see ref. [20, 21, 22]) a different treatment of the Δ_{1232} resonance is now included. In general the model uses mass dependent widths $\Gamma(M)$ for the resonance decay, but the lifetime is treated mass-independent as $\tau = 1/\Gamma(m_{pole})$, with m_{pole} being the pole mass, to avoid unphysically high lifetimes for low masses below the pole mass. Especially for far off-shell Delta resonances this might lead to the problem of overestimating the dilepton spectra at high masses and transverse momentum. In the Δ_{1232} case, we therefore now employ a mass-dependent lifetime for masses higher than the pole mass.

For further details on dilepton calculations within the UrQMD approach see ref. [20].

3. Results

In Figure 1 dilepton invariant mass spectra for p+p (top figure) and A+A collisions (bottom figure) are shown. At the top the UrQMD results for elementary collisions at $E_{lab} = 2.2$ and 3.5 GeV are plotted together with the experimental data [6, 7, 8, 9]. At 2.2 GeV the energy is sufficient to reach the pole mass of the ω and ρ while at 3.5 GeV the full phase space is open for vector meson production. In general we see a good agreement of our calculations for p+p with the HADES data. However, around the ρ pole mass a significant overestimation of the yield is

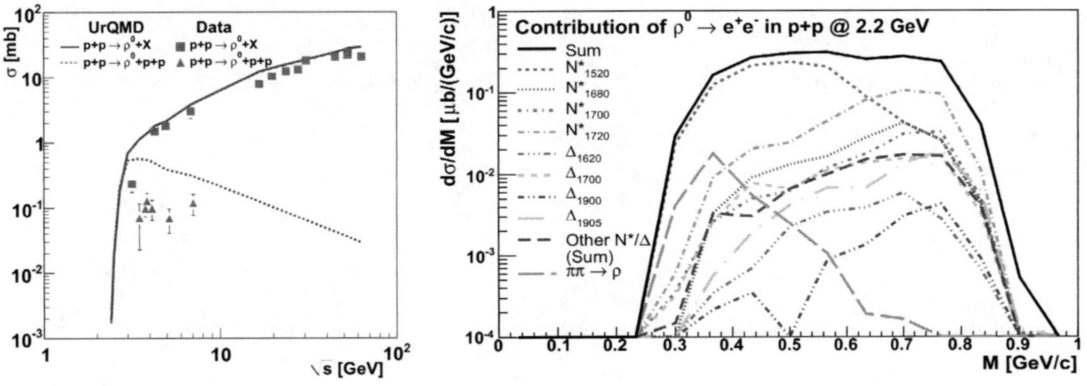

Figure 3. (Color online.) Left: Cross sections for inclusive and exclusive ρ^0 production in $p+p$ reactions within the UrQMD model. The results are compared to the experimental data from [9, 23]. Right: Different contribution channels to the $\rho^0 \to e^+e^-$ production via resonance decay and two pion reactions in the present calculations.

observed at 2.2 GeV, which it is not that prominent at 3.5 GeV. In contrast, the region between 150 and 500 MeV/c^2 invariant mass is underestimated in both cases, more significantly at 3.5 GeV.

In nucleus-nucleus collisions we expect additional effects compared to elementary reactions. Here we have not only p+p but also p+n and n+n collisions and secondary interactions, depending on system size and energy. Also the Fermi momentum of the nucleus and Pauli blocking are an issue. Such effects can be treated by the UrQMD model, while no further in-medium modifications are included in the current calculations. Looking at the results for Ar+KCl at 1.76 AGeV and C+C at 2 AGeV at the bottom of Figure 1, we observe (as in elementary reactions) too many dileptons that stem from the ρ^0 resonance, especially in the high-mass tail. Comparing the three cases for reactions at similar energies around ~2 AGeV, we see that the Delta Dalitz contribution increases with the system size. It is larger in C+C than in p+p and becomes dominant at low masses for Ar+KCl. Furthermore, there is an excess in the mass region dominated by the η Dalitz decay that increases with system size. However, it is not clear from the present investigation whether this might be related to additional in-medium modifications, as we see a similar effect in p+p at 3.5 GeV if compared to 2.2 GeV.

The transverse momentum distributions shown in Figure 2 describe the data at 3.5 GeV very well, especially in the low mass region dominated by the π^0 Dalitz decay. Here we see a clear improvement by the changed Delta lifetime treatment if we compare the spectra with previous results [9] where the Delta contribution is above the data at high p_t and high masses. Also the ρ meson contribution from our calculation fits the data, with only a slight overestimation at low transverse momentum. The underestimation at intermediate energies is again visible here in the mass range from 150 to 470 MeV/c^2.

As our rho contribution in parts overestimates the data significantly, we studied the production cross-sections for ρ^0 production in elementary reactions in Figure 3 (left). The comparison with experimental data [9, 23] shows a good description for the relevant inclusive channel, especially at higher energies. At threshold the recent HADES measurement at 3.5 GeV ($\hat{=}\ \sqrt{s} = 3.17$ GeV) indicates that the production rate is too high in the present model, leading to the observed overestimation of the invariant mass spectra. The same comparison for the η (which is not shown here) exhibits that this cross-section, in contrast, is underestimated at the threshold. Considering the different channels that add to the ρ^0 dilepton production, e.g. for p+p at 2.2 GeV (see Figure 3, right), the main contribution below the pole mass stems from the N^*_{1520}, while around the pole mass there is a multitude of different resonances that contribute. Other processes than resonance decay (e.g $\pi\pi \rightarrow \rho$) are mostly negligible in the cases considered here. This indicates one of the difficulties of the present hadronic resonance approach: The ρ cross-section is not implemented explicitly but one uses intermediate baryonic resonances. Unfortunately, for the production of these resonances and the subsequent decay the uncertainties concerning branching ratios and production cross-sections are quite high, due to little available data that could constrain these model parameters. Therefore the shape of the ρ mass distribution and the resonances contributing to it differ among transport models (e.g. compare our results with recent GiBUU calculations [24]), depending largely on the implementation of intermediate resonant states and their branchings.

4. Summary and Conclusion

Dilepton production in elementary and nucleus-nucleus collisions at SIS energies has been studied within the UrQMD approach and compared to HADES data. We achieve a good description of the experimental results for invariant mass and transverse momentum spectra. Particularly the changed treatment of the Delta resonance lifetime leads to an improved description of the p_t distribution. However, the comparison to data and experimental cross-sections from the HADES Collaboration also shows that the ρ production rate close to the threshold might be too high while the η cross-section seems to be underestimated. For an accurate and detailed study of possible in-medium modifictions in the future further detailed investigations of the elementary reactions as a baseline are necessary.

Acknowledgments

We thank the HADES Collaboration, especially T. Galatyuk and C. Sturm, for providing their acceptance filters and data. This work was supported by BMBF, HIC for FAIR and H-QM.

References

[1] Xia L H, Ko C M, Xiong L and Wu J Q 1988 *Nucl. Phys.* A **485** 721
[2] Rapp R and Wambach J 2000 *Adv. Nucl. Phys.* **25** 1 (*Preprint* hep-ph/9909229)
[3] Brown G E and Rho M 1991 *Phys. Rev. Lett.* **66** 2720
[4] Rapp R, Chanfray G and Wambach J 1997 *Nucl. Phys.* A **617** 472 (*Preprint* hep-ph/9702210)
[5] Porter R J *et al.* (DLS Collab.) 1997 *Phys. Rev. Lett.* **79** 1229 (*Preprint* nucl-ex/9703001)
[6] Agakichiev G *et al.* (HADES Collab.) 2007 *Phys. Rev. Lett.* **98** 052302 (*Preprint* nucl-ex/0608031)
[7] Agakishiev G *et al.* (HADES Collab.) 2011 *Phys. Rev.* C **84** 014902 (*Preprint* arXiv:1103.0876 [nucl-ex])
[8] Agakishiev G *et al.* (HADES Collab.) 2012 *Phys. Rev.* C **85** 054005 (*Preprint* arXiv:1203.2549 [nucl-ex])
[9] Agakishiev G *et al.* (HADES Collab.) 2012 *Eur. Phys. J.* A **48** 64 (*Preprint* arXiv:1112.3607 [nucl-ex])
[10] Aichelin J and Stoecker H 1986 *Phys. Lett.* B **176** 14
[11] Beringer J *et al.* (Particle Data Group Collab.) 2012 *Phys. Rev.* D **86** 010001
[12] Bass S A *et al.* 1998 *Prog. Part. Nucl. Phys.* **41** 255 (*Preprint* nucl-th/9803035)
[13] Bleicher M *et al.* 1999 *J. Phys.* G **25** 1859 (*Preprint* hep-ph/9909407)
[14] Petersen H, Bleicher M, Bass S A and Stoecker H 2008 *Preprint* arXiv:0805.0567 [hep-ph]
[15] Li G Q, Ko C M, Brown G E and Sorge H 1996 *Nucl. Phys.* A **611** 539 (*Preprint* nucl-th/9611037)
[16] Landsberg L G 1985 *Phys. Rept.* **128** 301
[17] Wolf G, Batko G, Cassing W, Mosel U, Niita K and Schaefer M 1990 *Nucl. Phys.* A **517** 615
[18] Bratkovskaya E L *et al.* 1999 *Nucl. Phys.* A **653** 301 (*Preprint* nucl-th/9903009)
[19] Li G Q and Ko C M 1995 *Nucl. Phys.* A **582** 731 (*Preprint* nucl-th/9407016)
[20] Schmidt K, Santini E *et al.* 2009 *Phys. Rev.* C **79** 064908 (*Preprint* arXiv:0811.4073 [nucl-th])
[21] Vogel S, Petersen H, *et al.* 2008 *Phys. Rev.* C **78** 044909 (*Preprint* arXiv:0710.4463 [hep-ph])
[22] Santini E, Steinheimer J *et al.* 2011 *Phys. Rev.* C **84** 014901 (*Preprint* arXiv:1102.4574 [nucl-th])
[23] Flaminio V, Moorhead W G, Morrison D R O and Rivoire N 1984 *CERN-HERA-84-01*
[24] Weil J *et al.* 2012 *Eur. Phys. J.* A **48** 111; Erratum-ibid. 2012 A **48** 150 (*Preprint* arXiv:1203.3557 [nucl-th])

Journal of Physics: Conference Series **426** (2013) 012034 doi:10.1088/1742-6596/426/1/012034

Proton induced dielectron radiation off Nb: P_t and Y distributions

M. Lorenz[7], M. Weber[9], G. Agakishiev[6], C. Behnke[7], D. Belver[16], A. Belyaev[6], J.C. Berger-Chen[8], A. Blanco[1], C. Blume[7], M. Böhmer[9], P. Cabanelas[16], S. Chernenko[6], C. Dritsa[10], A. Dybczak[2], E. Epple[8], L. Fabbietti[8], O. Fateev[6], P. Fonte[1,a], J. Friese[9], I. Fröhlich[7], T. Galatyuk[4,b], J. A. Garzón[16], K. Gill[7], M. Golubeva[11], D. González-Díaz[4], F. Guber[11], M. Gumberidze[14], S. Harabasz[4], T. Hennino[14], R. Holzmann[3], P. Huck[9], C. Hhne[10], A. Ierusalimov[6], A. Ivashkin[11], M. Jurkovic[9], B. Kämpfer[5,c], T. Karavicheva[11], I. Koenig[3], W. Koenig[3], B. W. Kolb[3], G. Korcyl[2], G. Kornakov[16], R. Kotte[5], A. Krása[15], E. Krebs[7], F. Krizek[15], H. Kuc[2,14], A. Kugler[15], A. Kurepin[11], A. Kurilkin[6], P. Kurilkin[6], V. Ladygin[6], R. Lalik[8], S. Lang[3], K. Lapidus[8], A. Lebedev[12], L. Lopes[1], L. Maier[9], A. Mangiarotti[1], J. Markert[7], V. Metag[10], J. Michel[7], C. Müntz[7], R. Münzer[8], L. Naumann[5], M. Palka[2], Y. Parpottas[13,d], V. Pechenov[3], O. Pechenova[7], J. Pietraszko[7], W. Przygoda[2], B. Ramstein[14], L. Rehnisch[7], A. Reshetin[11], A. Rustamov[7], A. Sadovsky[11], P. Salabura[2], T. Scheib[7], H. Schuldes[7], J. Siebenson[8], Yu.G. Sobolev[15], S. Spataro[e], H. Ströbele[7], J. Stroth[7,3], P Strzempek[2], C. Sturm[3], O. Svoboda[15], A. Tarantola[7], K. Teilab[7], P. Tlusty[15], M. Traxler[3], H. Tsertos[13], T. Vasiliev[6], V. Wagner[15], C. Wendisch[5,c], J. Wüstenfeld[5], S. Yurevich[3], Y. Zanevsky[6]

(HADES collaboration)

[1]LIP-Laboratório de Instrumentação e Física Experimental de Partículas , 3004-516 Coimbra, Portugal
[2]Smoluchowski Institute of Physics, Jagiellonian University of Cracow, 30-059 Kraków, Poland
[3]GSI Helmholtzzentrum für Schwerionenforschung GmbH, 64291 Darmstadt, Germany
[4]Technische Universität Darmstadt, 64289 Darmstadt, Germany
[5]Institut für Strahlenphysik, Helmholtz-Zentrum Dresden-Rossendorf, 01314 Dresden, Germany
[6]Joint Institute of Nuclear Research, 141980 Dubna, Russia
[7]Institut für Kernphysik, Goethe-Universität, 60438 Frankfurt, Germany
[8]Excellence Cluster 'Origin and Structure of the Universe' , 85748 Garching, Germany
[9]Physik Department E12, Technische Universität München, 85748 Garching, Germany
[10]II.Physikalisches Institut, Justus Liebig Universität Giessen, 35392 Giessen, Germany
[11]Institute for Nuclear Research, Russian Academy of Science, 117312 Moscow, Russia
[12]Institute of Theoretical and Experimental Physics, 117218 Moscow, Russia
[13]Department of Physics, University of Cyprus, 1678 Nicosia, Cyprus
[14]Institut de Physique Nucléaire-Université Paris Sud, F-91406 Orsay Cedex, France
[15]Nuclear Physics Institute, Academy of Sciences of Czech Republic, 25068 Rez, Czech Republic
[16]LabCAF. Dpto. Física de Partículas, 15706 Santiago de Compostela, Spain
[a] also at ISEC Coimbra, Coimbra, Portugal
[b] also at ExtreMe Matter Institute EMMI, 64291 Darmstadt, Germany
[c] also at Technische Universität Dresden, 01062 Dresden, Germany
[d] also at Frederick University, 1036 Nicosia, Cyprus

[e] also at Dipartimento di Fisica Generale and INFN, Università di Torino, 10125 Torino, Italy

E-mail: m.lorenz@gsi.de

Abstract. Following our recent letter [1] on inclusive e^+e^- pair production in proton induced reactions at E_{kin} = 3.5 GeV on the nucleus Nb, we present here in addition the transverse and rapidity distributions for various e^+e^- invariant mass bins and compare them to reference data measured in p+p reactions.

1. Introduction

The prospect to relate hadron properties inside a strongly interacting medium with chiral symmetry restoration has motivated plenty of work, both in theoretical and experimental physics. The spontaneous breaking of chiral symmetry leads to a non vanishing value of the two-quark condensate, which is related to the dynamical generation of hadron masses. Focussing mainly on vector mesons, various models predict relatively strong changes of particle masses and/or widths already at normal nuclear matter density ρ_0 [2, 3, 4, 5]. However a consistent picture of in-medium hadron properties has not yet emerged and more experimental input is needed. Moreover such modifications are expected to be most pronounced for particles with small relative momenta to the surrounding medium according to hadronic models [6], a region which is challenging to access in experiment.

Experimentally, in-medium properties can be directly studied using the decay of vector mesons into an e^+e^- pair, since electrons and positrons are not affected by strong final state interactions. Although medium modifications are expected to be stronger in heavy-ion collisions, compared proton-, pion- or photon- induced reactions on nuclei, measured observables represent an average over the complete space-time evolution of the temperature and the density of the system and are hence more complicated to model. On the other hand, in induced reactions on nuclei the system does not undergo a noticeable density and temperature evolution in time, hence conditions of the system are better defined. The experimental drawback is that, for a measurement sensitive to the in-medium spectral shape, the decay to an e^+e^- pair has to take place inside the nucleus. Therefore good acceptance for decays of low momentum vector mesons is of crucial importance, in particular, for the relatively long living ω and ϕ mesons.

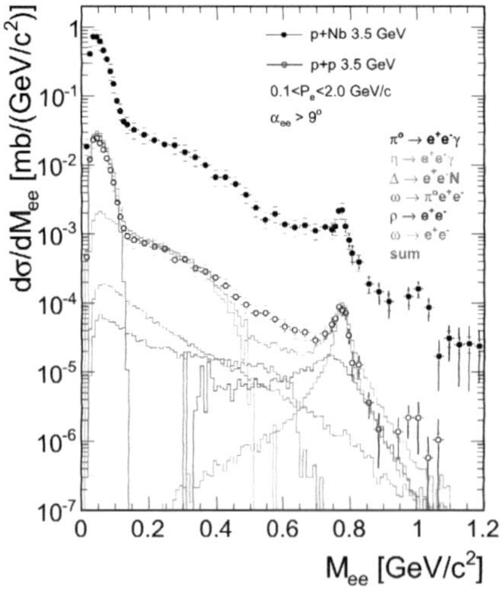

Figure 1. Dielectron cross sections measured in p+Nb and p+p reactions at a kinetic beam energy of 3.5 GeV. In case of the p+p data a comparison to a calculated PYTHIA/PLUTO dielectron cocktail is shown in addition (the different sources are listed in the legend).

FAIRNESS 2012: FAIR NExt Generation of ScientistS 2012 IOP Publishing
Journal of Physics: Conference Series **426** (2013) 012034 doi:10.1088/1742-6596/426/1/012034

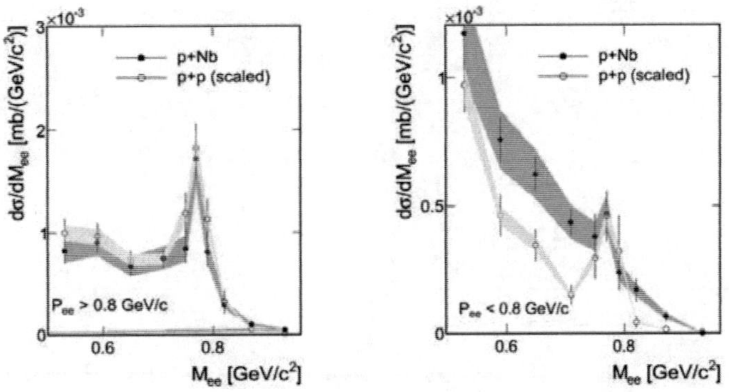

Figure 2. Left: Comparison of the dielectron invariant mass spectra in the vector meson mass region for pairs with $P_{ee} > 0.8$ GeV/c from p+p (scaled) and p+Nb at a kinetic beam energy of 3.5 GeV. Right: For pairs with $P_{ee} < 0.8$ GeV/c.

In fact, most experiments focusing on the spectral distribution of dielectrons produced off nuclei in photon and proton induced reactions are restricted to relatively high momenta ($P_{ee} > 0.8$ GeV/c) and are not conclusive yet. For the ρ meson, the CLAS experiment at JLab [7] reports a slight broadening and no shift of the ρ pole postion in photon induced reactions, while the E325 experiment at KEK [8] deduced a shift but no broadening in proton induced reactions.
In this paper we follow our recent letter [1] on inclusive e^+e^- pair production in proton induced reactions at $E_{kin} = 3.5$ GeV on the nucleus Nb, representing the first high statistics measurement with small momenta of e^+e^- pairs relative to the medium ($P_{ee} < 0.8$ GeV/c). We present here in addition the transverse momentum and rapidity distributions for various e^+e^- invariant mass bins. These data are compared to our reference data measured in p+p reactions at the same incident beam energy [9] [1].

2. Experimental data

The efficiency corrected invariant mass distributions of e^+e^- pairs within the HADES acceptance are displayed in Fig. 1. The systematic uncertainties are represented by colored horizontal, which are the quadratic sum of errors estimated from different particle identification methods (10%), from consistency checks of the efficiency correction (10%), including the uncertainty due to combinatorial background subtraction, and from the normalization to cross sections (15%). The total systematic error amounts to 21% in case of the p+Nb data, and to 20% for the p+p data. Only the systematic errors of the normalization are taken into account for the comparison of the shape of the invariant mass spectra as the other systematic errors should cancel at first order. For the p+p data, a dielectron cocktail, generated using an adapted version of the event generator PYTHIA (see [9] for details) is displayed in addition. We distinct between four different mass regions: M_{ee} [GeV/c^2] < 0.15 (dominated by neutral pion decays), $0.15 < M_{ee}$ [GeV/c^2] < 0.47 (η Dalitz decay dominated), $0.47 < M_{ee}$ [GeV/c^2] < 0.7 (dominated by direct ρ decays and Dalitz decays of baryonic resonances and ω mesons) and $0.7 < M_{ee}$ [GeV/c^2] (ρ and ω dominated) as can be seen from the cocktail. Around 1 GeV/c^2 a low statistics ϕ signal is

[1] Note that, according to the systematics on dielectron emission obtained in p+p and d+p collisions at various beam energies by the DLS collaboration [10], it is save to conclude that at the kinetic beam energy of 3.5 GeV isospin effects play only a secondary role. Therefore, in order to extract medium effects in p+Nb, the above discussed p+p data at the same kinetic beam energy of 3.5 GeV/c is a valuable reference.

Journal of Physics: Conference Series **426** (2013) 012034 doi:10.1088/1742-6596/426/1/012034

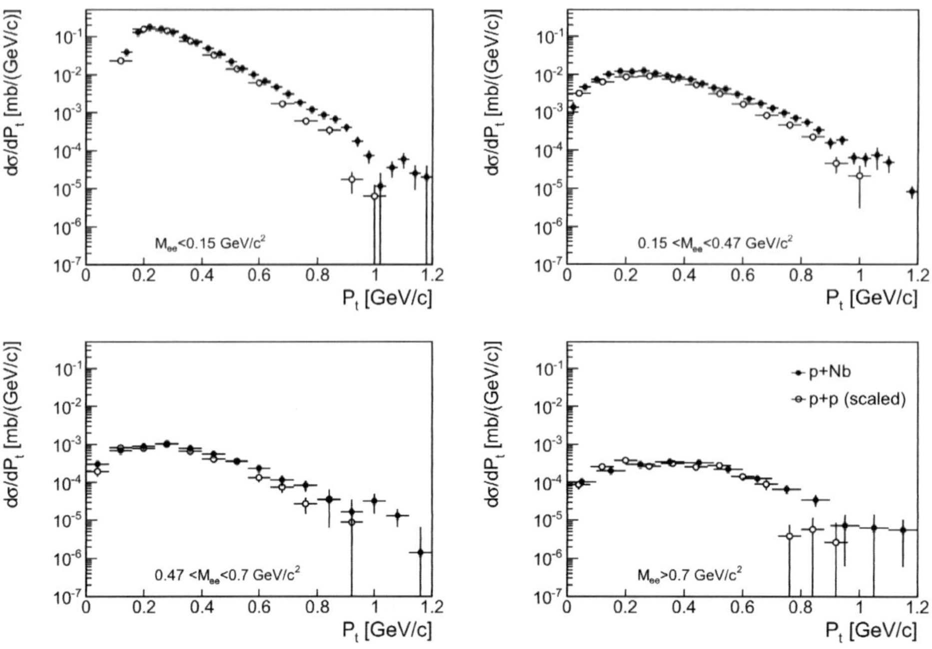

Figure 3. Transverse momentum distributions of dielectrons in p+p (scaled) and p+Nb collisions.

visible, which will be further discussed in future, making use of additional information from its hadronic decay channel to charged kaons. The underestimation of the dielectron yield in the mass region from $0.47 < M_{ee}$ $[GeV/c^2] < 0.7$ is addressed in e.g. [12, 13], where the authors improve the description by the introduction of a coupling between ρ mesons and baryonic resonances.

Comparing the shape of the invariant mass spectra separately for pairs with momenta larger and smaller 0.8 GeV/c to pairs from p+p, scaled to the number of participants and the total reaction cross section, we observe a strong e^+e^- excess yield below the ω pole mass in the small momentum sample displayed in the right panel of Fig. 2, while for pairs with $P_{ee} > 0.8$ GeV/c no significant difference in the vector meson mass region within the systematic uncertainties indicated by the colored bands in the panel of Fig. 2, is visible. For pairs with $P_{ee} < 0.8$ GeV/c (right panel of Fig. 2) the e^+e^- yield at the ω pole mass is not reduced, but as the underlying smooth distribution is enhanced, the yield in the peak is reduced to almost zero within errors. This additional yield we attribute to ρ-like channels as it is supposed to be the dominating source for radiation from the medium due to its large total width. For more details, see [1]. Comparing the P_t and Y distributions for the four previously discussed invariant mass regions the very same pattern is observed: In the p+Nb system the distributions reach out to slightly higher transverse momentum and the rapidity distributions shift towards target rapidity, see Fig. 3 and Fig. 4. The latter one is much stronger pronounced and corresponds to second generation particles produced from a slower source compared to the initial one. Although this behavior is to some extent expected (stronger transformation of longitudinal energy to transverse degrees of freedom in p+Nb) the defined values are important constraints to phenomenological models in order to correctly model the experimental conditions.

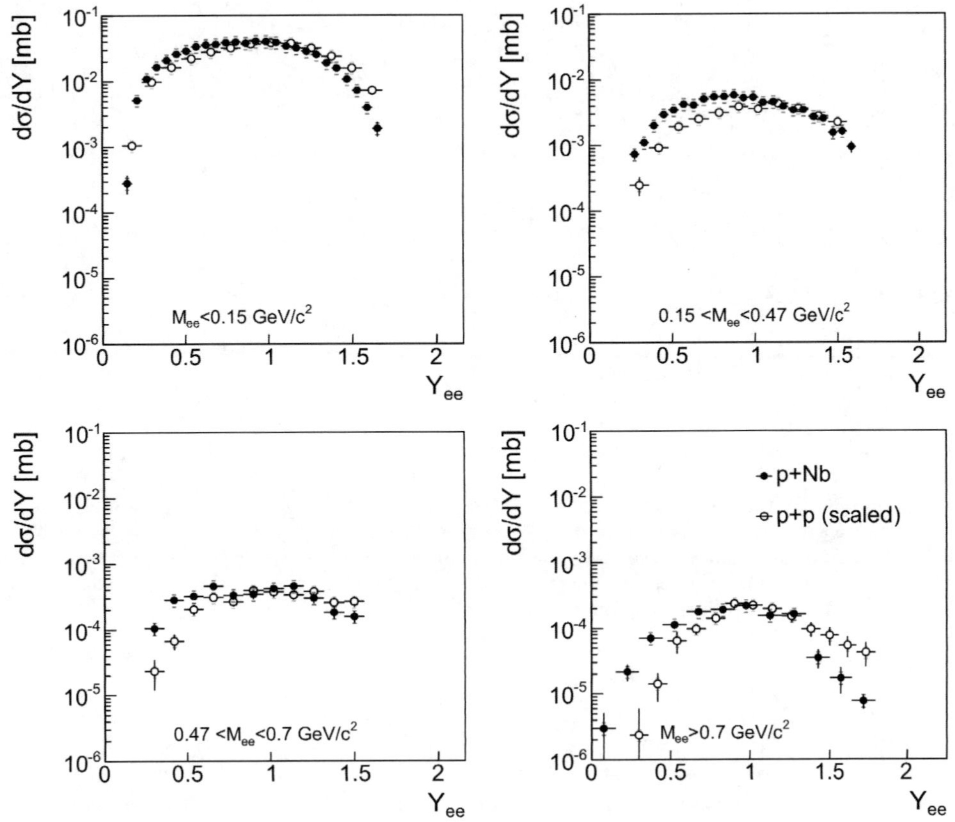

Figure 4. Rapidity distributions of dielectrons in p+p (scaled) and p+Nb collisions.

3. Summary

In addition to our recent letter [1] on inclusive e^+e^- pair production in proton induced reactions at E_{kin} = 3.5 GeV on the nucleus Nb, we presented the transverse and rapidity distributions for various e^+e^- invariant mass bins. Comparing these distributions to reference data from p+p collisions, the following pattern is observed: In the p+Nb system the distributions reach out to slightly higher transverse momentum and the rapidity distributions are shifted towards target rapidity. Especially the defined shift in rapidity gives important constraints to phenomenological models.

[1] G. Agakishiev *et al.* (HADES Collaboration]), Phys. Lett. B **715** 304 (2012).
[2] R. Brockmann and W. Weise, Phys. Lett. B **367**, 40 (1996).
[3] G. E. Brown and M. Rho, Phys. Rev. Lett. **66**, 2720 (1991).
[4] T. Hatsuda and S. H. Lee, Phys. Rev. C **46**, R34 (1992).
[5] R. Rapp et al., Nucl. Phys. A **617**, 472 (1997).
[6] S. Leupold, V. Metag, and U. Mosel, Int. J. Mod. Phys. **E 19**, 147 (2010).
[7] R. Nasseripour et al. (CLAS Collaboration), Phys. Rev. Lett. **99**, 262302 (2007).
[8] M. Naruki et al., Phys. Rev. Lett. **96**, 092301 (2006).
[9] G. Agakishiev et al. (HADES Collaboration), Eur. Phys.J. **A 48** 64 (2012).
[10] W.K. Wilson et al. (DLS Collaboration), Phys. Rev. **C57** 1865 (1998).
[11] G. Agakishiev et al. (HADES Collaboration), Eur. Phys. J. **A 41**, 243 (2009).
[12] J. Weil, K. Gallmeister and U. Mosel, arXiv:1105.0314 [nucl-th].
[13] J. Weil, H. van Hees, and U. Mosel, arXiv:1203.3557 [nucl-th].

Journal of Physics: Conference Series **426** (2013) 012035 doi:10.1088/1742-6596/426/1/012035

Dilepton production at SIS energies with the GiBUU transport model

Janus Weil and Ulrich Mosel

Institut für Theoretische Physik, Universität Giessen, D-35392 Giessen, Germany

E-mail: janus.weil@theo.physik.uni-giessen.de

Abstract. We present dilepton spectra from nucleus-nucleus collisions at SIS energies, which were simulated with the GiBUU transport model in a resonance-model approach. These spectra are compared to the data published by the HADES collaboration. We argue that the interpretation of dilepton spectra at SIS energies critically depends on the couplings between the ρ meson and the baryonic resonances.

1. Introduction

While the vacuum properties of most hadrons are known to good accuracy nowadays, it is a heavily debated question how these properties change in nuclear matter. In particular, various theoretical predictions regarding the in-medium properties of the light vector mesons have been suggested. For recent reviews on in-medium physics, see [1–3].

Among the expected in-medium effects, a so-called "collisional broadening" of the meson spectral function, due to collisions with the hadronic medium, is expected. A second class of predictions claims that the vector-meson masses are shifted in the medium due to the partial restoration of chiral symmetry [4]. QCD sum rules can constrain these effects, but do not provide definitive predictions [5].

The more prominent hadronic decay modes of the vector mesons are unfavorable for studying in-medium effects, since they are affected by strong final-state interactions with the hadronic medium – in contrast to the rare dilepton decay modes. Since the leptons only interact electromagnetically, they are ideally suited to carry the in-medium information outside to the detector, nearly undisturbed by the hadronic medium.

Early measurements of dilepton spectra from heavy-ion collisions in the low-energy regime were conducted by the DLS collaboration [6], showing a clear excess over the expected yield. A similar excess was also observed in experiments at higher energies [7, 8], where it could be attributed to an in-medium broadening of spectral functions [9–11]. For the DLS data such in-medium effects never provided a convincing explanation - a problem that was soon known as the "DLS puzzle" [12–15].

More recently, the HADES collaboration at GSI has set up an extensive program for measuring dilepton spectra from p+p, p+A and A+A reactions [16–21], in order to systematically check the old DLS data with improved statistics and to finally resolve the DLS puzzle. Up to now this endeavor has fully confirmed the validity of the DLS data and shifted the puzzle into the theory sector.

2. The model

In this paper, we apply the Gießen Boltzmann-Uehling-Uhlenbeck transport model (GiBUU) [22, 23] to the nucleus-nucleus reactions studied by the HADES collaboration. The same model was used earlier to describe the elementary HADES and DLS data [24]. GiBUU is a hadronic transport model, which relies on a resonance-model description of NN and πN collisions at low energies.

In addition to the features described in [24], the model has received a few extensions in the meantime: For the production of baron resonances via $NN \to NR$, we use angular distributions according to $d\sigma/dt \propto 1/t$, inspired by one-boson exchange, in order to improve upon the isotropic phase-space approach used earlier. At low energies, these angular distributions can become important due to the geometrical acceptance of the detector. In the Δ Dalitz decay, we now use the form factor from [25], which gives only moderate deviations from the assumption of a constant form factor and overall seems more reasonable than the one used previously. However, one should keep in mind that the issue of the Δ form factor has still not been settled in any way.

For the elementary dilepton spectra, of course only NN collisions played a role. In heavy-ion collisions, other types of secondary collisions can occur, such as πN and $\pi\pi$ collisions. Details about the treatment of these processes in GiBUU can be found in [22]. For πN Bremsstrahlung, we use the soft-photon approximation, in the same way as previously applied to NN Bremsstrahlung [24].

3. Dilepton spectra from heavy-ion collisions

We present here preliminary results, obtained with the GiBUU transport model, of dilepton spectra from nucleus-nucleus collisions, namely the light C+C system at beam kinetic energies of 1.0 and 2.0 AGeV and the intermediate Ar+KCl system at 1.756 AGeV, all of which have been measured by the HADES collaboration in a fixed-target setup at the SIS18 accelerator at GSI.

Table 1. Kinematic conditions of the collision systems measured by HADES and corresponding cuts on the single lepton momenta (all in GeV), and the maximum impact parameters in fm.

system	E_{kin}/A	$\sqrt{s}_{NN}$	$p_{\text{lep}}^{\text{min}}$	$p_{\text{lep}}^{\text{max}}$	b_{max}
C + C	1.0	2.32	0.05	1.8	5.0
C + C	2.0	2.70	0.05	1.8	5.0
Ar + KCl	1.756	2.61	0.10	1.1	6.5

Table 1 gives an overview over the different reactions, including kinematic conditions, cuts and impact parameters used. All reactions have been simulated with impact parameters restricted to the range $0 \leq b \leq b_{\text{max}}$ Furthermore, all spectra have been filtered through the HADES acceptance filter [26, 27] in order to account for the geometrical acceptance and resolution of the detector. In addition, a dilepton opening angle cut of $\theta_{ee} > 9°$ is applied in all cases, as well as the single-lepton momentum cuts listed in table 1, matching the experimental analysis procedure.

3.1. C + C

We start by showing in fig. 1 the dilepton mass spectrum for ^{12}C+^{12}C collisions at beam energies of 1.0 and 2.0 AGeV, compared to the HADES data. Since the published data are normalized relative to the π^0 yield only, but the simulation is given in absolute units, the data were scaled by

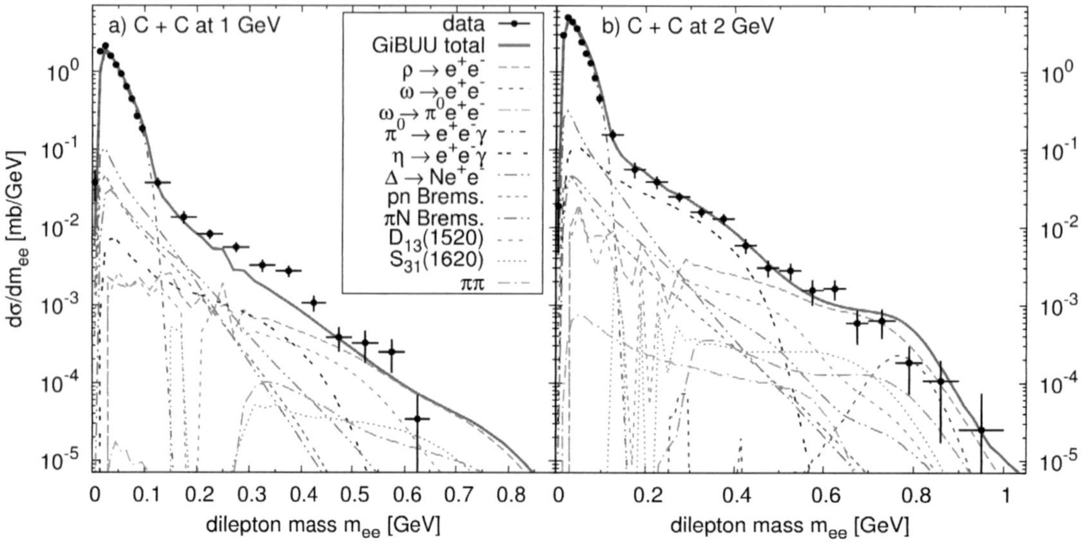

Figure 1. Dilepton mass spectrum for C+C at 1 and 2 AGeV in comparison to the data from [16, 17].

an arbitrary factor to match the simulation in the pion channel. As has been argued in [18], C+C is a sufficiently light system, so that it can be regarded, in first order, as a simple superposition of elementary NN collisions, without significant secondary effects or modifications due to the hadronic medium. It has also been shown that both C+C data sets are compatible with a superposition of the NN data (when subtracting the η component, which has a different beam energy dependence) [18].

From this point of view, we can expect our results for the C+C collisions to reflect the quality of our elementary NN results. And indeed we observe an excellent description of the C+C data at 2 AGeV, due to the good description of pp collisions at 2.2 and 3.5 GeV [24]. The largest deviation in the C+C spectrum at 2 AGeV is a slight overestimation of the data in the ω mass region, which is barely significant since the data points already have quite large error bars. Most other models showed a much more severe overestimation in this mass regime [16, 28].

In the C+C spectrum at 1 AGeV one can observe more significant deviations, namely an underestimation of the data around 300 - 400 MeV. This is probably connected to the fact that our model's agreement to the elementary data is not as good at lower energies (in particular in d+p at 1.25 GeV). Improving the description of the elementary collisions would certainly also improve the agreement with the C+C data at 1 AGeV.

We stress that, in contrast to other models [29, 30], the Δ Dalitz channel plays only a minor role in our cocktail, while the ρ production via baryon resonances is much more important. The effects of the ρ meson coupling to the $D_{13}(1520)$ have previously been discussed e.g. in [31]. Fig. 1 shows separately the dominant sources of ρ mesons, which are given by decays of the $D_{13}(1520)$ and $S_{31}(1620)$ resonances and the process $\pi\pi \to \rho$. We also note that the latter is the only production process of the ρ which has a physical threshold at $m = 2m_\pi$. All other processes yield dilepton contributions below the 2π threshold (but unfortunately suffer from poor statistics there, due to our numerical treatment).

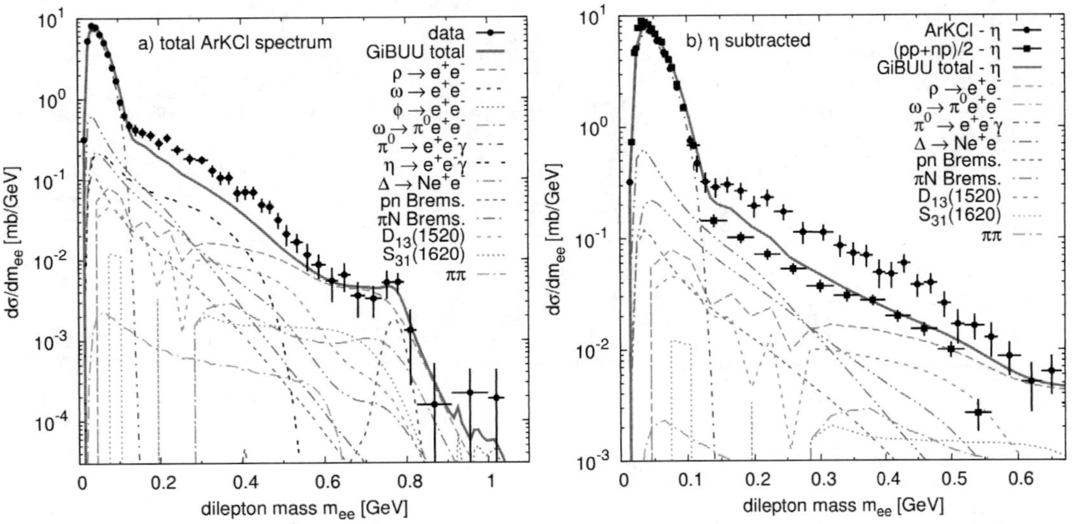

Figure 2. Dilepton mass spectrum for Ar+KCl at 1.756 AGeV in comparison to the data from [19]. Left: Total spectrum; right: η component subtracted.

3.2. Ar + KCl

The data for this reaction have been obtained by shooting ^{40}Ar ions on a natural KCl target. In the simulation we use ^{37}Ar (i.e. the average of ^{35}Cl and ^{39}K) to approximate the composite target. Fig. 2a shows a comparison of our simulation results to the data, which clearly exhibits larger deviations than in the C+C case. The discrepancies mostly concern the intermediate mass range of 150 to 500 MeV.

As in the C+C case at 1 AGeV, part of the missing yield might be due to disagreement with elementary data. However, it is unlikely that this can account for all of the missing yield, since our agreement with the elementary HADES and DLS data is quite good in the energy regime relevant for the Ar+KCl system [24]. Instead, it seems more plausible that the missing yield represents an actual medium effect. This is supported by the fact that the ArKCl data apparently shows an excess over the elementary reference cocktail [19], which is of similar magnitude as the discrepancy to our model. To illustrate this, fig. 2b shows the ArKCl spectrum with the η Dalitz component subtracted, compared to the NN reference spectrum (both normalized to the pion channel, for details see [19]). While our model underestimates the ArKCl data, it lies close above the elementary reference spectrum.

The excess yield could in principle come from two distinct sources:

(i) Dilepton emission from secondary collisions (e.g. πN), which is in principle covered by our model but could be underestimated. Here also πN Bremsstrahlung could play a role, which is treated in our model only in soft-photon approximation (SPA).

(ii) An in-medium modification of the spectral functions, e.g. a broadening of the ρ meson or modifications of the nucleon resonances, which are involved in the production dynamics of the ρ, most prominently the $D_{13}(1520)$. Such spectral modifications are not included in our current simulation, which fully relies on vacuum spectral functions.

In order to decide if any of these two effects can contribute to the missing yield, both should be studied more closely in future investigations.

4. Conclusions

We have shown that the GiBUU transport model provides a reasonable description of dilepton spectra from C+C collisions, both at 1 and 2 AGeV. At the lower beam energy, minor discrepancies show up, which might be connected to the underestimation of the d+p reaction at 1.25 GeV. At the higher beam energy of 2 AGeV, the high-mass region of the spectrum is well-described by a dominant contribution from ρ mesons produced via baryonic resonances. Most other models showed a significant overestimation of the data in this region.

Regardless of the good agreement with the elementary and C+C data, our model significantly underestimates the Ar+KCl data in the intermediate mass region when using vacuum spectral functions. This excess in the data could be an indication of non-trivial effects of the hadronic medium and gives rise to the hope that even larger effects can be observed in the Au+Au system.

Acknowledgments

We thank Tetyana Galatyuk and Romain Holzmann for providing the data. This work was supported by HGS-HIRe, BMBF and the Hessian LOEWE initiative through HIC for FAIR.

References

[1] S. Leupold, V. Metag, U. Mosel, Int. J. Mod. Phys. E **19**, 147 (2010), `arXiv:0907.2388[nucl-th]`
[2] R.S. Hayano, T. Hatsuda, Rev. Mod. Phys. **82**, 2949 (2010), `arXiv:0812.1702[nucl-ex]`
[3] R. Rapp, J. Wambach, H. van Hees, Landolt-Börnstein **I/23**, 4 (2010), `arXiv:0901.3289[hep-ph]`
[4] T. Hatsuda, S.H. Lee, Phys. Rev. C **46**, 34 (1992)
[5] S. Leupold, W. Peters, U. Mosel, Nucl. Phys. A **628**, 311 (1998), `arXiv:9708016[nucl-th]`
[6] R. Porter et al. (DLS Collaboration), Phys. Rev. Lett. **79**, 1229 (1997), `arXiv:9703001[nucl-ex]`
[7] D. Adamova et al. (CERES/NA45 Collaboration), Phys. Rev. Lett. **91**, 042301 (2003), `arXiv:0209024[nucl-ex]`
[8] R. Arnaldi et al. (NA60 Collaboration), Phys. Rev. Lett. **96**, 162302 (2006), `arXiv:0605007[nucl-ex]`
[9] H. van Hees, R. Rapp, Phys. Rev. Lett. **97**, 102301 (2006), `arXiv:0603084[hep-ph]`
[10] J. Ruppert, C. Gale, T. Renk, P. Lichard, J.I. Kapusta, Phys. Rev. Lett. **100**, 162301 (2008), `arXiv:0706.1934[hep-ph]`
[11] H. van Hees, R. Rapp, Nucl. Phys. A **806**, 339 (2008), `arXiv:0711.3444[hep-ph]`
[12] E. Bratkovskaya, W. Cassing, R. Rapp, J. Wambach, Nucl. Phys. A **634**, 168 (1998), `arXiv:9710043[nucl-th]`
[13] C. Ernst, S. Bass, M. Belkacem, H. Stoecker, W. Greiner, Phys. Rev. C **58**, 447 (1998), `arXiv:9712069[nucl-th]`
[14] E. Bratkovskaya, C. Ko, Phys. Lett. B **445**, 265 (1999), `arXiv:9809056[nucl-th]`
[15] K. Shekhter, C. Fuchs, A. Faessler, M. Krivoruchenko, B. Martemyanov, Phys. Rev. C **68**, 014904 (2003), `arXiv:0305015[nucl-th]`
[16] G. Agakichiev et al. (HADES Collaboration), Phys. Rev. Lett. **98**, 052302 (2007), `arXiv:0608031[nucl-ex]`
[17] G. Agakishiev et al. (HADES Collaboration), Phys. Lett. B **663**, 43 (2008), `arXiv:0711.4281[nucl-ex]`
[18] G. Agakishiev et al. (HADES Collaboration), Phys. Lett. B **690**, 118 (2010), `arXiv:0910.5875[nucl-ex]`
[19] G. Agakishiev et al. (HADES Collaboration), Phys. Rev. C **84**, 014902 (2011), `arXiv:1103.0876[nucl-ex]`
[20] G. Agakishiev et al. (HADES Collaboration), Eur. Phys. J. A **48**, 64 (2012), `arXiv:1112.3607[nucl-ex]`
[21] G. Agakishiev et al. (HADES Collaboration), Phys. Rev. C **85**, 054005 (2012), `arXiv:1203.2549[nucl-ex]`
[22] O. Buss, T. Gaitanos, K. Gallmeister, H. van Hees, M. Kaskulov et al., Phys. Rep. **512**, 1 (2012), `arXiv:1106.1344[hep-ph]`
[23] *GiBUU website*, `http://gibuu.physik.uni-giessen.de`
[24] J. Weil, H. van Hees, U. Mosel, Eur. Phys. J. A **48**, 111 (2012), `arXiv:1203.3557[nucl-th]`
[25] G. Ramalho, M. Pena, Phys. Rev. D **85**, 113014 (2012), `arXiv:1205.2575[hep-ph]`
[26] *HADES website*, `http://www-hades.gsi.de`
[27] T. Galatyuk, *private communications*
[28] H. Barz, B. Kampfer, G. Wolf, M. Zetenyi (2009), `arXiv:0910.1541[nucl-th]`
[29] E. Bratkovskaya, W. Cassing, Nucl. Phys. A **807**, 214 (2008), `arXiv:0712.0635[nucl-th]`
[30] K. Schmidt, E. Santini, S. Vogel, C. Sturm, M. Bleicher et al., Phys. Rev. C **79**, 064908 (2009), `arXiv:0811.4073[nucl-th]`
[31] S. Vogel, M. Bleicher, Phys.Rev. **C74**, 014902 (2006), `arXiv:0509105[nucl-th]`